AF411313

Hedgehog Signaling in Organogenesis and Tumor Microenvironment

Hedgehog Signaling in Organogenesis and Tumor Microenvironment

Special Issue Editor

Tsuyoshi Shimo

MDPI • Basel • Beijing • Wuhan • Barcelona • Belgrade • Manchester • Tokyo • Cluj • Tianjin

Special Issue Editor
Tsuyoshi Shimo
Health Sciences University of Hokkaido
Japan

Editorial Office
MDPI
St. Alban-Anlage 66
4052 Basel, Switzerland

This is a reprint of articles from the Special Issue published online in the open access journal *International Journal of Molecular Sciences* (ISSN 1422-0067) (available at: https://www.mdpi.com/journal/ijms/special_issues/hedgehog_signaling).

For citation purposes, cite each article independently as indicated on the article page online and as indicated below:

LastName, A.A.; LastName, B.B.; LastName, C.C. Article Title. *Journal Name* **Year**, *Article Number*, Page Range.

ISBN 978-3-03936-260-8 (Hbk)
ISBN 978-3-03936-261-5 (PDF)

Contents

About the Special Issue Editor

Tsuyoshi Shimo is a professor in the Division of Reconstructive Surgery for the Oral and Maxillofacial Region at the Health Sciences University of Hokkaido. He received his medical education at Hiroshima University, and he obtained his Ph.D. from the Okayama University in Japan, in cancer and developmental biology. He received further postdoctoral education at University of Pennsylvania, Philadelphia, and he received his surgical training in oral cancer and orthognathic surgery from Okayama University Hospital. His research interests include cancer-induced bone destruction and development.

International Journal of
Molecular Sciences

Editorial

Hedgehog Signaling in Organogenesis and the Tumor Microenvironment

Tsuyoshi Shimo

Division of Reconstructive Surgery for Oral and Maxillofacial Region, Department of Human Biology and
Pathophysiology, School of Dentistry, Health Sciences University of Hokkaido, Hokkaido 061-0293, Japan;
shimotsu@hoku-iryo-u.ac.jp; Tel./Fax: +81-133-23-1429

Received: 8 April 2020; Accepted: 13 April 2020; Published: 17 April 2020

The Hedgehog signaling pathway was first discovered in 1980 during a large-scale genetic screening seeking to find mutations that affect larval body segment development in the fruit fly, *Drosophila melanogaster* [1]. The Hedgehog signaling pathway is an evolutionarily conserved pathway that governs complex developmental processes including stem cell maintenance, proliferation, differentiation, and patterning. Several recent studies have shown that the aberrant activation of Hedgehog signaling is associated with neoplastic transformation, cancer cell proliferation, metastasis, multiple cancers' drug resistance, and survival rates. This Special Issue focuses on several aspects of Hedgehog signaling in organogenesis and the tumor microenvironment, and we called for reviews and original papers on the recent efforts in the field of Hedgehog signaling.

This Special Issue of the *International Journal of Molecular Sciences*, entitled "Hedgehog Signaling in Organogenesis and the Tumor microenvironment", thus includes four original articles and five reviews that provide new insights regarding the roles of Hedgehog signaling in organogenesis and the tumor microenvironment.

Tarulli et al., report on "Discrete Hedgehog Factor Expression and Action in the Developing Phallus", and they describe a potential developmental interaction involved in urethral closure that mimics bone differentiation and incorporates discrete Hedgehog activity within the developing phallus and phallic urethra [2].

Takebe et al., examined Gli-CreERT2; tdTomato mice, and they demonstrate that the SHH-Gli1 signaling pathway is involved in intramembranous and endochondral ossification during the fracture healing process [3].

Takabatake et al., describe "The Role of Sonic Hedgehog Signaling in the Tumor Microenvironment of Oral Squamous Cell Carcinoma", and their findings revealed that (1) autocrine effects of SHH induce cancer invasion and (2) paracrine effects of SHH govern parenchyma–stromal interactions of oral squamous cell carcinoma [4].

El Shahawy et al., propose that "Sonic Hedgehog Signaling Is Required for Cyp26 Expression during Embryonic Development", and they explain that rigidly calibrated Hedgehog and retinoic acid activities are required for normal organogenesis and tissue patterning [5].

Hosoya et al., provide an overview of recent advances related to the role of SHH signaling in tooth development, homeostasis, regeneration, and the regulatory mechanism of stem cell properties in the dental mesenchyme from experiments using tamoxifen administration in iGli1/Tomato mice [6].

Jeng et al., extensively review the recent progress made in the field of "Sonic Hedgehog Signaling in Organogenesis, Tumors, and Tumor Microenvironments", focusing on the combined use of SHH signaling inhibitors and chemotherapy/radiation therapy/immunotherapy targeting cancer stem cells [7].

Hyuga et al., contribute a comprehensive overview of the "Hedgehog Signaling for Urogenital Organogenesis and Prostate Cancer: An Implication for the Epithelial-Mesenchyme Interaction (EMI)" and compare possible similarities and divergences in Hedgehog signaling functions and the interaction of this signaling with other local growth factors between organogenesis and tumorigenesis. They discuss two pertinent research aspects of Hedgehog signaling: (1) the potential signaling crosstalk between Hedgehog and androgen signaling and (2) the effect of Hedgehog signaling between the epithelia and the mesenchyme on the status of the basement membrane with extracellular matrix structures located on the epithelial–mesenchymal interface [8].

Bechtold et al., offer a thorough review of the recent progress made in studies on the roles of Indian Hedgehog signaling in temporomandibular joint (TMJ) formation, and they discuss important findings regarding the involvement of Hedgehog signaling in TMJ development during embryonic and early postnatal stages as well as in the establishment and postnatal maintenance of TMJs, plus the possible involvement of Hedgehog pathways in osteoarthritic conditions [9].

Haraguchi et al., provide a detailed discussion about "Recent Insights into Long Bone Development: Central Role of Hedgehog Signaling Pathway in Regulating Growth Plate", and they review the multiple roles of the Hedgehog pathway in the regulation of growth plate formation and differentiation, as well as longitudinal bone development and skeletal disorders [10].

The Editor hopes that these articles will help readers update their knowledge about the role of Hedgehog signaling in physiology and pathology. The efforts of the authors who contributed their excellent articles to this Special Issue are greatly appreciated.

Funding: The author received no funding for this editorial.

Conflicts of Interest: The author declares no conflict of interest.

References

1. Varjosalo, M.; Taipale, J. Hedgehog: Functions and mechanisms. *Genes Dev.* **2008**, *22*, 2454–2472. [CrossRef]
2. Tarulli, G.A.; Pask, A.J.; Renfree, M.B. Discrete Hedgehog Factor Expression and Action in the Developing Phallus. *Int. J. Mol. Sci.* **2020**, *21*, 1237. [CrossRef] [PubMed]
3. Takebe, H.; Shalehin, N.; Hosoya, A.; Shimo, T.; Irie, K. Sonic Hedgehog Regulates Bone Fracture Healing. *Int. J. Mol. Sci.* **2020**, *21*, 677. [CrossRef]
4. Takabatake, K.; Shimo, T.; Murakami, J.; Anqi, C.; Kawai, H.; Yoshida, S.; Wathone Oo, M.; Haruka, O.; Sukegawa, S.; Tsujigiwa, H.; et al. The Role of Sonic Hedgehog Signaling in the Tumor Microenvironment of Oral Squamous Cell Carcinoma. *Int. J. Mol. Sci.* **2019**, *20*, 5779. [CrossRef] [PubMed]
5. El Shahawy, M.; Reibring, C.G.; Hallberg, K.; Neben, C.L.; Marangoni, P.; Harfe, B.D.; Klein, O.D.; Linde, A.; Gritli-Linde, A. Sonic Hedgehog Signaling Is Required for Cyp26 Expression during Embryonic Development. *Int. J. Mol. Sci.* **2019**, *20*, 2275. [CrossRef] [PubMed]
6. Hosoya, A.; Shalehin, N.; Takebe, H.; Shimo, T.; Irie, K. Sonic Hedgehog Signaling and Tooth Development. *Int. J. Mol. Sci.* **2020**, *21*, 1587. [CrossRef] [PubMed]
7. Jeng, K.S.; Chang, C.F.; Lin, S.S. Sonic Hedgehog Signaling in Organogenesis, Tumors, and Tumor Microenvironments. *Int. J. Mol. Sci.* **2020**, *21*, 758. [CrossRef] [PubMed]
8. Hyuga, T.; Alcantara, M.; Kajioka, D.; Haraguchi, R.; Suzuki, K.; Miyagawa, S.; Kojima, Y.; Hayashi, Y.; Yamada, G. Hedgehog Signaling for Urogenital Organogenesis and Prostate Cancer: An Implication for the Epithelial-Mesenchyme Interaction (EMI). *Int. J. Mol. Sci.* **2019**, *21*, 58. [CrossRef] [PubMed]
9. Bechtold, T.E.; Kurio, N.; Nah, H.D.; Saunders, C.; Billings, P.C.; Koyama, E. The Roles of Indian Hedgehog Signaling in TMJ Formation. *Int. J. Mol. Sci.* **2019**, *20*, 6300. [CrossRef] [PubMed]
10. Haraguchi, R.; Kitazawa, R.; Kohara, Y.; Ikedo, A.; Imai, Y.; Kitazawa, S. Recent Insights into Long Bone Development: Central Role of Hedgehog Signaling Pathway in Regulating Growth Plate. *Int. J. Mol. Sci.* **2019**, *20*, 5840. [CrossRef] [PubMed]

International Journal of
Molecular Sciences

Article

Sonic Hedgehog Signaling Is Required for Cyp26 Expression during Embryonic Development

Maha El Shahawy [1,2,†], Claes-Göran Reibring [1,†], Kristina Hallberg [1], Cynthia L. Neben [3], Pauline Marangoni [3], Brian D. Harfe [4], Ophir D. Klein [3,5], Anders Linde [1] and Amel Gritli-Linde [1,*]

[1] Department of Oral Biochemistry, Sahlgrenska Academy at the University of Gothenburg, SE-40530 Göteborg, Sweden; maha.el.shahawy@odontologi.gu.se (M.E.S.); claes-goran.reibring@gu.se (C.-G.R.); kristina.hallberg@odontologi.gu.se (K.H.); linde@odontologi.gu.se (A.L.)

[2] Department of Oral Biology, Minia University, Minia 51161, Egypt

[3] Program in Craniofacial Biology and Department of Orofacial Sciences, University of California San Francisco, San Francisco, CA 94143, USA; cynthianeben@gmail.com (C.L.N.); Pauline.Marangoni@ucsf.edu (P.M.); Ophir.Klein@ucsf.edu (O.D.K.)

[4] Department of Molecular Genetics and Microbiology, University of Florida College of Medicine, Gainesville, FL 32610, USA; bharfe@UFL.EDU

[5] Department of Pediatrics and Institute for Human Genetics, University of California San Francisco, San Francisco, CA 94143, USA

[*] Correspondence: amel@odontologi.gu.se; Tel.: +46-31-7863392

[†] These authors contributed equally to this work.

Received: 1 April 2019; Accepted: 3 May 2019; Published: 8 May 2019

Abstract: Deciphering how signaling pathways interact during development is necessary for understanding the etiopathogenesis of congenital malformations and disease. In several embryonic structures, components of the Hedgehog and retinoic acid pathways, two potent players in development and disease are expressed and operate in the same or adjacent tissues and cells. Yet whether and, if so, how these pathways interact during organogenesis is, to a large extent, unclear. Using genetic and experimental approaches in the mouse, we show that during development of ontogenetically different organs, including the tail, genital tubercle, and secondary palate, Sonic hedgehog (SHH) loss-of-function causes anomalies phenocopying those induced by enhanced retinoic acid signaling and that SHH is required to prevent supraphysiological activation of retinoic signaling through maintenance and reinforcement of expression of the *Cyp26* genes. Furthermore, in other tissues and organs, disruptions of the Hedgehog or the retinoic acid pathways during development generate similar phenotypes. These findings reveal that rigidly calibrated Hedgehog and retinoic acid activities are required for normal organogenesis and tissue patterning.

Keywords: Cyp26 enzymes; congenital anomalies; CRE/LoxP; hedgehog signaling; mouse models; retinoic acid; smoothened; sonic hedgehog

---

1. Introduction

Development and homeostasis of multicellular organisms crucially rely on concerted functions of a multitude of proteins and small molecules that operate within signaling pathways. Understanding how signaling pathways interact to ensure normal embryonic development and maintenance of proper shape, size, cellular organization and function of tissues and organs is requisite to decipher the etiopathogenesis of congenital malformations and diseases.

The Hedgehog and retinoic acid (RA) signaling pathways play key roles during embryogenesis, organogenesis, and tissue homeostasis [1–13], and genetic disruption of Hedgehog signaling can lead to neoplasia [1,14–18]. Mammals produce three Hedgehog ligands, Desert hedgehog, Indian hedgehog (IHH) and Sonic hedgehog (SHH) [1,19]. Hedgehog ligands, notably SHH and IHH proteins, emanating from producing cells, can signal both short and long-range [1,4]. The Hedgehog signaling cascade is regulated by several factors at different levels, from ligand modifications and release to ligand reception and signal transduction [4,6]. In the absence of Hedgehog ligands, the Hedgehog receptor PTCH1 accumulates predominantly in the primary cilium and inhibits Smoothened (SMO), an obligatory factor for the transduction of all Hedgehog signals, leading to the formation of repressor forms of GLI transcription factors that repress Hedgehog target genes. Upon ligand binding to PTCH1, the activated SMO protein translocates to the cilium and initiates a signaling cascade that reaches its acme in the nucleus, where the activator forms of GLI proteins activate transcription of Hedgehog target genes. The principal GLI activator function derives primarily from GLI2, whereas the GLI repressor function largely emanates from GLI3 [1,4,6,18,20,21].

All-trans retinoic acid (RA), the predominant active metabolite of the dietary-derived vitamin A, is a small, highly diffusible and biologically potent lipophilic molecule. During embryonic development, RA is produced from maternally-derived vitamin A. Experimental studies in rodents and avians established the importance of vitamin A for proper development, as vitamin A deficiency during embryogenesis and early organogenesis engenders a wide range of congenital anomalies [22]. However, exposure of embryos to excess vitamin A or RA is teratogenic. Direct evidence for the crucial role of RA during development emanated from genetic gain and loss-of-function studies in mice and zebrafish, which demonstrated that proper tissue patterning and cell fate specification require well-calibrated spatio-temporal RA activity [2,3,5,23,24].

RA synthesis from retinol, the alcohol form of vitamin A, is a stepwise process catalyzed by various dehydrogenases. First, retinol is oxidized into retinaldehyde by alcohol dehydrogenases and retinol dehydrogenases. Thereafter, oxidation of retinaldehyde to RA is catalyzed by retinaldehyde dehydrogeneases, including RALDH1, RALDH2, and RALDH3, encoded by *Aldh1a1*, *Aldh1a2* and *Aldh1a3*, respectively [2,22,24]. RA is degraded by the cytochrome P450 isoenzymes CYP26A1 [25], CYP26B1 [26], and CYP26C1 [27]. Thus, cells expressing CYP26 enzymes are protected from physiological RA activity. RA signaling is mediated by heterodimers of two classes of DNA-binding nuclear receptors that bind to RA response elements (*RARE*) to regulate target gene transcription: (1) the retinoic acid receptors (RARα, RARβ, and RARγ encoded by *RARa*, *RARb* and *RARg*, respectively) which bind to *all-trans* RA and (2) the retinoid X receptors (RXRα, RXRβ, and RXRγ encoded by *RXRa*, *RXRb* and *RXRg*, respectively) that bind to 9-cis-RA. In the absence of ligand, RAR/RXR dimers recruit co-repressors to inhibit transcription of RA target genes, whereas ligand-bound RAR/RXR dimers recruit co-activators to activate the same targets [2,22,24].

Previous studies have shown that cells can respond to both SHH and RA signaling, and that coordinated functions of these pathways are required for normal development. In this respect, during patterning of the spinal cord, SHH and RA exhibit complementary roles in specification of motor neuron progenitor identity [28–30]. Likewise, the SHH and RA pathways converge to influence other developmental processes, including patterning and differentiation of the forebrain, early specification of neuronal and mesodermal derivatives, and the establishment of left-right asymmetry [1,31–36]. RA and Hedgehog activities may also directly control expression of the same target genes, as exemplified by the existence of functional GLI and RAR-RXR binding sites in the *Ngn2* enhancer [37]. However, in other biological settings SHH has been shown to oppose RA activity. In the developing limb for example, SHH operates within a signaling network to promote proximal-distal growth by enhancing CYP26B1-mediated RA degradation [38]. In the human bone marrow, multiple myeloma cells modify their microenvironment to escape differentiation and reinforce chemoprotection by inhibiting RA activity in the stroma through SHH-mediated upregulation of *CYP26A1* expression [39].

Recently, we showed that in the developing tongue antagonistic activities of SHH and RA control patterning, growth and epithelial cell fate specification and that SHH inhibits RA inputs through maintenance and enhancement of *Cyp26a1* and *Cyp26c1* expression in the lingual epithelium [40].

While reviewing the literature pertaining to the RA and Hedgehog signaling pathways, we noticed that in several tissues and organs loss of Hedgehog signaling generates malformations that are strikingly similar to those engendered by supraphysiological activation of RA signaling. We therefore sought to determine whether in murine tissues known to depend on SHH for normal development, SHH antagonizes RA signaling through CYP26. To this end, we used mutant mice lacking SHH signaling and complementary experimental approaches in vitro. We found that loss of SHH signaling causes indeed loss of expression of *Cyp26* genes and enhancement of RA signaling during ontogeny of organs as disparate as craniofacial structures, genital tubercle and tail, and generates anomalies mimicking those engendered by genetically or pharmacologically induced activation of RA signaling. These findings show that in different developing organs SHH signaling uses a common strategy to antagonize RA activity. Our findings provide a concept to further the understanding of the pathogenesis of congenital malformations caused by altered Hedgehog signaling and the mechanisms underlying Hedgehog-dependent tumorigenesis.

2. Results and Discussion

To determine whether, as in the developing tongue [40], SHH signaling also impinges upon RA activity in other embryonic structures, we generated and studied *K14-CRE/Shh^{f/f}* mutant embryos, in which the *Shh* gene is disabled in Keratin-14 expressing cells and their progeny [40,41], as well as *ShhGFPCRE/Smo^{f/f}* and *ShhCreER^{T2}/Shh^f* mutant embryos, which lack the function of the *Smo* and *Shh* genes, respectively, in cells that express *Shh* and their progeny [40–43]. In the *ShhGFPCRE/Smo^{f/f}* mutants, only cells that express or have expressed SHH are unable to respond to SHH signaling. In the *ShhCreER^{T2}/Shh^f* mutants exposure to tamoxifen (TAM) abrogates SHH production, leading to loss of both autocrine and paracrine SHH signaling. Similary, in the *K14-CRE/Shh^{f/f}* mutants, both autocrine and paracrine SHH signaling are disabled. Embryos not expressing the CRE gene and/or the floxed *Smo* and *Shh* alleles were phenotypically normal; they were thus used as controls [40–42].

2.1. SHH Signaling Antagonizes RA Activity through CYP26A1 to Ensure Proper Development of the Tail

Experimental and genetic studies have demonstrated that SHH emanating from the notochord, a mesodermal midline rod-like structure, and the neural floor plate is required for survival and expansion of the sclerotomes, somite-derived structures that form the vertebral column [1,44]. Homozygous *Shh* null (*Shh^{n/n}*) mutant embryos, in which *Shh* is disabled in the germ line exhibit severe axial defects with nearly total absence of sclerotomal derivatives, including the entire vertebral column [44]. In the *Shh^{n/n}* mutants, the notochord differentiates, but is subsequently lost, indicating that autocrine SHH signaling is essential for maintenance of this important structure [44]. After fulfilling its function in patterning adjacent tissues, the notochord persists only in prospective intervertebral discs, where it develops into the *nucleus pulposus*. *ShhGFPCRE/Smo^{f/f}* and TAM-induced *ShhCreER^{T2}/Shh^f* mutants, in which abrogation of SHH signaling occurs shortly after formation of the notochord and floor plate, exhibit an abnormally thin notochord and lack intervertebral discs in the thoracic and lumbar regions. The latter anomaly is due to loss of notochordal integrity, leading to failure of development of the *nucleus pulposus* [42].

Shh^{n/n}, *ShhGFPCRE/Smo^{f/f}* and TAM-induced *ShhCreER^{T2}/Shh^f* mutants all display a severely truncated and abnormally thin tail totally lacking vertebrae [42,44] (see also Figure 1A–G). Furthermore, immunostaining for SHH and Keratin 8, molecular markers of the notochord and *nucleus pulposus* [42,45,46], showed that in contrast to control tails which exhibited a notochord, the mutants tails were devoid of this structure, except rostrally, where an abnormally thin Keratin 8-positive notochord was detectable (Figure 1H–O). Development of vertebrae is heralded by condensation of sclerotome-derived chondrogenic mesenchymal cells. These structures failed to develop in the mutant tails (Figure 1H–O), consistent with failure of development of tail vertebrae upon loss of SHH signaling [42,44].

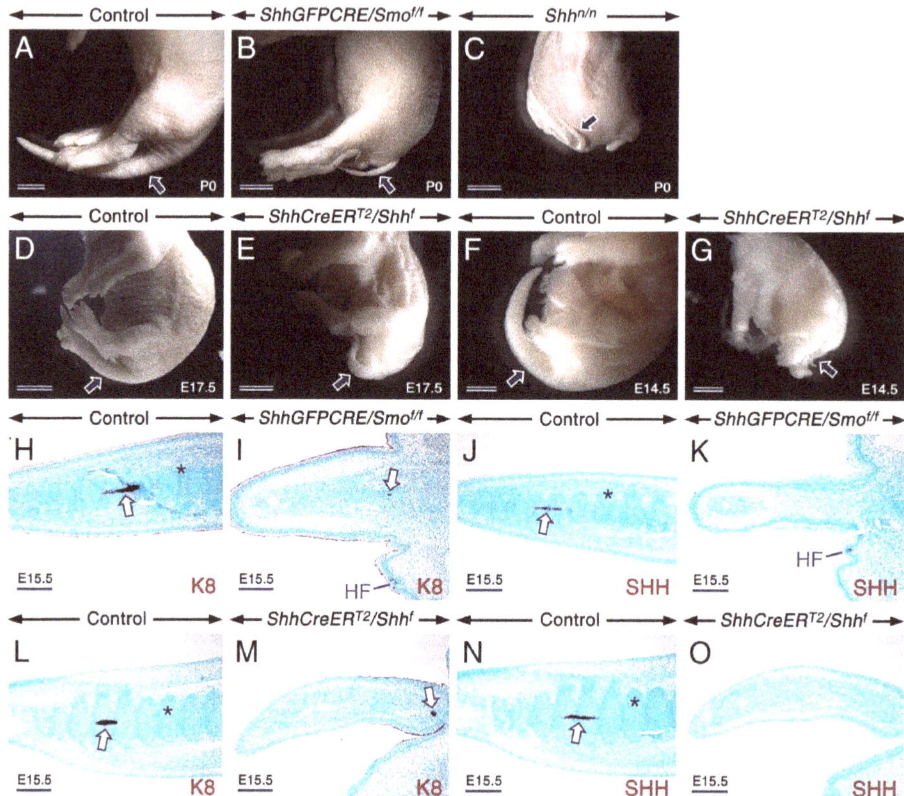

Figure 1. Loss of sonic hedgehog (SHH) signaling generates an abnormally thin and truncated tail lacking the notochord and vertebral chondrogenic condensations. (**A–G**). Representative external tail phenotype (arrows) of mutants relative to controls. Control (**A**; $n = 15$), *ShhGFPCRE/Smo^{f/f}* mutant (**B**; $n = 11$), and *Shh^{n/n}* mutant (**C**; $n = 2$) newborns (P0). E17.5 control (**D**; $n = 8$) and *ShhCreER^{T2}/Shh^f* mutant (**E**; $n = 9$) embryos first exposed to tamoxifen (TAM) at E11.5. E14.5 control (**F**; $n = 5$) and *ShhCreER^{T2}/Shh^f* mutant (**G**; $n = 6$) embryos first exposed to TAM at E10.5. The mutants exhibit severe tail defects. (**H–O**) Tail sections from E15.5 mutants and controls immunostained (dark purple) for Keratin 8 (K8) and Sonic hedgehog (SHH) to visualize the notochord. Tails from a control embryo (**H,J**) and a *ShhGFPCRE/Smo^{f/f}* embryo (**I,K**). Tails from a control embryo (**L,N**) and a *ShhCreER^{T2}/Shh^f* mutant embryo (**M,O**) first exposed to TAM at E10.5. The control tails display chondrogenic mesenchymal condensations of presumptive vertebrae (asterisks) and a notochord (arrows) in the caudal region, whereas the mutant tails lack these structures. K8-positive (arrows in **I** and **M**) remnants of the notochord are visible in the rostral region of the mutant tails. HF, hair follicle. Scale bars: 2 mm (**A–C**), 1 mm (**D–G**) and 200 μm (**H–O**).

Tail development initiates in the future lumbosacral region and coincides with the closure of the posterior neuropore. Tail tissues, including the neural tube, notochord and somites, originate from the tail bud mesenchyme, a progenitor zone located at the tip of the embryonic tail. The hindgut extends a short distance into the elongating tail after closure of the posterior neuropore [47]. The developing tail expresses components of the SHH and RA pathways. SHH is produced in the notochord and neural floor plate and elicits responses in the notochord, neuroepithelium, as well as in somites and sclerotomes [1,42]. *Aldh1a2* is expressed in presomitic and somitic mesoderm anterior to the tail bud [22,48,49], whereas *RARs* are expressed in presomitic and somitic mesoderm, sclerotomes, and tail bud [22,47,50–52]. RA signaling is tightly controlled by the activities of RALDHs and CYP26s, and loss-of-function of CYP26s during development leads to supraphysiological activation of RA signaling with entailing congenital malformations [53,54]. In the embryonic tail, *Cyp26a1* is expressed at high levels in the tail bud mesoderm, the neuroepithelium and hindgut endoderm [25,55–58].

Remarkably, the tail phenotype characterized by formation of a truncated and thin tail lacking vertebrae in the *Shh$^{n/n}$*, *ShhGFPCRE/Smo$^{f/f}$* and TAM-induced *ShhCreER^{T2}/Shh^{f}* mutants is strikingly similar to that in *Cyp26a1$^{n/n}$* mice [25,59,60] and rodent embryos exposed to teratogenic doses of vitamin A or RA [61–65]. Furthermore, exposure of hamster embryos to exogenous RA causes degeneration of the notochord and alters the formation of axial chondrogenic condensations [66], mimicking the anomalies caused by loss of SHH signaling. *Cyp26b1* and *Cyp26c1* are not expressed during the critical, SHH-dependent stages of tail formation [57,67] and embryos with loss-of-function of *Cyp26b1* [26,68] and *Cyp26c1* [27] do not exhibit tail truncation. *Cyp26b1* transcripts become detectable at later developmental stages concomitantly with the formation of chondrogenic mesenchymal condensations prefiguring vertebrae [69]. These become visible in the proximal part of the caudal region of mouse embryos at E12.5-E13 [70]. It is noteworthy that chondrogenic mesenchymal condensations express *Indian Hedgehog* [71,72]. These observations may be taken to suggest that the tail defects engendered by loss of SHH signaling are caused, at least in part, by abnormal activation of RA signaling owing to loss CYP26A1-mediated RA degradation.

To explore this possibility, we assessed the expression levels of *RARb* and *RARg*, well-established direct transcriptional targets of RA signaling [23], as well as the expression patterns of *Cyp26a1* in control and mutant tails. Reverse transcription quantitative PCR (RT-qPCR) revealed significant upregulation of *RARb* and *RARg* transcripts in tails from *ShhGFPCRE/Smo$^{f/f}$* and TAM-induced *ShhCreER^{T2}/Shh^{f}* mutants (Figure 2J,K). Furthermore, *Cyp26a1* in situ hybridization signals were either abolished or dramatically diminished in the mutant tails (Figure 2A–I).

RA activity can be visualized in tissues from mice carrying the *RAREhsplacZ* transgene [73]. Although this transgene fails to accurately reveal RA activity in several tissues and organs, including the developing tongue [10,73,76] and a large part of the palatal shelves of the secondary palate [77], it is able to visualize abnormal activation of RA signaling in the developing tail [59,60]. We thus took advantage of this possibility by examining tails from controls and *ShhGFPCRE/Smo$^{f/f}$* mutants carrying the *RAREhsplacZ* transgene and found that similar to *Cyp26a1$^{n/n}$* embryos [60] the *ShhGFPCRE/Smo$^{f/f}$* mutants exhibited expansion of *RAREhsplacZ* activity in the developing tail (Figure 2L–O), indicating ectopic activation of RA signaling. Taken together, these findings show that loss of SHH signaling in the developing tail causes a decrease of *Cyp26a1* expression and enhancement of RA signaling.

Recently, we showed that in the developing tongue, SHH activity is required for maintenance and reinforcement of *Cyp26a1* and *Cyp26c1* expression but not for the initiation of their expression [40]. This phenomenon occurs also in the developing tail, since in vitro treatment of tails with SAG, a SMO agonist enhanced the intensity of *Cyp26a1* hybridization signals in tails but failed to induce ectopic *Cyp26a1* expression in adjacent tissues (Figure 2P,Q).

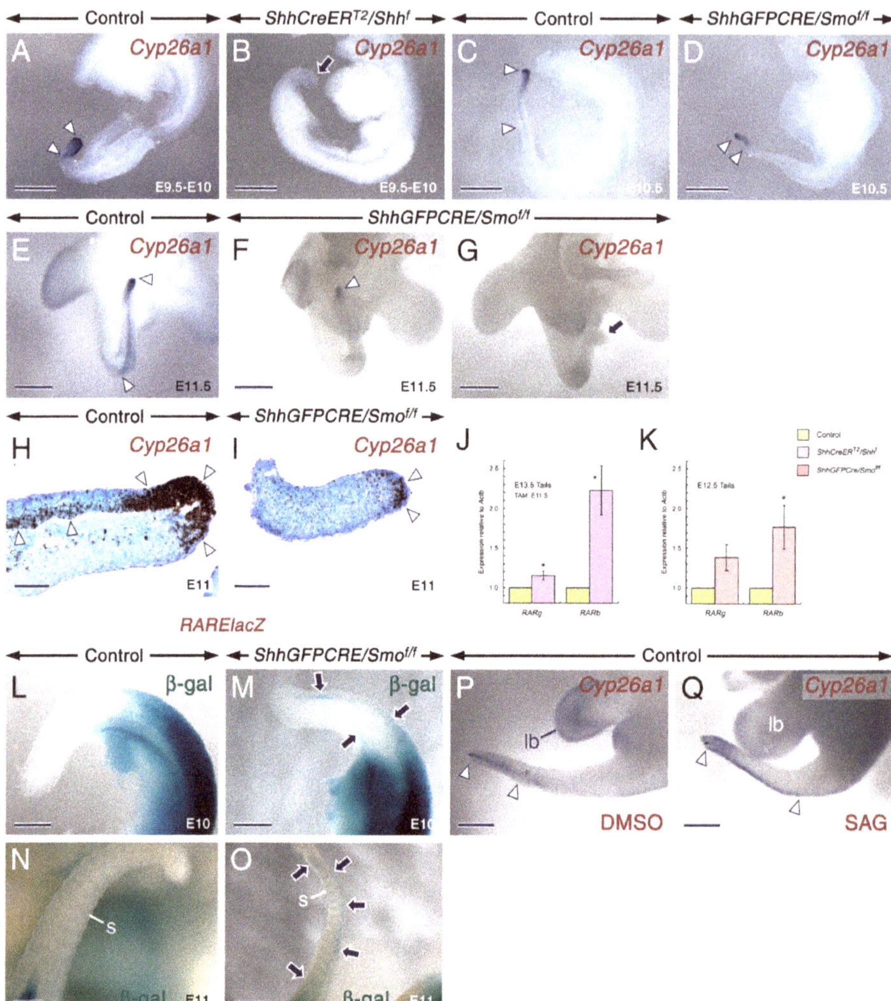

Figure 2. Loss of SHH signaling in the developing tail causes loss of *Cyp26a1* expression and ectopic activation of retinoic acid signaling. (**A–G**) Representative whole-mount in situ hybridization (ISH) with riboprobes showing *Cyp26a1* expression (purple) in developing tails. E9.5-E10 control (**A**; $n = 3$) and *ShhCreER^{T2}/Shh^f* mutant (**B**; $n = 4$) embryos first exposed to tamoxifen (TAM) at E8-E8.5. Control (**C,E**) and *ShhGFPCRE/Smo^{f/f}* mutant (**D,F,G**) embryos at E10.5 (**C,D**; $n = 4$ controls and $n = 4$ mutants) and E11.5 (**E–G**; $n = 4$ controls and $n = 3$ mutants). In the control tails, the *Cyp26a1* expression domain extends from the tail bud to more rostral levels of the tail (arrowheads in **A,C** and **E**). The mutant tails exhibit either a severely reduced domain of *Cyp26a1* expression (arrowheads in **D** and **F**) or abolished *Cyp26a1* expression (arrows in **B** and **G**). (**H,I**) Representative tail sections from E11 control embryos (**H**; $n = 2$) and a *ShhGFPCRE/Smo^{ff}* mutant embryo (**I**) after ISH for *Cyp26a1* with oligonucleotide probes (black). Decreased *Cyp26a* hybridization signals in the mutant tail as compared to the control tail (arrowheads in **H** and **I**). (**J,K**) RT-qPCR analysis showing the expression levels of *RARb* and *RARg* relative to *Actb* (β-actin). Upregulation of *RARb* ($p = 0.0162$) and *RARg* ($p = 0.0261$) levels in tails from E13.5 *ShhCreER^{T2}/Shh^f* mutant ($n = 3$ and $n = 4$ for *RARb* and *RARg* analyses, respectively) as compared to tails from control ($n = 3$ and $n = 4$ for *RARb* and *RARb* analyses, respectively) embryos first exposed to TAM at E11.5 (**J**). Upregulation of *RARb* ($p = 0.0476$) and *RARg* ($p = 0.0610$) levels in tails from E12.5

ShhGFPCRE/Smo^{f/f} mutants (*n* = 3 and *n* = 4 for *RARb* and *RARg* analyses, respectively) as compared to tails from controls (*n* = 3 and *n* = 4 for *RARb* and *RARg* analyses, respectively) (**K**). Data are mean values ± standard deviation; *: $p < 0.05$. (**L–O**) Representative β-galactosidase (β-gal) histochemistry visualizing retinoic acid activity (blue) in control (**L,N**) and *ShhGFPCRE/Smo^{f/f}* mutant (**M,O**) embryos carrying the *RAREhsplacZ* transgene (*RARElacZ*) at E10 (**L,M**; *n* = 3 controls and *n* = 3 mutants) and at E11 (**N,O**; *n* = 7 controls and *n* = 3 mutants). The mutants exhibit ectopic retinoic acid activity (arrows in **M** and **O**) in tail tissues. s, somite. (**P,Q**) Representative tail explants from E11.5 control embryos treated for 24 h with DMSO (**P**; *n* = 5) and 0.2 μM SAG (**Q**; *n* = 4) showing expansion of *Cyp26a1* expression domain (arrowheads in **P** and **Q**) and increased *Cyp26a1* hybridization signals in the SAG-treated tail and failure of SAG to induce ectopic *Cyp26a1* expression in adjacent structures, including the hindlimb bud (lb). Scale bars: 300 μm (**A–G,L–Q**) and 100 μm (**H,I**).

To determine whether increased RA signaling is indeed involved in the genesis of tail anomalies upon loss of SHH signaling, we cultured tails from TAM-treated *ShhCreER^{T2}/Shh^f* mutant and control embryos (Figure 3A) in the presence of BMS493, a RA signaling inhibitor, or DMSO (control vehicle). Compared to tails from control embryos, the DMSO-treated mutant tails exhibited an abnormally thin notochord in the rostral region and were devoid of notochord in the posterior region (Figure 3B–3D'). However, the BMS493-treated mutant tails exhibited an intact notochord (Figure 3E,E'), indicating that degeneration of the notochord was prevented upon inhibition of RA signaling. These data suggest that RA signaling participates in the degeneration of the caudal notochord upon loss of SHH signaling. However, compared to tails from control embryos treated with DMSO or BMS493 (Figure 3B–C'), the BMS493-treated mutant tails failed to show chondrogenic mesenchymal condensations flanking the notochord (Figure 3D,D'), indicating that the inhibition of RA signaling only partially rescued the mutant tails. This finding was not surprising, as survival and expansion of sclerotomal cells, which form axial chondrogenic condensations, are SHH-dependent [1,44].

RA activity is required for apoptosis mediated removal of the interdigital mesenchyme [78], and genetic or teratogenic overactivation of RA signaling is known to induce apoptosis in developing organs, including the testes, limb mesenchyme, chondrogenic mesenchymal condensations [54], and the developing tail [65,79]. Loss of SHH signaling in the developing tail causes enhanced apoptosis [42]. Accordingly, the TAM-induced *ShhCreER^{T2}/Shh^f* mutant tails treated with DMSO exhibited increased numbers of apoptotic cells, as compared to the DMSO-treated tails from control embryos (Figure 3F,H,J). We also found that BMS493 treatment significantly reduced the number of apoptotic cells in the mutant tails (Figure 3H–J). These findings strongly suggest that enhanced apoptosis in the mutant tails is at least partly caused by ectopic activation of RA signaling.

Altogether, our data reveal a hitherto unknown mechanism behind abnormal tail development upon loss of SHH signaling and strongly suggest involvement of ectopic RA activation in the genesis of this anomaly. The fact that loss of SHH signaling [42,44] (this study) and ectopic activation of RA signaling [25,59–65] during tail development generates strikingly similar tail defects further supports our conclusion.

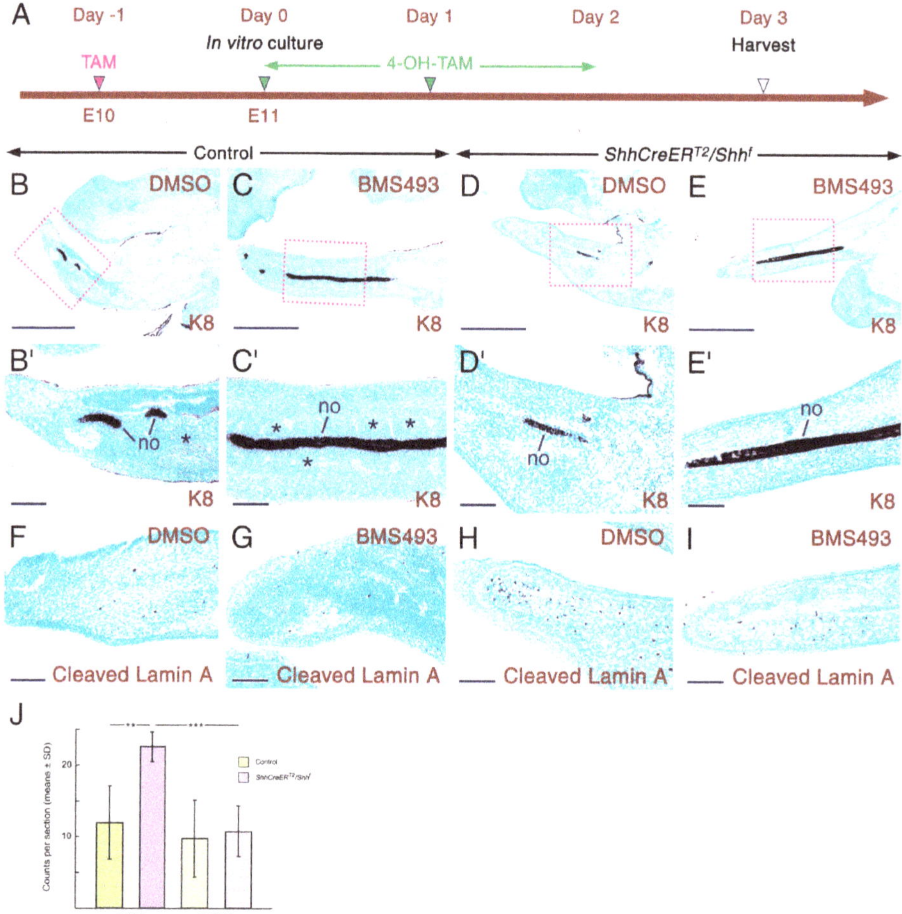

Figure 3. In vitro inhibition of retinoic acid signaling partially rescues the tail phenotype of SHH-deficient embryos. (**A**) Timeline representing the induction of CRE-mediated deactivation of *Shh* in embryos and in tail explants. The tails are from E11 control and *ShhCreER^T2/Shh^f* mutant embryos first exposed in utero to tamoxifen (TAM) at E10 (red arrowhead). All tail explants were cultivated in vitro for two days in the presence of 4-hydroxytamoxifen (4-OH-TAM; green arrowheads). During the in vitro cultivation period (three days), the tails were treated with DMSO or 12.5 μM BMS493. The time of harvest of the explants is indicated by a white arrowhead. (**B–E**) Representative Keratin 8 (K8; dark purple) immunostaining visualizing the notochord (no) in sections of tail explants from control and mutant embryos. The tails were treated with DMSO (n = 5 controls and n = 6 mutants) or BMS493 (n = 13 controls and n = 8 mutants). **B′–E′** are magnified images of the boxed areas in **B–E**. All the control tails treated with DMSO (**B,B′**) or BMS493 (**C,C′**) exhibit a notochord and chondrogenic mesenchymal condensations (asterisks in **B′** and **C′**). All the DMSO-treated mutant tails lack a notochord in the posterior region, while in the rostral region they display an abnormally thin notochord (**D,D′**). The BMS493-treated mutant tails (**E,E′**) display a notochord (n = 6/8), but fail to exhibit chondrogenic mesenchymal condensations (n = 8/8). (**F–I**) Representative sections of tail explants from control and mutant embryos were immunostained for cleaved Lamin A (dark purple) to visualize apoptotic cells. Massive apoptosis in the DMSO-treated mutant tails (**H**; n = 6) as compared to the BMS493-treated mutant tails (**I**; n = 6) and the DMSO-treated (**F**; n = 3) and BMS493-treated (**G**; n = 7) control tails. (**J**) Quantitation of apoptosis in tail explants (the number of explants assessed is

described above). The number of apoptotic cells in the DMSO-treated mutant tails is significantly higher than in the DMSO-treated ($p < 0.005$) and BMS493-treated ($p = 0.002$) control tails. The BMS493-treated mutant tails show a significant decrease in apoptosis, as compared to the DMSO-treated mutant tails ($p < 0.001$). BMS493 had no effects on the extent of apoptosis in the control tails ($p = 0.59$). Data are mean values ± standard deviation; **: $p < 0.01$; ***: $p < 0.001$. Scale bars: 500 μm (**B–E**) and 100 μm (**B′–I**).

2.2. SHH Signaling in the Developing Secondary Palate Is Required to Prevent Enhancement of RA Activity

Development of the secondary palate depends on complex spatio-temporal cellular and molecular events, and genetic mutations and/or environmental factors that alter these events cause cleft palate, the most common congenital malformation in humans, with debilitating consequences [80–84]. In mice, development of the secondary palate begins at E11.5. First, bilateral paired palatal shelves (PS) arise from the oral side of the maxillary processes and grow downwards while flanking the developing tongue (E11.5-E14.5). Thereafter, the PS elevate to a horizontal position (E14.5-E15) above the tongue, and further growth enables the opposing PS to adhere to each other and form a median epithelial seam which eventually disappears, allowing fusion of the PS [80–84]. The developing palate exhibits molecular and histological heterogeneity along its anterior-posterior and oral-nasal axes. Along the oral-nasal axis, this heterogeneity is translated into formation of ciliated respiratory epithelium that differentiates on the nasal side of the PS, and development of periodic epithelial ridges known as *rugae palatinae* in the oral surface of the PS [80,84].

SHH signaling plays a crucial role during PS growth and patterning of *rugae palatinae* through regulation of the expression of signaling molecules and transcription factors, and loss of SHH signaling causes cleft palate [6,80,84–86] and mispatterning of palatal *rugae* [86]. During palatogenesis, SHH is produced by the PS epithelium and signals within the PS epithelium and to the PS mesenchyme. At E11.5 *Shh* is expressed in the entire epithelium of the emerging PS, and from E12 onwards *Shh* expression is restricted to the developing *rugae palatinae* [86–91].

In the developing palate, *RARs* are expressed in both the epithelium and mesenchyme [92], whereas, *Cyp26a1* [56] and *Cyp26b1* [69,77] are expressed in the epithelium and mesenchyme, respectively. Interestingly, *Cyp26a1* expression is restricted to the oral epithelium of the PS [56], overlapping with *Shh* expression [86–91]. Exposure of rodent embryos to excess RA or vitamin A causes cleft palate [61–64,93], and *Cyp26b1*$^{n/n}$ mice exhibit cleft palate [68,77] due to failure of elevation of PS [77].

To determine whether ablation of SHH signaling causes enhancement of RA signaling during palate development, we generated *ShhCreER*^{T2}/*Shh*^{f} mutant and control embryos that had been first exposed to TAM at E10.5-E11 (Figure S1). We found that all the E10.5-E11 TAM-induced *ShhCreER*^{T2}/*Shh*^{f} mutants assessed displayed cleft palate, a defect that was not observed in control embryos (Figure S2A–F). RT-qPCR revealed that *RARg* and *RARb* expression levels were significantly enhanced in the mutant PS (Figure 4M), indicating enhanced RA activity.

To explore putative sources of RA in the developing palate, we assessed the expression patterns of RALDH1-3 proteins in TAM-treated control and *ShhCreER*^{T2}/*Shh*^{f} mutant embryos. We found that all three RALDHs were expressed in the developing palate at E13.5 and that their expression patterns were not altered in the mutant palate (Figure S3). These findings show that RA synthesis occurs in the developing palate and that enhanced RA signaling in the *ShhCreER*^{T2}/*Shh*^{f} mutant palate is not caused by increased RA synthesis. Interestingly, *Cyp26b1* loss-of-function generates cleft palate owing to enhanced RA signaling and abnormal mesenchymal proliferation in the bend region of the PS [77], a site that we found to be enriched in RALDH1-3 expression (Figure S3).

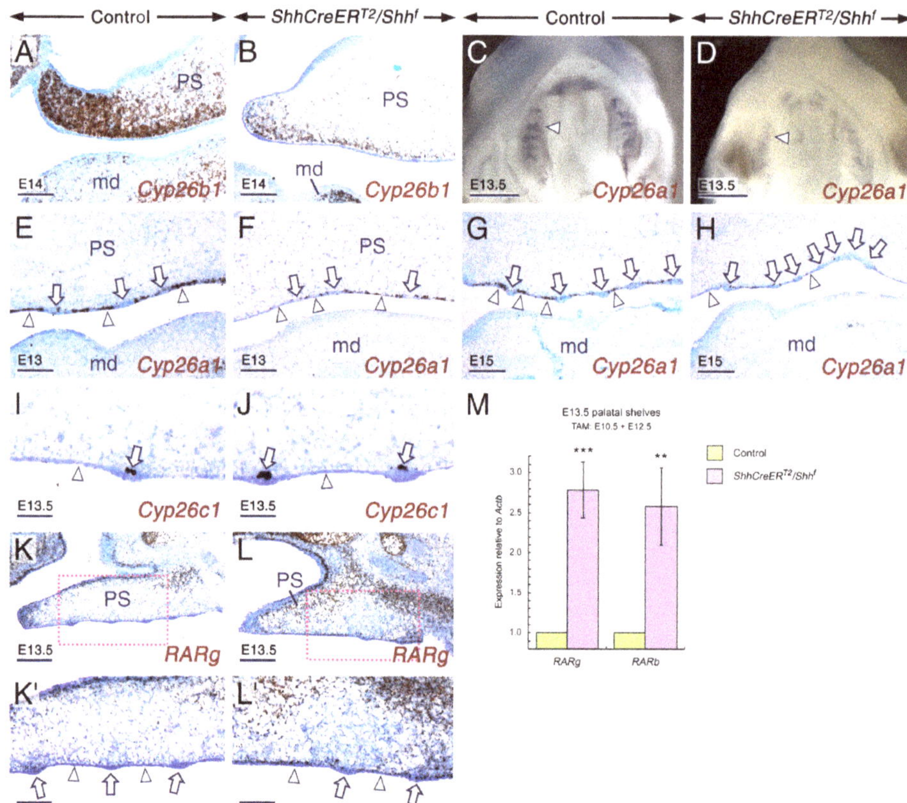

Figure 4. SHH signaling in the developing secondary palate is required for expression of *Cyp26a1* and *Cyp26b1* to prevent enhancement of retinoic acid signaling. (**A–L**) Representative developing palates from control and *ShhCreER^T2/Shh^f* mutant embryos first exposed to tamoxifen (TAM) at E10-5-E11. The developmental stages are indicated on the panels. Whole-mount in situ hybridization (WMISH) with Dig-labelled riboprobes (**C,D**) and in situ hybridization in parasagittal sections (anterior palatal region towards the left of the panels) with oligonucleotide probes (**A,B,E–L**). The inter-rugal epithelium and *rugae palatinae* are indicated by arrowheads and arrows, respectively. (**A,B**) *Cyp26b1* expression in sections of palates (see also Figure S4) from control (**A**, *n* = 2) and mutant (**B**; *n* = 2) embryos. The mutant palate shows decreased *Cyp26b1* hybridization signals (brown) as compared to the control palate. (**C–H**) The mutant palates (**D,F,H**; *n* = 3 for WMISH and *n* = 4 for ISH in sections) show decreased *Cyp26a1* hybridization signals (dark purple in whole-mounts and black in sections) as compared to control palates (**C,E,G**; *n* = 3 for WMISH and *n* = 4 for ISH in sections). In control palates *Cyp26a1* transcripts are enriched in the inter-rugal epithelium. (**I,J**) *Cyp26c1* (black) is expressed in subsets of cells within the basal layer of *rugae palatinae* (arrows in **I** and **J**) in control (**I**; *n* = 2) and mutant (**J**; *n* = 2) palates. (**K,L**) The mutant palate (**L**; *n* = 3) shows increased *RARg* hybridization signals (brown) in the mesenchyme and inter-rugal epithelium as compared to the control palate (**K**; *n* = 3). **K′** and **L′** are magnified views of the boxed areas in **K** and **L**, respectively. (**M**) RT-qPCR assay for *RARb* and *RARg* relative to *Actb* (β-actin) in paired palatal shelves from E13.5 controls (*n* = 7) and *ShhCreER^T2/Shh^f* mutants (*n* = 7) first exposed to TAM at E10.5 showing upregulation of *RARb* (*p* = 0.004) and *RARg* (*p* = 0.000) in the mutant palatal shelves as compared to the control palatal shelves. Data are mean values ± standard deviation; **: *p* < 0.01; ***: *p* < 0.001. md, mandible; PS, palatal shelf. Scale bars: 500 μm (**C,D**), 200 μm (**K,L**), 100 μm (**A,B,E–H,K′,L′**) and 50 μm (**I,J**).

To determine whether increased RA signaling in the *ShhCreER^T2^/Shh^f* mutant palate is due to decreased CYP26-mediated RA catabolism, we assessed the expression patterns of *Cyp26a1*, *Cyp26b1*, and *Cyp26c1* transcripts in the palate of TAM-treated control and *ShhCreER^T2^/Shh^f* mutant embryos. In both control and mutant embryos, *Cyp26b1* displayed a gradient of hybridization signals along the anterior-posterior axis of the PS, with highest and lowest intensities seen anteriorly and posteriorly, respectively (Figure 4A,B and Figure S4). However, compared to control PS, the *ShhCreER^T2^/Shh^f* mutant PS showed diminished *Cyp26b1* hybridization signals (Figure 4 and Figure S4). Furthermore, *Cyp26a1* hybridization signals were diminished in the epithelium of the mutant PS (Figure 4C–H). By contrast, *Cyp26c1* expression, which we found to be restricted to subsets of cells in *rugae palatinae*, was unaltered in the mutant PS (Figure 4I,J). Notably, in the control PS, *Cyp26a1* transcripts were enriched in the inter-rugal epithelium (Figure 4C,E,G). Consistent with *RARg* RT-qPCR analysis (Figure 4M), the mutant PS displayed increased *RARg* hybridization signals in the palatal mesenchyme and in the inter-rugal epithelium (Figure 4K–L'). Thus, loss of SHH signaling in the developing palate leads to enhanced RA signaling in both the palatal epithelium and palatal mesenchyme as a result of loss of *Cyp26a1* and *Cyp26b1* expression.

Abrogation of SHH signaling in *K14-Cre/Shh^f/f* mutant mice causes mispatterning of *rugae palatinae* manifested as furcations, fusions, and formation of supernumerary *rugae* [86] similar to those observed in rat embryos exposed to excess RA [94]. To determine whether, like the *K14-Cre/Shh^f/f* mutants, the TAM-induced *ShhCreER^T2^/Shh^f* embryos exhibit mispatterning of *rugae palatinae*, we immunostained sections of control and *ShhCreER^T2^/Shh^f* mutant palates for FOXA1 whose encoding gene is expressed in *rugae palatinae* [95] (see also Figure 5A). In the oral epithelium of control palates, FOXA1 was expressed in the periderm and *rugae palatinae* (Figure 5B,B'). However, in the *ShhCreER^T2^/Shh^f* mutant palates, FOXA1 expression was expanded (Figure 5C,C'), indicating development of supernumerary *rugae*. Taken together, these findings show that loss of SHH signaling during palatogenesis leads to enhanced RA signaling and suggest involvement of enhanced RA signaling in the genesis of cleft palate and mispatterning of *rugae palatinae* upon loss of SHH inputs in the developing palate.

Our study revealed a new function for SHH signaling during growth of the PS, which is to keep RA activity in check in both the palatal epithelium and palatal mesenchyme. It is possible that elevated RA availability in the palatal epithelium of the TAM-induced *ShhCreER^T2^/Shh^f* mutant embryos (as a result of diminished RA degradation by CYP26A1) not only causes enhanced RA signaling within the palatal epithelium, but also contributes in enhancing RA signaling in the palatal mesenchyme, since RA is a highly potent and diffusible small molecule. Vice versa, in the TAM-induced *ShhCreER^T2^/Shh^f* mutant palates, RA overproduced in the palatal mesenchyme (as a result of diminished RA degradation by CYP26B1) may also contribute to enhancement of RA signaling in the palatal epithelium.

Nature is replete with repeating, regularly spaced structures such as *rugae palatinae*, feather and hair follicles, lingual fungiform papillae, and tracheal cartilage rings. We have shown recently that antagonistic SHH and RA activities are involved in patterning of the lingual epithelium, whereby SHH inhibits while RA promotes the formation of taste placodes and lingual glands [40]. Patterning of *rugae palatinae* has been shown to involve Turing-based mechanisms, where FGF and SHH function as activator and inhibitor, respectively [86]. However, how SHH inhibits *rugae* formation is unknown. In addition, besides the SHH-FGF signaling pair, other signaling pathways have been incriminated in patterning of *rugae palatinae* [86]. In the present study, we confirmed that SHH inhibits the formation of *rugae palatinae*, since in the TAM-induced *ShhCreER^T2^/Shh^f* mutants the palate forms supernumerary *rugae*. Furthermore, our findings suggest that RA signaling is involved in patterning of *rugae palatinae*. Several lines of evidence support this notion: (1) components of the RA signaling pathway are expressed in developing *rugae palatinae* and in the inter-rugal epithelium; (2) *Cyp261a1* expression is enriched in the inter-rugal epithelium; (3) in the inter-rugal epithelium of the *ShhCreER^T2^/Shh^f* mutant palates *Cyp26a1* expression is severely diminished and *RARg* expression is enhanced; and (4) loss of SHH signaling causes abnormal patterning of *rugae palatinae* [86], this study similar to that engendered by exposure of the developing palate to excess RA [94]. These findings suggest that RA signaling promotes the formation of *rugae palatinae* and that one mechanism by which SHH inhibits *rugae* formation is through the attenuation of RA signaling.

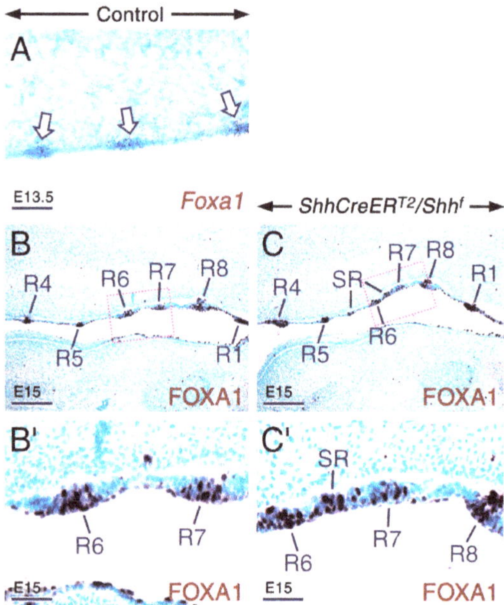

Figure 5. Loss of SHH signaling in the *ShhCreER^T2/Shh^f* mutant palate causes mispatterning of *rugae palatinae* (**A**) Bright-field view of a frontal section across the palate of an E13.5 control embryo after in situ hybridization with a ^{35}S-UTP-labelled *Foxa1* riboprobe showing *Foxa1* expression (black dots) in *rugae palatinae* (arrows). (**B,C**) Representative FOXA1 immunostaining (dark purple) of para-sagittal sections of palates (anterior palatal region towards the left of the panels) from E15 control (**B**; *n* = 3) and *ShhCreER^T2/Shh^f* mutant (**C**; *n* = 3) embryos first exposed to tamoxifen at E10.5-E11. **B′** and **C′** are magnified views of the boxed areas in **B** and **C**, respectively. In the control palate FOXA1 is detected in the palatal periderm and in a subset of cells of *rugae palatinae*. The orthotopic *rugae* (R) are labelled with arabic numerals according to the order of their formation as described previously [91]. In the mutant palate supernumerary *rugae* (SR) develop between *rugae* R5 and R6 and between *rugae* R6 and R7 (**C,D**). Scale bars: 200 μm (**B,C**) and 50 μm (**A,B′,C′**).

2.3. SHH Signaling Is Required for Cyp26 Expression in Other Developing Structures

To explore whether *Cyp26* expression requires SHH inputs in other SHH-dependent developing structures, such as the genital tubercle and embryonic teeth known to express factors involved in RA signaling, we analyzed these organs in SHH-deficient and control embryos.

The genital tubercle (GT), primordium of the penis and clitoris, consists of a mesenchyme covered by ectoderm and a ventral midline structure, the urethral plate epithelium. The urethral plate epithelium derives from the endoderm of the cloaca and generates the entire penile urethra [96]. Previous work established a crucial role for SHH signaling for normal development of the genitourinary system, including proximal-distal outgrowth of the GT and formation of the urethral tube [1,97–100]. *Shh* expression begins in the cloacal membrane before the onset of GT development, and during GT outgrowth SHH is produced by the urethral plate epithelium and signals to the mesenchyme and ventral ectoderm of the tubercle [96–98]. Loss of SHH signaling in the developing GT generates various anomalies, including developmental arrest, hypoplasia due to stunted proximal-distal outgrowth, and/or hypospadias [97–100].

Strikingly, rodent embryos exposed to teratogenic doses of vitamin A or RA exhibit GT anomalies [61], mimicking those caused by loss of SHH signaling [97–100], including GT agenesis and stunted outgrowth of the GT. Components of the RA signaling cascade are expressed before and during outgrowth of the GT. *Aldh1a2* is expressed in the cloacal membrane and urethral plate epithelium [48,49,101], all three *RARs* are expressed in the urethral plate epithelium and in the mesenchyme of the GT [50,51,102,103], and *Cyp26b1* is expressed in the GT mesenchyme [69]. Furthermore, RA activity is readily detectable in the urethral plate epithelium and proximal GT mesenchyme (Figure 6F).

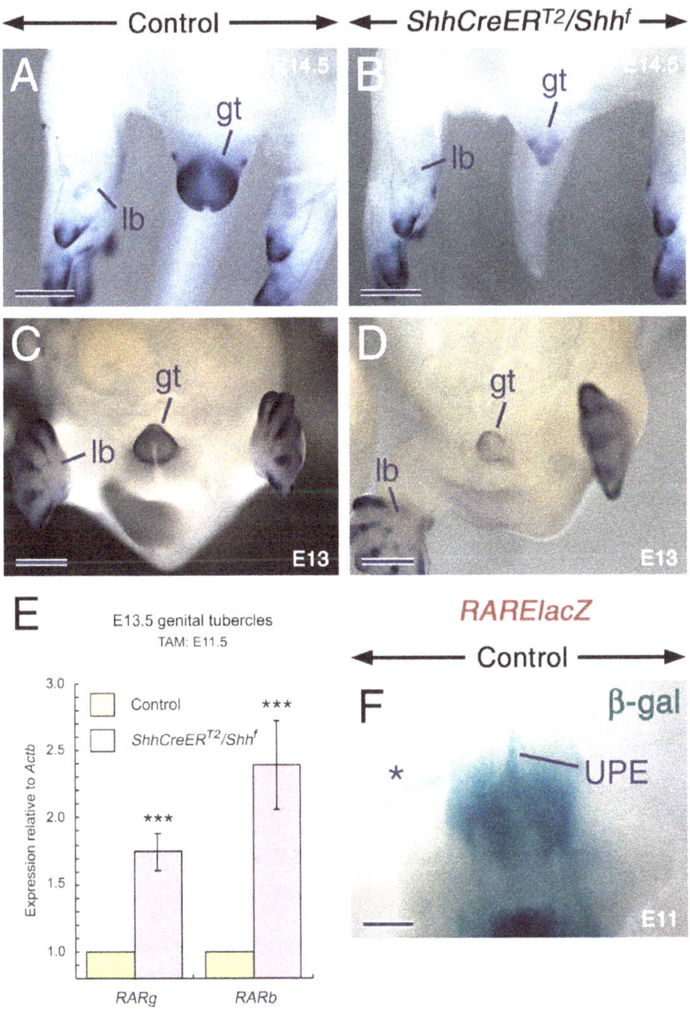

Figure 6. Loss of SHH signaling in the developing genital tubercle causes downregulation of *Cyp26b1* expression and enhancement of retinoic acid signaling. (**A–D**) Representative *Cyp26b1* whole-mount in situ hybridization with riboprobes (purple). E14.5 control (**A**; *n* = 2) and *ShhCreER^T2/Shh^f* mutant (**B**; *n* = 2) embryos first exposed to tamoxifen (TAM) at E12. E13 control (**C**; *n* = 2) and *ShhCreER^T2/Shh^f* mutant (**D**; *n* = 2) embryos first exposed to TAM at E11.5. Diminished *Cyp26b1* hybridization signals in the genital tubercle (gt) of the mutants. Note that *Cyp26b1* signals are not altered in chondrogenic condensations within limb buds (lb) of the mutant as these cellular condensations do not express *Shh*.

(**E**) RT-qPCR analysis for *RARb* and *RARg* relative to *Actb* (β-actin) in genital tubercles from E13.5 control and *ShhCreER^{T2}/Shh^f* mutant embryos first exposed to TAM at E11.5. Upregulation of *RARb* ($p = 0.0009$) and *RARg* ($p = 0.0002$) in the mutant ($n = 8$ and $n = 7$ for *RARb* and *RARg*, respectively) as compared to the control ($n = 8$ and $n = 7$ for *RARb* and *RARg*, respectively) genital tubercles. Data are mean values ± standard deviation; ***: $p < 0.001$. (**F**) β-galactosidase (β-gal) histochemistry revealing retinoic acid activity in the genital tubercle of control embryos carrying the *RAREhsplacZ* transgene ($n = 7$). Asterisk in **F** indicates an artefact due to tissue detachment. UPE, urethral plate epithelium. Scale bars: 300 μm (**F**) and 500 μm (**A–D**).

To determine whether *Cyp26b1* expression in the GT requires SHH signaling, we compared the expression of *Cyp26b1* in TAM-treated control and *ShhCreER^{T2}/Shh^f* mutant embryos and found down-regulation of *Cyp26b1* expression in the GT of the mutants (Figure 6A–D). As previously described [97–99], the mutants exhibited hypoplastic GT (Figure 6A–D). Furthermore, RA signaling was enhanced in the mutant GT as shown by significant enhancement of *RARb* and *RARg* expression levels (Figure 6E). Thus, like in the limb bud [38], in the developing GT SHH inputs are required for modulating RA activity through maintenance of proper levels of *Cyp26b1* expression.

Cyp26b1^{n/n} embryos display enhanced RA signaling in the GT and exhibit a range of anomalies of external genitalia, including enlarged width due to increased proliferation of the GT mesenchyme [101]. Yet, unlike mouse embryos deficient in SHH signaling [97–100] (this study), the *Cyp26b1^{n/n}* mutants have intact proximal-distal outgrowth of the GT [101]. The major function of SHH in the GT mesenchyme is to maintain proper rates of mesenchymal cell proliferation required for proximal-distal outgrowth [96]. The lack of abnormal proximal-distal outgrowth of the *Cyp26b1^{n/n}* mutant GT is likely due to that *Cyp26b1* ablation occurs in the presence of a functional *Shh* gene, a condition that differs from that of the TAM-induced *ShhCreER^{T2}/Shh^f* mutant GT, in which diminished *Cyp26b1* expression occurs in the absence of SHH inputs. In fact, in *Cyp26b1^{n/n}* embryos *Shh* expression in the urethral epithelium and SHH signaling in the GT mesenchyme were found to be upregulated, as a result of increased RA signaling [101]. The phenotype of the GT in the *Cyp26b1^{n/n}* mutants is also different from that of rodent embryos exposed to teratogenic doses of vitamin A or RA [61], as in the latter the GT fails to form or is truncated, mimicking the anomalies induced by loss of SHH signaling. A likely explanation for these differences is that in the *Cyp26b1^{n/n}* mutants, the GT is exposed to RA emanating from endogenous sources, leading to upregulation of *Shh* expression in this organ [101]. By contrast, in embryos exposed to excess exogenous retinoids the GT is exposed to overwhelming levels of RA. Since teratogenic levels of RA are known to abolish *Shh* expression [32,104], it is possible that in the GT of embryos overexposed to exogenous retinoids, SHH signaling is reduced or lost. Thus, in these embryos, combined loss of SHH signaling and enhanced RA signaling may lead to conditions resembling those that occur upon the genetic loss of SHH signaling in the GT. Taken together, these observations suggest that RA bio-availability must be precisely controlled to ensure normal development of the GT.

Mouse models revealed the importance of SHH signaling during odontogenesis. Loss of SHH signaling in developing teeth of *K14-Cre/Shh^{n/f}*, *K14-Cre/Smo^{n/f}* and *Evc^{n/n}* mutant embryos generates tooth anomalies, including abnormally small and misshapen teeth with enamel defects [41,105,106] and failure of differentiation of enamel-producing ameloblasts, epithelial cells that differentiate from the inner dental epithelium [41,105]. SHH, is produced by the dental epithelium and signals within the dental epithelium and to the dental mesenchyme [41,105]. Developing teeth express genes encoding components of the RA pathway, including RALDHs [48,107], CYP26A1 [69], CYP26C1 [67], and RARs [108]. Remarkably, exposure of mice to excess RA generates enamel defects and abnormal ameloblast differentiation [109], and in vitro exposure of embryonic mouse teeth to supraphysiological levels of RA leads to formation of misshapen teeth [110]. However, physiological levels of RA seem to be required for normal tooth formation, since vitamin A deficiency in rats causes a range of defects, including enamel hypoplasia, abnormal dentine formation, and metaplasia of dental epithelia [111–113].

We found that developing teeth from *K14-CRE/Shh^{f/f}* mutant mice exhibit severe downregulation of *Cyp26a1* and *Cyp26c1* expression levels in the inner dental epithelium (Figure 7). However, compared to control teeth that showed an absence of *Cyp26b1* expression in the dental papilla mesenchyme (Figure 7C,C'), consistent with previous findings [69], the mutant teeth exhibited ectopic *Cyp26b1* hybridization signals in this tissue (Figure 7D,D'). By contrast, *Cyp26b1* expression in osteoblast progenitors, including in the developing alveolar bone at the periphery of developing teeth [69], was as expected unaltered in the mutants (Figure 7C,D) as these cells do not express Keratin 14 and SHH.

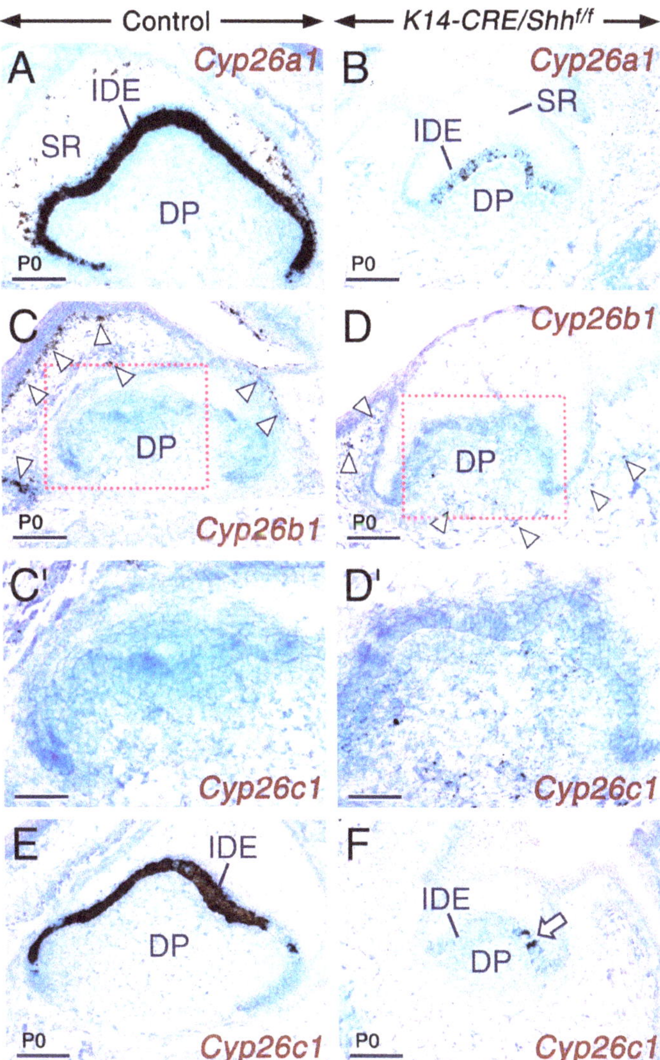

Figure 7. SHH signaling is required for maintenance of the expression of *Cyp26a1* and *Cyp26c1* in the developing tooth. (**A–F**). Representative *Cyp26a1* (**A,B**; *n* = 2 controls and *n* = 2 mutants), *Cyp26b1* (**C,D**; one control and one mutant) and *Cyp26c1* (**E,F**; *n* = 2 controls and *n* = 2 mutants) in situ hybridization (black) with oligonucleotide probes in frontal sections across developing first molars from control (**A,C,E**) and *K14-CRE/Shh^{f/f}* mutant (**B,D,F**) newborn (P0) mice. **C'** and **D'** are magnified views of the boxed areas in **C** and **D**, respectively. The mutant molars show severely diminished *Cyp26a1* hybridization signals in

the inner dental epithelium (IDE) and abolished *Cyp26a1* expression in cells of the stellate reticulum (SR). Note also the severely reduced domain of *Cyp26c1* expression in the IDE of the mutant tooth (arrow in **F**). The mutant molars (**D**) exhibit ectopic expression of *Cyp26b1* in the dental papilla mesenchyme (DP). *Cyp26b1* expression in cells of the developing alveolar bone (arrowheads in **C** and **D**) is unaltered in the mutant. Scale bars: 100 μm (**A–D,E,F**) and 50 μm (**C′,D′**).

The dependence of *Cyp26b1* expression on SHH signaling seems to be context-dependent. While loss of SHH signaling in the developing limb [38], genital tubercle, and palate (this study) leads to diminished levels of *Cyp26b1* expression, *Cyp26b1* transcript levels are enhanced in tongue mesenchyme of SHH-deficient embryos [40] (see also Figure S4), and *Cyp26b1* is expressed ectopically in the dental papilla mesenchyme of *K14-CRE/Shh^{f/f}* mutant molars.

2.4. Conclusions

Previous studies in the embryonic limb [38] and tongue [40] together with our present findings in several developing structures showed that SHH signaling abates RA signaling through the maintenance/reinforcement of *Cyp26* expression. Thus, during development of various organs, SHH uses this same strategy to antagonize RA signaling. Furthermore, loss of SHH signaling in the developing tongue [40], tail, secondary palate, genital tubercle, and tooth (this study) causes these structures to develop defects that are remarkably akin to those engendered by genetically or pharmacologically induced overactivation of RA signaling. A literature search revealed strikingly similar congenital anomalies caused by deregulation of Hedgehog and RA signaling (Table S1), suggesting that antagonism between the two pathways may be a common phenomenon.

It is unlikely that SHH directly induces the initial expression of *Cyp26* transcripts, since experiments showed that the Smoothened agonist SAG reinforces *Cyp26a1* expression in *Cyp26a1*-expressing tissues but fails to induce *de novo* expression of this gene in *Cyp26a1*-non-expressing tissues [40] (this study). Which factor(s) whose activities are modulated by SHH signaling directly regulate *Cyp26* expression, and thus RA activity, remain to be identified.

Delineating how cell signaling cascades interact to control tissue patterning, cell fate specification and organogenesis is key to understanding the etiopathogenesis of congenital malformations and malignancies, and knowledge of developmental pathway interactions constitutes a basis for regenerative medicine. Our findings in the developing tail provide a probable, mechanistic explanation for the tail anomalies engendered by loss of SHH signaling, that is, involvement of aberrant enhancement of RA signaling in the genesis of these malformations.

Human embryos develop a tail bud and a transient tail, the latter being normally fated to regress [114,115] through apoptosis [116]. A congenital midline malformation known as "human tail" [114,117–119] has been suggested to result from failure of regression of the embryonic tail [115]. Currently, the etiology of "human tail" and underlying molecular mechanisms leading to this anomaly are unknown. Our findings in mouse embryos not only provide insights into the interplay between signaling pathways in the control of development of the caudal region of the embryo, but also provide valuable information for future work aiming at deciphering the etiopathogenesis of "human tail".

Besides their role in tissue patterning, the Hedgehog and RA pathways play crucial roles during organogenesis and postnatal tissue homeostasis. Compelling evidence suggests that Hedgehog signaling promotes cell proliferation and cell survival [1,19], whereas RA inhibits cell proliferation and induces differentiation and/or apoptosis [3,54,65,78,79,120]. In human multiple myeloma, a B cell malignancy, SHH derived from myeloma plasma cells has been shown to antagonize RA activity through upregulation of CYP26A1 [39]. Altogether these observations prompt the question of whether in other tumor types in which Hedgehog signaling is pathologically upregulated [17,18,121], RA signaling is mitigated through Hedgehog-dependent CYP26-mediated clearance of RA, providing favorable conditions for growth and survival of tumor cells.

3. Materials and Methods

3.1. Ethics Statement

The procedures involving the use of mice were reviewed and approved by the Animal Research Ethics Committee in Göteborg, Sweden (Dnr. 230-2010 (29 September 2010), 174-2013 (12 November 2013) and 40-2016 (27 April 2016)). Mouse experiments were also carried out under approved protocols in strict accordance with the policies and procedures established by the University of California, San Francisco (UCSF) Institutional Animal Care and Use Committees (UCSF protocol AN084146 re-approved on 05 March 2019).

3.2. Mouse Lines

The *K14-CRE/Shh^{f/f}* mutant, the *ShhGFPCRE/Smo^{f/f}* mutant, the tamoxifen (TAM)-inducible *ShhCreER^{T2}/Shh^{f}* mutant, and the *Shh^{n/n}* mutant embryos as well as their control littermates were generated and identified as described previously [40–43]. Control and *ShhGFPCRE/Smo^{f/f}* mutant embryos carrying the *RAREhsplacZ* transgene [73] were generated as described [40]. For CRE-mediated ablation of *Shh* in *ShhCreER^{T2}/Shh^{f}* embryos, pregnant females were treated with intraperitoneal injections of TAM every other day (excluding the day of embryo harvest) as described [40].

3.3. Histology, Immunohistochemistry, In Situ Hybridization, β-Galactosidase Histochemistry and RT-qPCR

Tissues and organs were processed for histology (Alcian blue van Gieson staining), immunohistochemistry, in situ hybridization, and β-galactosidase histochemistry as described previously [40]. Rabbit antibody targeting cleaved Lamin A (small subunit; 1:1000 dilution) was obtained from Cell Signaling Technology (Danvers, MA, USA). Rabbit monoclonal (MAB) antibody against FOXA1 (1:5000 dilution) was from Abcam (Cambridge, UK). For detection of *Cyp26b1* and *Cyp26c1* transcripts in tissue sections, oligonucleotide probes targeting *Mm-Cyp26b1* (NM_175475.3; target sequence. 460-1308) and *Mm-Cyp26c1* (NM_001105201.1; target sequences: 21-1134) were used. For RT-qPCR assays, the entire tail per embryo, the entire genital tubercle per embryo, and a pair of palatal shelves per embryo were analyzed. *Actb* (β-actin) was used as a reference gene for RT-qPCR data. Primers and conditions for RT-qPCR, other probes used for in situ hybridization, and other antibodies have been described [40].

3.4. In Vitro Explant Cultures and Quantification of Apoptosis

Shh^{f/f} females were mated with *ShhCreER^{T2}* males. The pregnant females received an intraperitoneal injection of TAM [40] to induce in utero CRE-mediated *Shh* deactivation at embryonic day 10 (E10). The next day (E11), tails/pelvic girdles were dissected from control and *ShhCreER^{T2}/Shh^{f}* mutant embryos and cultivated in vitro in an organ culture system as described previously [40]. The medium contained 2.5 and 1.25 µM 4-OH-TAM (4-Hydroxytamoxifen, Sigma-Aldrich, Stockholm, Sweden) during the first and second days of culture, respectively, to enable continuation of CRE-mediated ablation of *Shh* in vitro. The explants were cultivated for a total period of 3 days in the presence of vehicle control (DMSO) or 12.5 µM BMS493, a pan-RAR inverse agonist (Tocris Bioscience, Abingdon, UK). Thereafter, the explants treated with DMSO or BMS493 were processed for Keratin 8 immunohistochemistry. For quantification of apoptosis, sections of tail explants cultivated as described above were processed for immunostaining for cleaved Lamin A to visualize apoptotic cells. Apoptotic epithelial and mesenchymal cells in sections of tail explants from controls and *ShhCreER^{T2}/Shh^{f}* mutants were counted in the caudal portion of the tail through a ×20 objective. Student's *t*-test was used for statistical analysis.

For *Cyp26a1* whole-mount in situ hybridization, tails/pelvic girdles were dissected from E11.5 control embryos and cultivated for 24 h in vitro under conditions described previously [40] in the presence of DMSO or 0.2 µM SAG, a small molecule agonist of Smoothened [122].

Supplementary Materials: Supplementary materials can be found at http://www.mdpi.com/1422-0067/20/9/2275/s1. Figure S1. Loss of SHH signaling in the developing palate of *ShhCreER*^{T2}*/Shh^f* mutant embryos. Figure S2. The *ShhCreER*^{T2}*/Shh^f* mutant embryos display cleft palate. Figure S3. RALDH1-3 proteins are produced in the developing palate of control and *ShhCreER*^{T2}*/Shh^f* mutant embryos. Figure S4. Diminished *Cyp26b1* hybridization signals in the palatal mesenchyme upon loss of SHH signaling. Table S1. Phenotypes caused by loss of Hedgehog signaling and enhancement of retinoic acid signaling in animal models. References [123–190] are cited in the Supplementary Materials files.

Author Contributions: Conceptualization, A.G.-L.; Data curation, A.L., A.G.-L., B.D.H., C.-G.R., C.L.N., K.H., M.E.S., O.D.K. and P.M.; Formal analysis, A.L., A.G.-L., B.D.H., C.-G.R., C.L.N., K.H., M.E.S., O.D.K. and P.M.; Funding acquisition, A.G.-L., C.L.N. and O.D.K; Investigation, A.L., A.G.-L., B.D.H., C.-G.R., C.L.N., K.H., M.E.S., O.D.K. and P.M.; Methodology, A.G.-L., B.D.H., C.-G.R., C.L.N., K.H., M.E.S. and P.M.; Project administration, A.G.-L. and O.D.K.; Supervision, A.G.-L., Validation, A.L., A.G.-L., B.D.H., C.-G.R., C.L.N., K.H., M.E.S., O.D.K. and P.M; Visualization, A.L., A.G.-L., C.-G.R., C.L.N., K.H., M.E.S., and P.M.; Writing-original draft, A.G.-L., A.L., C.-G.R. and M.E.S.; Writing-review and editing, B.D.H., C.L.N., K.H., O.D.K., and P.M.

Funding: This work was funded by The Swedish Research Council-Medicine (grant 20614- www.vr.se), The Thurèus Foundation, and TUA Västra Götaland Region to A.G.-L.; The Institute of Odontology, Sahlgrenska Academy at The University of Gothenburg to C-G.R.; TL1 Postdoctoral Fellowship Training Program (PRESCIENT; National Institutes of Health 1TL1TR001871-01- www.nih.gov) to C.L.N.; and National Institutes of Health (Grant R35-DE026602- www.nih.gov) to O.D.K.

Acknowledgments: The authors are grateful to P. Chambon, P. Dollé, A.P. McMahon and P. Oh for providing plasmids used to generate riboprobes. We thank A.P. McMahon and C.J. Tabin for providing mouse lines, A.P. McMahon for SHH antibody, and K. Nobelius for valuable assistance.

Conflicts of Interest: The authors declare no conflict of interest.

Abbreviations

BMS493	Pan Retinoic acid receptor antagonist
CRE	(Cyclization recombination) DNA recombinase
CYP26A1,B1,C1	Cytochrome P450 isoenzymes A1, B1 and C1
f	Floxed allele
FGF	Fibroblast growth factor
FNP	Frontonasal process
FOXA1/*Foxa1*	Forkhead box protein A1 protein/gene
GLI1-3	Glioma-associated oncogene family members 1, 2 and 3
IHH	Indian hedgehog
ISH	In situ Hybridization
K8	Keratin 8
K14	Keratin 14
LacZ	Gene encoding *E. Coli* β-galactosidase
n	Null allele
SAG	Smoothened agonist
SHH/*Shh*	Sonic Hedgehog protein/gene
SMO/*Smo*	Smoothened protein/gene
PS	Palatal shelf/shelves
Ptch1	Patched 1
RA	Retinoic acid
RALDH/*Aldh1a*	Retinaldehyde dehydrogenase protein/gene
RAR	Retinoic acid receptor
RARE	Retinoic acid response element
RT-qPCR	Reverse transcription quantitative polymerase chain reaction
RXR	Retinoid X receptor
TAM	Tamoxifen
WMISH	Whole-mount in situ hybridizationWingless/integrated 3a
4-OH-TAM	4-hydroxytamoxifen

References

1. McMahon, A.P.; Ingham, P.W.; Tabin, T.J. Developmental roles and clinical significance of hedgehog signaling. *Curr. Top. Dev. Biol.* **2003**, *53*, 1–114.
2. Duester, G. Retinoic acid synthesis and signaling during early organogenesis. *Cell* **2008**, *134*, 921–931. [CrossRef] [PubMed]
3. Niederreither, K.; Dollé, P. Retinoic acid in development: Towards an integrated view. *Nat. Rev. Genet.* **2008**, *9*, 541–553. [CrossRef] [PubMed]
4. Briscoe, J.; Thérond, P. The mechanisms of hedgehog signaling and its roles in development and disease. *Nat. Rev. Mol. Cell Biol.* **2013**, *14*, 416–429. [CrossRef]
5. Mark, M.; Teletin, M.; Vernet, N.; Ghyselinck, N. Role of retinoic acid receptor (RAR) signaling in postnatal male germ cell differentiation. *Biochim. Biophys. Acta* **2015**, *1849*, 84–93. [CrossRef] [PubMed]
6. Xavier, M.G.; Seppala, M.; Barrell, W.; Birjandi, A.A.; Geoghegan, F.; Cobourne, M. Hedgehog receptor function during craniofacial development. *Dev. Biol.* **2016**, *415*, 198–215. [CrossRef]
7. Cotton, J.L.; Li, Q.; Ma, L.; Wang, J.; Park, J.S.; Ou, J.; Zhu, L.J.; YT, I.P.; Johnson, R.L.; Mao, J. YAP/TAZ and Hedgehog coordinate growth and patterning in gastrointestinal mesenchyme. *Dev. Cell* **2017**, *43*, 35–47. [CrossRef]
8. Hibsher, D.; Epshtein, A.; Oren, N.; Landsman, L. Pancreatic mesenchyme regulates islet cellular composition in a Patched/Hedgehog-dependent manner. *Sci. Rep.* **2016**, *28*, 38008. [CrossRef]
9. Wang, Q.; Yang, X.; Li, Y.; Zhang, X.; Zhang, Z. Supressor of fused restraint of Hedgehog activity level is critical for osteogenic proliferation and differentiation during calvarial bone development. *J. Biol. Chem.* **2017**, *292*, 1514–1525.
10. Millington, G.; Elliott, K.H.; Chang, Y.T.; Chang, C.F.; Dlugosz, A.; Brugmann, S.A. Cilia-dependent GLI processing in neural crest cells is required for tongue development. *Dev. Biol.* **2017**, *424*, 124–137. [CrossRef]
11. Shimo, T.; Koyama, E.; Okui, T.; Kunisada, Y.; Ibaragi, S.; Yoshioka, N.; Yoshida, S.; Sasaki, A.; Masui, M.; Kurio, N.; et al. Retinoic receptor signaling regulates hypertrophic chondrocyte-specific gene expression. *In Vivo* **2019**, *33*, 85–91. [CrossRef]
12. Smith, J.N.; Walker, H.M.; Thompson, H.; Collinson, J.M.; Vargesson, N.; Erskine, L. Lens-regulated retinoic acid signalling controls expansion of the developing eye. *Development* **2018**, *145*. [CrossRef] [PubMed]
13. Marchwicka, A.; Macinkowska, E. Regulation of expression of CEBP genes by variably expressed vitamin D and retinoic acid receptor in human acute myeloid leukemia cell lines. *Int. J. Mol. Sci.* **2018**, *19*, 918. [CrossRef]
14. Akhavan-Sigari, R.; Schulz-Schaeffer, W.; Harcej, A.; Rohde, V. The importance of the hedgehog signaling pathway in tumorigenesis of spinal and cranial chordoma. *J. Clin. Med.* **2019**, *8*, 248. [CrossRef] [PubMed]
15. Park, K.S.; Martelloto, L.G.; Peifer, M.; Sos, M.L.; Karnesis, A.N.; Mahjoub, M.R.; Bernard, K.; Conklin, J.F.; Szczepny, A.; Yuan, J.; et al. A crucial requirement for Hedgehog signaling in small cell lung cancer. *Nat. Med.* **2011**, *17*, 1504–1508. [CrossRef] [PubMed]
16. Han, M.E.; Lee, Y.S.; Baek, S.Y.; Kim, B.S.; Kim, J.B.; Oh, S.O. Hedgehog signaling regulates the survival of gastric cancer cells by regulating the expression of Bcl-2. *Int. J. Mol. Sci.* **2009**, *6*, 3033–3043. [CrossRef]
17. Teglund, S.; Toftgård, R. Hedgehog beyond medulloblastoma and basal cell carcinoma. *Biochim. Biophys. Acta* **2010**, *1805*, 181–208. [CrossRef] [PubMed]
18. Hui, C.C.; Angers, S. Gli proteins in development and disease. *Ann. Rev. Cell Dev. Biol.* **2011**, *27*, 513–537. [CrossRef]
19. Bitgood, M.J.; McMahon, A.P. Hedgehog and Bmp genes are coexpressed at many sites of cell-cell interaction in the mouse embryo. *Dev. Biol.* **1995**, *172*, 126–138. [CrossRef]
20. Corbit, K.C.; Aanstad, P.; Norman, A.R.; Stainier, D.Y.; Reiter, J.F. Vertebrate Smoothened functions at the primary cilium. *Nature* **2005**, *437*, 1018–1021. [CrossRef]
21. Rohatgi, R.; Milenkovic, L.; Scott, M. Patched regulates hedgehog signaling at the primary cilium. *Science* **2007**, *317*, 372–376. [CrossRef]
22. Rhinn, M.; Dollé, P. Retinoic acid signaling during development. *Development* **2012**, *139*, 843–858. [CrossRef] [PubMed]
23. Blomhoff, R.; Blomhoff, H.K. Overview of retinoid metabolism and function. *J. Neurobiol.* **2006**, *66*, 606–630. [CrossRef]

24. Mark, M.; Ghyselinck, N.B.; Chambon, P. Function of retinoic acid receptors during embryonic development. *Nucl. Recept. Signal.* **2009**, *7*, e002. [CrossRef]

25. Abu-Abed, S.; Dollé, P.; Metzger, D.; Beckett, B.; Chambon, P.; Petkovitch, M. The retinoic acid-metabolizing enzyme, Cyp26A1, is essential for normal hindbrain patterning, vertebral identity, and development of posterior structures. *Genes Dev.* **2001**, *15*, 226–240. [CrossRef]

26. Yashiro, K.; Zhao, X.; Uehara, M.; Yamashita, J.; Nishijima, M.; Nishino, J.; Saijoh, Y.; Sakai, Y.; Hamada, H. Regulation of retinoic acid distribution is required for proximodistal patterning and outgrowth of the developing mouse limb. *Dev. Cell* **2004**, *6*, 411–422. [CrossRef]

27. Uehara, M.; Yashiro, K.; Mamiya, S.; Nishino, J.; Chambon, P.; Dollé, P.; Sakai, Y. CYP26A1 and CYP26C1 cooperatively regulate anterior-posterior patterning of the developing brain and the production of migratory neural crest cells in the mouse. *Dev. Biol.* **2007**, *302*, 399–411. [CrossRef]

28. Sockanathan, S.; Jessell, T.M. Motor neuron-derived retinoic acid specifies the subtype identity of spinal motor neurons. *Cell* **1998**, *94*, 503–514. [CrossRef]

29. Novitch, B.G.; Wichterle, H.; Jessell, T.M.; Stockanathan, S. A requirement for retinoic acid-mediated transcriptional activation in ventral neural patterning and motor neuron specification. *Neuron* **2003**, *40*, 81–95. [CrossRef] [PubMed]

30. Ghaffari, L.; Starr, A.; Nelson, A.T.; Sattler, R. Representing diversity in the dish: Using patient-derived in vitro models to recreate the heterogeneity of neurological disease. *Front. Neurosci.* **2018**, *12*, 56. [CrossRef]

31. Tsukui, T.; Capdevila, J.; Tamura, K.; Ruiz-Lozano, P.; Rodriguez-Esteban, C.; Yonei-Tamura, S.; Magallón, J.; Chandraratna, R.A.; Chien, K.; Blumberg, B.; et al. Multiple left-right asymmetry defects in Shh$^{-/-}$ mutant mice unveil a convergence of the Shh and retinoic acid pathways in the control of Lefty-1. *Proc. Natl. Acad. Sci. USA* **1999**, *96*, 11376–11381. [CrossRef]

32. Schneider, R.A.; Hu, D.; Rubenstein, J.L.R.; Maden, M.; Helms, J.A. Local retinoid signaling coordinates forebrain and facial morphogenesis by maintaining FGF8 and SHH. *Development* **2001**, *128*, 2755–2767. [PubMed]

33. Tanaka, K.; Okada, Y.; Hirokawa, N. FGF-induced vesicular release of Sonic hedgehog and retinoic acid in leftward nodal flow is critical for left-right detemination. *Nature* **2005**, *435*, 172–177. [CrossRef]

34. Ribes, V.; Wang, Z.; Dollé, P.; Niederreither, K. Retinaldehyde dehydrogennase 2 (RALDH2)-mediated retinoic acid synthesis regulates early mouse embryonic forebrain development by controlling FGF and sonic hedgehog signaling. *Development* **2006**, *133*, 351–361. [CrossRef]

35. Ribes, V.; Le Roux, I.; Rhinn, M.; Schuhbaur, B.; Dollé, P. Early mouse caudal development relies on crosstalk between retinoic acid, Shh and Fgf signaling pathways. *Development* **2009**, *136*, 665–676. [CrossRef]

36. Mich, J.K.; Chen, J.K. Hedgehog and retinoic acid signaling cooperate to promote motoneurogenesis in zebrafish. *Development* **2011**, *138*, 5113–5119. [CrossRef] [PubMed]

37. Ribes, V.; Stutzmann, F.; Bianchetti, L.; Guillemot, F.; Dollé, P.; Le Roux, I. Combinatorial signaling controls Neurogenin2 expression at the onset of spinal neurogenesis. *Dev. Biol.* **2008**, *321*, 470–481. [CrossRef]

38. Probst, S.; Kraemer, C.; Demougin, P.; Sheth, R.; Martin, G.R.; Shiratori, H.; Hamada, H.; Iber, D.; Zeller, R.; Zuniga, A. SHH propagates distal limb bud development by enhancing CYP26B1-mediated retinoic acid clearance via AER-FGF signaling. *Development* **2011**, *138*, 1913–1923. [CrossRef]

39. Alonso, S.; Hernandez, D.; Chang, Y.T.; Gocke, C.B.; McCray, M.; Varadham, R.; Matsui, W.H.; Jones, R.J.; Ghiaur, G. Hedgehog and retinoic acid signaling alters multiple myeloma microenvironment and generates bortezomib resistance. *J. Clin. Investig.* **2016**, *126*, 4460–4468. [CrossRef] [PubMed]

40. El Shahawy, M.; Reibring, C.G.; Neben, C.L.; Hallberg, K.; Marangoni, P.; Harfe, B.D.; Klein, O.D.; Linde, A.; Gritli-Linde, A. Cell fate specification in the lingual epithelium is controlled by antagonistic activities of Sonic hedgehog and retinoic acid. *PLoS Genet.* **2017**, *13*, e1006914. [CrossRef]

41. Dassule, H.; Lewis, P.; Bei, M.; McMahon, A.P. Sonic hedgehog regulates growth and morphogenesis of the tooth. *Development* **2000**, *127*, 4775–4785.

42. Choi, K.S.; Harfe, B.D. Hedgehog signaling is required for formation of the notochord sheath and patterning of nuclei pulposi within the intervertebral discs. *Proc. Natl. Acad. Sci. USA* **2011**, *108*, 9484–9489. [CrossRef] [PubMed]

43. Harfe, B.D.; Scherz, P.J.; Nissim, S.; Tian, H.; McMahon, A.P.; Tabin, C. Evidence for expansion-based temporal Shh gradient in specifying vertebrate digit identities. *Cell* **2004**, *118*, 517–528. [CrossRef]

44. Chiang, C.; Litingtun, Y.; Lee, E.; Young, K.E.; Corden, J.L.; Westphal, H.; Beachy, P.A. Cyclopia and defective axial patterning in mice lacking Sonic Hedgehog gene function. *Nature* **1996**, *383*, 407–413. [CrossRef]

45. Götz, W.; Kasper, M.; Fischer, G.; Herken, R. Intermediate filament typing of the human embryonic and fetal notochord. *Cell Tissue Res.* **1995**, *280*, 455–462. [CrossRef]

46. Richardson, S.M.; Ludwinski, F.E.; Gnanalingham, K.K.; Atkinson, R.A.; Freemont, A.J.; Hoyland, J.A. Notochordal and nucleus pulposus marker expression is maintained by subpopulations of adult human nucleosus pulposus cells through aging and regeneration. *Sci. Rep.* **2017**, *7*, 1501–1511. [CrossRef]

47. Gofflot, F.; Hall, M.; Morriss-Kay, G.M. Genetic patterning of the developing mouse tail at the time of posterior neuropore closure. *Dev. Dyn.* **1997**, *210*, 431–445. [CrossRef]

48. Niederreither, K.; McCaffery, P.; Dräager, U.C.; Chambon, P.; Dollé, P. Restricted expression and retinoic acid-induced downregulation of the retinaldehyde dehydrogenenase type 2 (RALDH-2) gene during mouse development. *Mech. Dev.* **1997**, *62*, 67–78. [CrossRef]

49. Haselbeck, R.J.; Hoffmann, I.; Duester, G. Distinct functions for Aldh1 and Raldh2 in the control of ligand production for embryonic retinoid signaling. *Dev. Genet.* **1999**, *25*, 353–364. [CrossRef]

50. Dollé, P.; Ruberte, E.; Leroy, P.; Morris-Kay, G.; Chambon, P. Retinoic acid receptors and cellular retinoid binding proteins. I. A systematic study of their differential patterns of transcription during mouse organogenesis. *Development* **1990**, *110*, 1133–1151.

51. Mollard, R.; Viville, S.; Ward, S.J.; Décimo, D.; Chambon, P.; Dollé, P. Tissue-specific expression of retinoic acid receptor isoform transcripts in the mouse embryo. *Mech. Dev.* **2000**, *94*, 223–232. [CrossRef]

52. Abu-Abed, S.; Dollé, P.; Metzger, D.; Wood, C.; MacLean, G.; Chambon, P.; Petkovich, M. Developing with lethal RA levels: Genetic ablation of RARg can restore the viability of mice lacking Cyp26a1. *Development* **2003**, *130*, 1449–1459. [CrossRef]

53. White, R.J.; Schilling, T.F. How degrading: Cyp26s in hindbrain development. *Dev. Dyn.* **2008**, *237*, 2775–2790. [CrossRef]

54. Pennimpede, T.; Cameron, D.A.; MacLean, G.A.; Li, H.; Abu-Abed, S.; Petkovitch, M. The role of CYP26 enzymes in defining appropriate retinoic acid exposure during embryogenesis. *Birth Defects Res.* **2010**, *88*, 883–894. [CrossRef] [PubMed]

55. Fujii, H.; Sato, T.; Kanedo, S.; Gotoh, O.; Fujii-Kuriyama, Y.; Osawa, K.; Kato, S.; Hamada, H. Metabolic inactivation of retinoic acid by a novel P450 differentially expressed in developing mouse embryos. *EMBO J.* **1997**, *16*, 4163–4173. [CrossRef] [PubMed]

56. De Roos, K.; Sonneveld, E.; Compaan, B.; ten Berge, D.; Durston, A.J.; van der Saag, P.T. Expression of retinoic acid 4-hydroxylase (CYP26) during mouse and Xenopus laevis embryogenesis. *Mech. Dev.* **1999**, *82*, 205–211. [CrossRef]

57. MacLean, G.; Abu-Abed, S.; Dollé, P.; Tahayato, A.; Chambon, P.; Petkovitch, M. Cloning of a novel retinoic acid metabolizing cytochrome P450 Cyp26B1, and comparative expression analysis with Cyp26A1 during early murine development. *Mech. Dev.* **2001**, *107*, 195–201. [CrossRef]

58. Molotkova, N.; Molotkov, A.; Sirbu, I.O.; Duester, G. Requirement of mesodermal retinoic acid generated by Raldh2 for posterior neural transformation. *Mech. Dev.* **2005**, *122*, 145–155. [CrossRef]

59. Sakai, Y.; Meno, C.; Fujii, H.; Nishino, J.; Shiratori, H.; Saijoh, Y.; Rossant, J.; Hamada, H. The retinoic acid-inactivating enzyme Cyp26 is essential for establishing an uneven distribution of retinoic acid along the anterio-posterior axis within the mouse embryo. *Genes Dev.* **2001**, *15*, 213–225. [CrossRef]

60. Niederreither, K.; Abu-Abed, S.; Schuhbar, B.; Petkovich, M.; Chambon, P.; Dollé, P. Genetic evidence that oxidative derivatives of retinoic acid are not involved in retinoid signaling during mouse development. *Nat. Genet.* **2002**, *31*, 84–88. [CrossRef]

61. Shenefelt, R.E. Morphogenesis of malformations in hamsters caused by retinoic acid: Relation to dose and stage at treatment. *Teratology* **1972**, *5*, 103–118. [CrossRef] [PubMed]

62. Geelen, J.A.G.; Peters, P.W.J. Hypervitaminosis A-induced teratogenesis. *CRC Crit. Rev. Toxicol.* **1979**, *6*, 351–375. [CrossRef] [PubMed]

63. Yasuda, Y.; Okamoto, M.; Konishi, H.; Matsuo, T.; Kihara, T.; Tanimura, T. Developmental anomalies induced by all-trans retinoic acid in fetal mice: Macroscopic findings. *Teratology* **1986**, *34*, 37–49. [CrossRef] [PubMed]

64. Elmazar, M.M.A.; Reichert, B.; Scroot, B.; Nau, H. Patterns of retinoid-induced teratogenenic effects: Possible relationships with relative selectivity for nuclear retinoid receptors RARa, RARb and RARg. *Teratology* **1996**, *53*, 158–167. [CrossRef]

65. Padmanabhan, R. Retinoic-acid-induced caudal regression syndrome in the mouse fetus. *Reprod. Toxicol.* **1998**, *12*, 139–151. [CrossRef]

66. Wiley, M.J. The pathogenesis of retinoic acid-induced vertebral anomalies in golden syrian hamster fetuses. *Teratology* **1983**, *28*, 341–353. [CrossRef] [PubMed]

67. Tahayato, A.; Dollé, P.; Petkovitch, M. Cyp26c1 encodes a novel retinoic acid-metabolizing enzyme expressed in the hindbrain, inner ear, first brachial arch and tooth buds during murine development. *Gene Exp. Patterns* **2003**, *3*, 449–454. [CrossRef]

68. MacLean, G.; Dollé, P.; Petkovitch, M. Genetic disruption of CYP26B1 severely affects development of the neural crest derived head structures but does not compromise hindbrain patterning. *Dev. Dyn.* **2009**, *238*, 732–745. [CrossRef]

69. Abu-Abed, S.; MacLean, G.; Fraulob, V.; Chambon, P.; Petkovitch, M.; Dollé, P. Differential expression of the retinoic acid-metabolizing enzymes CYP26A1 and CYP26B1 during murine organogenesis. *Mech. Dev.* **2002**, *110*, 173–177. [CrossRef]

70. Kaufman, M.H. *The Atlas of Mouse Development*, Revised ed.; Academic Press: London, UK, 2003; pp. 172–174.

71. Abzhanov, A.; Rodda, S.J.; McMahon, A.P.; Tabin, C.J. Regulation of skeletogenic differentiation in cranial dermal bone. *Development* **2007**, *134*, 3133–3144. [CrossRef]

72. Ferguson, C.; Alpern, E.; Miclau, T.; Helms, J.A. Does adult fracture repair recapitulate embryonic skeletal formation? *Mech. Dev.* **1999**, *87*, 57–66. [CrossRef]

73. Rossant, J.; Zirngibl, R.; Cado, D.; Shago, M.; Giguère, V. Expression of a retinoic acid response element-hsplacZ transgene defines specific domains of transcriptional activity during mouse embryogenesis. *Genes Dev.* **1991**, *5*, 1333–1344. [CrossRef] [PubMed]

74. Mendelsohn, C.; Larkin, S.; Mark, M.; LeMeur, M.; Clifford, J.; Zelent, A.; Chambon, P. RARb isoforms: Distinct transcriptional control by retinoic acid and specific spatial patterns of promoter activity during mouse embryonic development. *Mech. Dev.* **1994**, *45*, 227–241. [CrossRef]

75. Sigenthaler, J.A.; Ashique, A.M.; Kostantino, Z.; Patterson, K.P.; Hecht, J.H.; Jane, M.A.; Folias, A.E.; Choe, Y.; May, S.R.; Kume, T.; et al. Retinoic acid from meninges regulates cortical neuron generation. *Cell* **2009**, *139*, 597–609. [CrossRef]

76. Dollé, P.; Fraulob, V.; Gallego-Llamas, J.; Vermot, J.; Niederreither, K. Fate of retinoic acid-activated embryonic cell lineages. *Dev. Dyn.* **2010**, *239*, 3260–3274. [CrossRef] [PubMed]

77. Okano, J.; Kimura, W.; Papaionnou, V.E.; Miura, N.; Yamada, G.; Shiota, K.; Sakai, Y. The regulation of endogenous retinoic acid level through CYP26B1 is required for elevation of palatal shelves. *Dev. Dyn.* **2012**, *241*, 1744–1756. [CrossRef]

78. Dollé, P. Developmental expression of retinoic acid receptors (RARs). *Nucl. Recept. Signal.* **2009**, *7*, e006. [CrossRef] [PubMed]

79. Shum, A.S.; Poon, L.L.; Tang, W.W.; Koide, T.; Chan, B.W.; Leung, Y.C.; Shiroishi, T.; Copp, A.J. Retinoic acid induces down-regulation of Wnt-3a, apoptosis and diversion of tail bud cells to neural fate in the mouse embryo. *Mech. Dev.* **1999**, *84*, 17–30. [CrossRef]

80. Gritli-Linde, A. Molecular control of secondary palate development. *Dev. Biol.* **2007**, *301*, 309–326. [CrossRef]

81. Gritli-Linde, A. The etiopathogenesis of cleft lip and cleft palate: Usefulness and caveats of mouse models. *Curr. Top. Dev. Biol.* **2008**, *84*, 37–138.

82. Dixon, M.J.; Marazita, M.L.; Beaty, T.H.; Murray, J.C. Cleft lip and palate: Understanding genetic and environmental influences. *Nat. Rev. Genet.* **2011**, *12*, 167–178. [CrossRef]

83. Rahimov, F.; Jugessur, A.; Murray, J.C. Genetics of nonsyndromic orofacial clefts. *Cleft Palate Craniofac. J.* **2012**, *49*, 73–91. [CrossRef]

84. Lan, Y.; Xu, J.; Jiang, R. Cellular and molecular mechanisms of palatogenesis. *Curr. Top. Dev. Biol.* **2015**, *115*, 59–84.

85. Xu, J.; Liu, H.; Lan, Y.; Aronow, B.J.; Kalinichenko, V.V.; Jiang, R. A Shh-Foxf-Fgf18-Shh molecular circuit regulating palate development. *PLoS Genet.* **2016**, *12*, e1005769. [CrossRef] [PubMed]

86. Economou, A.D.; Ohazama, A.; Porntaveetus, T.; Sharpe, P.T.; Kondo, S.; Basson, M.A.; Gritli-Linde, A.; Cobourne, M.T.; Green, J.B. Periodic stripe formation by a Turing mechanism operating at growth zones in the mammalian palate. *Nat. Genet.* **2012**, *44*, 358–361. [CrossRef] [PubMed]

87. Jeong, J.; Mao, J.; Tenzen, T.; Kottman, A.H.; McMahon, A.P. Hedgehog signaling in the neural crest cells regulates the patterning and growth of facial primordia. *Genes Dev.* **2004**, *18*, 937–951. [CrossRef]

88. Rice, R.; Spencer-Dene, B.; Connor, E.C.; Gritli-Linde, A.; McMahon, A.P.; Dickson, C.; Thesleff, I.; Rice, D.P. Disruption of Fgf10/Fgfr2b-coordinated epithelial mesenchymal interactions causes cleft palate. *J. Clin. Investig.* **2004**, *113*, 1692–1700. [CrossRef] [PubMed]

89. Rice, R.; Connor, E.; Rice, D.P.C. Expression patterns of Hedgehog signaling pathway members during mouse palate development. *Gene Exp. Patterns* **2006**, *6*, 206–212. [CrossRef] [PubMed]

90. Pantalacci, S.; Prochazka, J.; Martin, A.; Rothova, M.; Lambert, A.; Bernard, L.; Charles, C.; Viriot, L.; Peterkova, R.; Laudet, V. Patterning of palatal rugae through sequential addition reveals an anterior-posterior boundary in palate development. *BMC Dev. Biol.* **2008**, *8*, 116. [CrossRef]

91. Welsh, I.C.; O'Brian, T.P. Signaling integration in the rugae growth zone directs sequential SHH signaling center formation during the rostral outgrowth of the palate. *Dev. Biol.* **2009**, *336*, 53–67. [CrossRef]

92. Naitoh, H.; Mori, C.; Nishimura, Y.; Shiota, K. Altered expression of retinoic acid (RA) receptor mRNAs in the fetal mouse secondary palate by all-trans and 13-cis Ras: Implications for RA-induced teratogenesis. *J. Craniofac. Genet. Dev. Biol.* **1998**, *18*, 202–210.

93. Padmanabhan, R.; Ahmed, I. Retinoic acid-induced asymmetric craniofacial growth and cleft palate in the mouse fetus. *Reprod. Toxicol.* **1997**, *11*, 843–860. [CrossRef]

94. Ikemi, N.; Kawata, M.; Yasuda, M. All-trans-retinoic acid-induced variant patterns of palatal rugae in CRJ:Sd rat fetuses and their potential as indicators for teratogenesis. *Reprod. Toxicol.* **1995**, *9*, 369–377. [CrossRef]

95. Vaziri Sani, F.; Kaartninen, V.; El Shahawy, M.; Linde, A.; Gritli-Linde, A. Developmental changes and extracellular structural molecules in the secondary palate and in the nasal cavity of the mouse. *Eur. J. Oral Sci.* **2010**, *118*, 212–236. [CrossRef]

96. Cohn, M.J. Development of the external genitalia: Conserved and divergent mechanisms of appendage patterning. *Dev. Dyn.* **2011**, *240*, 1108–1115. [CrossRef]

97. Seifert, A.W.; Bouldin, C.M.; Choi, K.S.; Harfe, B.D.; Cohn, M.J. Multiphasic and tissue-specific roles for sonic hedgehog in cloacal septation and external genitalia development. *Development* **2009**, *136*, 3949–3957. [CrossRef] [PubMed]

98. Lin, C.; Yin, Y.; Veith, G.M.; Fisher, A.V.; Long, F.; Ma, L. Temporal and spatial dissection of Shh signaling in genital tubercle development. *Development* **2009**, *136*, 3959–3967. [CrossRef] [PubMed]

99. Miyagawa, S.; Moon, A.; Haraguchi, R.; Inoue, C.; Harada, M.; Nakahara, C.; Suzuki, K.; Matsumaru, D.; Kaneko, T.; Matsuo, I.; et al. Dosage-dependent hedgehog signals integrated with Wnt/beta-catenin signaling regulate external genitalia formation as an appendicular program. *Development* **2009**, *136*, 3969–3978. [CrossRef]

100. Seifert, A.W.; Zheng, Z.; Ormerod, B.K.; Cohn, M.J. Sonic hedgehog controls growth of external genitalia by regulating cell cycle kinetics. *Nat. Commun.* **2010**, *1*, 23. [CrossRef]

101. Liu, L.; Suzuki, K.; Nakagata, N.; Mihara, K.; Matsumaru, D.; Ogino, Y.; Yashiro, K.; Hamada, H.; Liu, Z.; Evans, S.M.; et al. Retinoic acid signaling regulates Sonic hedgehog and Bone morphogenetic protein signaling during genital tubercle development. *Birth Defects Res.* **2012**, *95*, 79–88. [CrossRef]

102. Ruberte, E.; Dollé, P.; Krust, A.; Zelent, A.; Morris-Kay, G.; Chambon, P. Specific spacial and temporal distribution of retinoic acid receptor gamma transcripts during mouse embryogenesis. *Development* **1990**, *108*, 213–222. [PubMed]

103. Armfield, B.A.; Seifert, A.W.; Zheng, Z.; Merton, E.M.; Rock, J.R.; Lopez, M.C.; Baker, H.V.; Cohn, M.J. Molecular characterization of the genital organizer: Gene expression profile of the mouse urethral plate epithelium. *J. Urol.* **2016**, *196*, 1295–1302. [CrossRef] [PubMed]

104. Helms, J.A.; Kim, C.H.; Minkoff, R.; Thaller, C.; Eichele, G. Sonic hedgehog participates in craniofacial morphogenesis and is down-regulated by teratogenic doses of retinoic acid. *Dev. Biol.* **1997**, *187*, 25–35. [CrossRef]

105. Gritli-Linde, A.; Bei, M.; Maas, R.; Zhang, X.M.; Linde, A.; McMahon, A.P. Shh signaling within the dental epithelium is necessary for cell proliferation, growth and polarization. *Development* **2002**, *129*, 5323–5337. [CrossRef] [PubMed]

106. Nakatomi, M.; Hovorakova, M.; Gritli-Linde, A.; Blair, H.J.; MacArthur, K.; Peterka, M.; Lesot, H.; Peterkova, R.; Ruiz-Perez, V.L.; Goodship, J.A.; et al. Evc regulates a symmetrical response to Shh signaling in molar development. *J. Dent. Res.* **2013**, *92*, 222–228. [CrossRef]

107. Niederreither, K.; Fraulob, V.; Garnier, J.M.; Chambon, P.; Dollé, P. Differential expression of retinoic acid-synthesizing (RALDH) enzymes during fetal development and organ differentiation in the mouse. *Mech. Dev.* **2002**, *110*, 167–171. [CrossRef]
108. Bloch-Zupan, A.; Décimo, D.; Loriot, M.; Mark, M.P.; Ruch, J.V. Expression of nuclear retinoic acid receptors during mouse odontogenesis. *Differentiation* **1994**, *57*, 195–203. [CrossRef]
109. Morkmued, S.; Laugel-Haushalter, V.; Mathieu, E.; Schuhbaur, B.; Hemmerlé, J.; Dollé, P.; Bloch-Zupan, A.; Niederreither, K. Retinoic acid excess impairs amelogenesis inducing enamel defects. *Front. Physiol.* **2017**, *7*, 673. [CrossRef]
110. Mark, M.P.; Bloch-Zupan, A.; Rush, J.V. Effects of retinoids on tooth morphogenesis and cytodifferentiation in vitro. *Int. J. Dev. Biol.* **1992**, *36*, 517–526.
111. Irving, J.T. The effects of avitaminosis and hypervitamisosis A upon the incisor teeth and incisal alveolar bone of rats. *J. Physiol.* **1949**, *108*, 92–101. [CrossRef]
112. Schour, I.; Hoffman, M.M.; Smith, M.C. Changes in the incisor teeth of albino rats with vitamin A deficiency and the effects of replacement therapy. *Am. J. Pathol.* **1941**, *17*, 529–562. [PubMed]
113. McDowell, E.M.; Shores, R.L.; Spangler, E.F.; Wenk, M.L.; De Luca, L.M. Anomalous growth of rat incisor teeth during chronic intermittent vitamin A deficiency. *J. Nutr.* **1987**, *117*, 1265–1274. [CrossRef]
114. Tubbs, S.S.; Malefant, J.; Loukas, M.; Oakes, W.J.; Oskouian, R.J.; Fries, F.N. Enigmatic human tails: A review of their history, embryology, classification and clinical manifestations. *Clin. Anat.* **2016**, *29*, 430–438. [CrossRef]
115. Tojima, S.; Makishima, H.; Takakuwa, T.; Shigehito, Y. Tail reduction process during human embryonic development. *J. Anat.* **2018**, *232*, 806–811. [CrossRef]
116. Fallon, J.F.; Simandl, B.K. Evidence of a role for cell death in the disappearance of the embryonic human tail. *Am. J. Anat.* **1978**, *152*, 111–130. [CrossRef] [PubMed]
117. Dao, A.H.; Netsky, M.G. Human tails and pseudotails. *Hum. Pathol.* **1984**, *15*, 449–453. [CrossRef]
118. Cai, C.; Shi, O.; Shen, C. Surgical treatment of a patient with human tail and multiple abnormalities of the spinal cord and column. *Adv. Orthop.* **2011**, *2011*, 153797. [CrossRef] [PubMed]
119. Robinson, C.G.; Duke, T.C.; Allison, A.W. Incidental finding of a true human tail in an adult: A case report. *J. Cutan. Pathol.* **2017**, *44*, 75–78. [CrossRef] [PubMed]
120. Das, B.C.; Thapa, P.; Karki, R.; Das, S.; Mahapatra, S.; Liu, T.C.; Torregroza, I.; Wallace, D.P.; Kambhampati, S.; Van Veldhuizen, P.; et al. Retinoic acid signaling pathway in development and diseases. *Bioorg. Med. Chem.* **2014**, *22*, 673–683. [CrossRef] [PubMed]
121. Jiang, J.; Hui, C.C. Hedgehog signaling in development and cancer. *Dev. Cell* **2008**, *15*, 801–812. [CrossRef] [PubMed]
122. Chen, J.K.; Taipale, J.; Young, K.E.; Beachy, P.A. Small molecule modulation of Smoothened activity. *Proc. Natl. Acad. Sci. USA* **2002**, *99*, 14071–14076. [CrossRef]
123. Echelard, Y.; Epstein, D.J.; St-Jacques, B.; Shen, L.; Mohler, J.; McMahon, J.A.; McMahon, A.P. Sonic hedgehog, a member of a family of putative signaling molecules, is implicated in the regulation of CNS polarity. *Cell* **1993**, *75*, 1417–1430. [CrossRef]
124. Becker, S.; Wang, Z.J.; Massey, H.; Arauz, H.; Labosky, P.; Hammerschmidt, M.; St-Jacques, B.; Bumcrot, D.; McMahon, A.; Grabel, L. A role for indian hedgehog in F9 cells and the early mouse embryo. *Dev. Biol.* **1997**, *187*, 298–310. [CrossRef] [PubMed]
125. Zhang, X.M.; Ramalho-Santos, M.; McMahon, A.P. Smoothened mutants reveal redundant roles for Shh and Ihh signaling including regulation of L/R asymmetry by the mouse node. *Cell* **2001**, *105*, 781–792. [CrossRef]
126. Schachter, K.A.; Krauss, R.S. Murine models of holoprosencephaly. *Curr. Top. Dev. Biol.* **2008**, *84*, 139–170.
127. Seppala, M.; Xavier, G.M.; Fan, C.M.; Cobourne, M.T. Boc modifies the spectrum of holoprosencephaly in the absence of Gas1 function. *Biol. Open* **2014**, *3*, 728–740. [CrossRef] [PubMed]
128. Izzi, L.; Lévesque, M.; Morin, S.; Laniel, D.; Wilkes, B.C.; Krauss, R.S.; McMahon, A.P.; Allen, B.L.; Charron, F. Boc and Gas each form distinct Shh receptor complexes with Ptch1 and are required for Shh-mediated cell proliferation. *Dev. Cell* **2011**, *20*, 788–801. [CrossRef] [PubMed]
129. Seppala, M.; Depew, M.J.; Martinelli, D.C.; Fan, C.M.; Sharpe, P.T.; Cobourne, M.T. Gas1 is a modifier for holoprosencephaly and genetically interacts with Sonic hedgehog. *J. Clin. Investig.* **2007**, *117*, 1575–1784. [CrossRef] [PubMed]

130. Zhang, W.; Hong, M.; Bae, G.U.; Kang, J.S.; Krauss, R.S. Boc modifies the holoprosencephaly spectrum of Cdo mutant mice. *Dis. Models Mech.* **2011**, *4*, 368–380. [CrossRef] [PubMed]

131. Clagett-Dame, M.; DeLuca, H.F. The role of vitamin A in mammalian reproduction and embryonic development. *Ann. Rev. Nutr.* **2002**, *22*, 347–381. [CrossRef]

132. Niederreither, K.; Vermot, J.; Schuhbaur, B.; Chambon, P.; Dollé, P. Embryonic retinoic acid synthesis is required for fore limb growth and anteroposterior patterning in the mouse. *Development* **2002**, *129*, 3563–3574. [PubMed]

133. Mic, F.A.; Sirbu, I.O.; Duester, G. Retinoic acid synthesis controlled by Raldh2 is required early for limb bud initiation and then later as a proximodistal signal during apical ectodermal ridge formation. *J. Biol Chem.* **2004**, *279*, 26698–26706. [CrossRef] [PubMed]

134. Mic, F.A.; Molotkov, A.; Fan, X.; Cuenca, A.E.; Duester, G. RALDH3, a retinaldehyde dehydrogenase that generates retinoic acid, is expressed in the ventral retina, otic vesicle and olfactory pit during mouse development. *Mech. Dev.* **2000**, *97*, 227–230. [CrossRef]

135. Cunningham, T.J.; Duester, G. Mechanisms of retinoic acid signalling and its roles in organ and limb development. *Nat. Rev. Mol. Cell. Biol.* **2015**, *16*, 110–123. [CrossRef] [PubMed]

136. Ribes, V.; Fraulob, V.; Petkovich, M.; Dollé, P. The oxidizing enzyme CYP26A1 tightly regulates the availability of retinoic acid in the gastrulating mouse embryo to ensure proper head development and vasculogenesis. *Dev. Dyn.* **2007**, *236*, 644–653. [CrossRef]

137. Uehara, M.; Yashirom, K.; Takaokam, K.; Yamamoto, M.; Hamada, H. Removal of maternal retinoic acid by embryonic CYP26 is required for correct Nodal expression during early embryonic patterning. *Genes Dev.* **2009**, *23*, 1689–1698. [CrossRef]

138. Sulik, K.K.; Dehart, D.B.; Rogers, J.M.; Chernoff, N. Teratogenicity of low doses of all-trans retinoic acid in presomite mouse embryo. *Teratology* **1995**, *51*, 398–403. [CrossRef]

139. Nolen, G.A. Effect of high systemic background level of vitamin A on the teratogenicity of all-trans-retinoic acid given either acutely or subacutely. *Teratology* **1989**, *39*, 333–339. [CrossRef]

140. Lohnes, D.; Kastner, P.; Dierich, A.; Mark, M.; LeMeur, M.; Chambon, P. Function of retinoic acid receptor gamma in the mouse. *Cell* **1993**, *73*, 643–658. [CrossRef]

141. Knudsen, P.A. Congenital malformations of upper incisors in exencephalic mouse embryos, induced by hypervitaminosis A. II. Morphology of fused upper incisors. *Acta Odontol. Scand.* **1965**, *23*, 391–409. [CrossRef]

142. Kalter, H.; Warkany, J. Experimental production of congenital malformations in strains of inbred mice by maternal treatment with hypervitaminosis A. *Am. J. Pathol.* **1961**, *38*, 1–21.

143. Rutledge, J.C.; Shoubaji, A.G.; Hughes, L.A.; Polifka, J.E.; Cruz, Y.P.; Bishop, J.B.; Generoso, W.M. Limb and lower-body duplications induced by retinoic acid in mice. *Proc. Natl. Acad. Sci. USA* **1994**, *91*, 5436–5440. [CrossRef] [PubMed]

144. Kubota, Y.; Shimotake, T.; Yanagihara, I.; Iwai, N. Development of anorectal malformations using etretinate. *J. Pediatr. Surg.* **1998**, *33*, 127–129. [CrossRef]

145. Kubota, Y.; Shimotake, T.; Iwai, N. Congenital anomalies in mice induced by etretinate. *Eur. J. Pediatr. Surg.* **2000**, *10*, 248–251. [CrossRef]

146. Mo, R.; Freer, A.M.; Zinyk, D.L.; Crackower, M.A.; Michaud, J.; Heng, H.Q.; Chik, K.W.; Shi, X.M.; Tsui, L.C.; Cheng, S.H.; et al. Specific and redundant functions of Gli2 and Gli3 zinc finger genes in skeletal patterning and development. *Development* **1997**, *124*, 113–123.

147. Lan, Y.; Jiang, R. Sonic Hedgehog signaling regulates reciprocal epithelial-mesenchymal interactions controlling palatal growth. *Development* **2009**, *136*, 1387–1396. [CrossRef] [PubMed]

148. Lipinski, R.J.; Song, C.; Sulik, K.K.; Everson, J.L.; Gipp, J.J.; Yan, D.; Bushman, W.; Rowland, I.J. Cleft lip and palate from hedgehog signaling antagonism in the mouse: Phenotypic characterization and clinical implications. *Birth Defects Res. Clin. Mol. Teratol.* **2010**, *88*, 232–240. [CrossRef]

149. Heyne, G.W.; Melberg, C.G.; Doroodchi, P.; Parins, K.F.; Kietzman, H.W.; Everson, J.L.; Ansen-Wilson, L.J.; Lipinski, R.J. Definition of critical periods for Hedgehog signaling pathway antagonist-induced holoprosencephaly, cleft lip and cleft palate. *PLoS Genet.* **2015**, *10*, e0120517.

150. Dupé, V.; Matt, N.; Garnier, J.M.; Chambon, P.; Mark, M.; Ghyselinck, N.B. A newborn lethal defect due to inactivation of retinaldehyde dehydrogenase type 3 is prevented by maternal retinoic acid treatment. *Proc. Natl. Acad. Sci. USA* **2003**, *100*, 14036–14041. [CrossRef] [PubMed]

151. Molotkova, N.; Molotkov, A.; Duester, G. Role of retinoic acid during forebrain development begins late when Raldh3 generates retinoic acid in the ventral subventricular zone. *Dev. Biol.* **2007**, *303*, 601–610. [CrossRef]

152. Laue, K.; Pogoda, H.M.; Daniel, P.B.; van Haeringen, A.; Alany, Y.; von Ameln, S.; Rachwalski, M.; Morgan, T.; Gray, M.J.; Breuning, M.H.; et al. Craniosynostosis and multiple skeletal anomalies in humans and zebrafish result from a defect in the localized degradation of retinoic acid. *Am. J. Hum. Genet.* **2011**, *89*, 595–606. [CrossRef] [PubMed]

153. Cuervo, R.; Valencia, C.; Chandraratna, R.A.; Covarrubias, L. Programmed cell death is required for palate shelf fusion and is regulated by retinoic acid. *Dev. Biol.* **2002**, *245*, 145–156. [CrossRef]

154. Abe, M.; Maeda, T.; Wakisaka, S. Retinoic acid affects craniofacial patterning by changing Fgf8 expression in the pharyngeal ectoderm. *Dev. Growth Differ.* **2008**, *50*, 717–729. [CrossRef]

155. Litingtung, Y.; Lei, L.; Westphal, H.; Chiang, C. Sonic Hedgehog is essential to foregut development. *Nat. Genet.* **1998**, *20*, 58–61. [CrossRef]

156. Pepicelli, C.V.; Lewis, P.M.; McMahon, A.P. Sonic hedgehog regulates branching morphogenesis in the mammalian lung. *Curr. Biol.* **1998**, *8*, 1083–1096. [CrossRef]

157. Motoyama, J.; Liu, J.; Mo, R.; Ding, Q.; Post, M.; Hui, C.C. Essential function of Gli2 and Gli3 in the formation of lung, trachea and oesophagus. *Nat. Genet.* **1998**, *20*, 54–57. [CrossRef]

158. Malpel, S.; Mendelsohn, C.; Cardoso, W.V. Regulation of retinoic acid signaling during lung morphogenesis. *Development* **2000**, *127*, 3057–3067.

159. Mendelsohn, C.; Lohnes, D.; Décimo, D.; Lufkin, T.; Chambon, P.; Mark, M. Function of the retinoic acid receptors (RARs) during development. II. Multiple abnormalities at various stages of organogenesis in RAR double mutants. *Development* **1994**, *120*, 2749–2771.

160. Ramalho-Santos, M.; Melton, D.A.; McMahon, A.P. Hedgehog signals regulate multiple aspects of gastrointestinal development. *Development* **2000**, *127*, 2763–2772.

161. Mo, R.; Kim, J.H.; Zhang, J.; Chiang, C.; Hui, C.H.; Kim, P.C.W. Anorectal malformations caused by defects in Sonic Hedgehog signaling. *Am. J. Pathol.* **2001**, *159*, 765–773. [CrossRef]

162. Yu, J.; Carroll, T.J.; McMahon, A.P. Sonic Hedgehog regulates proliferation and differentiation of mesenchymal cells in the mouse metanephric kidney. *Development* **2002**, *129*, 5301–5312. [PubMed]

163. Rosselot, C.; Spraggon, L.; Chia, I.; Batourina, E.; Riccio, P.; Lu, B.; Niederreither, K.; Dolle, P.; Duester, G.; Chambon, P.; et al. Non-cell autonomous retinoid signaling is crucial for renal development. *Development* **2010**, *137*, 283–292. [CrossRef]

164. Bitgood, M.J.; Shen, L.; McMahon, A.P. Sertoli cell signaling by Desert hedgehog regulates the male germline. *Curr. Biol.* **1996**, *6*, 298–304. [CrossRef]

165. Vernet, N.; Dennefeld, C.; Rochette-Egly, M.; Ouled Abdelghani, M.; Chambon, P.; Ghyselinck, N.B.; Mark, M. Retinoic acid metabolism and signaling pathways in adult and developing mouse testis. *Endocrinology* **2006**, *147*, 96–110. [CrossRef] [PubMed]

166. Teletin, M.; Vernet, N.; Yu, J.; Klopfenstein, M.; Jones, J.W.; Féret, B.; Kane, M.A.; Ghyselinck, N.B.; Mark, M. Two functionally redundant sources of retinoic acid secure spermatogonia differentiation in the seminiferous epithelium. *Development* **2019**, *146*. [CrossRef] [PubMed]

167. Bowles, J.; Feng, C.W.; Ineson, J.; Miles, K.; Spiler, C.M.; Harley, V.R.; Sinclair, A.H.; Koopman, P. Retinoic acid antagonizes testis development in mice. *Cell Rep.* **2018**, *24*, 1330–1341. [CrossRef]

168. Li, H.; MacLean, G.; Cameron, D.; Clagett-Dame, M.; Petkovitch, M. Cyp26b1 expression in murine Sertoli cells is required to maintain germ cells in an undifferentiated state during embryogenesis. *PLoS ONE* **2009**, *4*, e7501. [CrossRef]

169. Biswas, N.M.; Deb, C. Testicular degeneration in rats during hypervitamisosis A. *Endokrinologie* **1965**, *49*, 64–69.

170. Long, F.; Chung, U.I.; Ohba, S.; McMahon, J.; Kronenberg, H.M.; McMahon, A.P. Ihh signaling is directly required for the osteoblast lineage in the endochondral skeleton. *Development* **2004**, *131*, 1309–1318. [CrossRef]

171. Long, F.; Zhang, X.M.; Karp, S.; Yang, Y.; McMahon, A.P. Genetic manipulation of hedgehog signaling in the endochondral skeleton reveals a direct role in the regulation of chondrocyte proliferation. *Development* **2001**, *128*, 5099–5108.

172. Minegishi, Y.; Sakai, Y.; Yahara, Y.; Akiyama, H.; Yoshikawa, H.; Hosokawa, K.; Tsumaki, N. Cyp26b1 within the growth plate regulates bone growth in juvenile mice. *Biochem. Biophys. Res. Commun.* **2014**, *454*, 12–18. [CrossRef] [PubMed]

173. Shimono, K.; Tung, W.E.; Macolino, C.; Chi, A.H.T.; Didizian, J.J.; Mundy, C.; Chandraratna, R.A.; Mishina, Y.; Enomoto-Iwamoto, M.; Pacifici, M.; et al. Potent inhibition of heterotopic ossification by nuclear retinoic acid receptor-γ agonists. *Nat. Med.* **2011**, *17*, 454–460. [CrossRef]

174. Gritli-Linde, A.; Hallberg, K.; Harfe, B.D.; Reyahi, A.; Kannius-Janson, M.; Nilsson, J.; Cobourne, M.T.; Sharpe, P.T.; McMahon, A.P.; Linde, A. Abnormal hair development and apparent follicular transformation into mammary glands in the absence of hedgehog signaling. *Dev. Cell* **2007**, *12*, 99–112. [CrossRef] [PubMed]

175. Everts, H.B.; Sundberg, J.P.; King, L.E., Jr.; Ong, D.E. Immunolocalization of enzymes, binding proteins and receptors sufficient for retinoic acid synthesis and signaling during the hair cycle. *J. Investig. Dermatol.* **2007**, *127*, 1593–1604. [CrossRef] [PubMed]

176. Okano, J.; Lichti, U.; Mamiya, S.; Aronova, M.; Zhang, G.; Yuspa, S.H.; Hamada, H.; Sakai, Y.; Morasso, M.I. Increased retinoic acid levels through ablation of Cyp26b1 determine the processs of embryonic skin barrier formation and peridermal development. *J. Cell Sci.* **2012**, *125*, 1827–1836. [CrossRef]

177. Hardy, M.H. Glandular metaplasia of hair follicles and other responses to vitamin A excess in cultures of rodent skin. *J. Embryol. Exp. Morph.* **1968**, *19*, 157–180.

178. Viallet, J.P.; Dhouailly, D. Retinoic acid and mouse skin morphogenesis. II. Role of epidermal competence in hair glandular metaplasia. *Dev. Biol.* **1994**, *166*, 277–288. [CrossRef]

179. Blanchet, S.; Favier, B.; Chevalier, G.; Kastner, P.; Michaille, J.J.; Chambon, P.; Dhouailly, D. Both retinoic acid receptors alpha (RARalpha) and gamma (RAR gamma) are able to initiate mouse upper-lip skin glandular metaplasia. *J. Investig. Dermatol.* **1998**, *111*, 206–212. [CrossRef] [PubMed]

180. Fell, H.B.; Mellanby, E. Metaplasia produced in cultures of chick ectoderm by high vitamin A. *J. Physiol.* **1953**, *119*, 470–488. [CrossRef]

181. Lawrence, D.J.; Bern, H.A. Mucous gland formation in keratinized adult epithelium in situ treated with vitamin A. *Exp. Cell Res.* **1960**, *21*, 443–446. [CrossRef]

182. Lawrence, D.J.; Bern, H.A.; Steadman, M.G. Vitamin A and keratinization. Studies on the hamster cheek pouch. *Ann. Otol. Rhinol. Laryngol.* **1960**, *69*, 645–661. [CrossRef]

183. Covant, H.A.; Hardy, M.H. Stability of the glandular morphogenesis produced by retinoids in the newborn hamster cheek pouch in vitro. *J. Exp. Zool.* **1988**, *246*, 139–149. [CrossRef]

184. So, P.L.; Lee, K.; Hebert, J.; Walker, P.; Lu, Y.; Hwang, J.; Kopelovich, L.; Athar, M.; Bickers, D.; Aszterbaum, M.; et al. Topical tazarotene chemoprevention reduces basal cell carcinoma number and size in Ptch$^{+/-}$ mice exposed to ultraviolet or ionizing radiation. *Cancer Res.* **2004**, *64*, 4385–4390. [CrossRef]

185. Riddle, R.D.; Johnson, R.L.; Laufer, E.; Tabin, C. Sonic hedgehog mediates the polarizing activity of the ZPA. *Cell* **1993**, *75*, 1401–1416. [CrossRef]

186. Tickle, C.; Towers, M. Sonic hedgehog signaling in limb development. *Front. Cell Dev. Biol.* **2017**, *5*, 14. [CrossRef] [PubMed]

187. Bouldin, C.M.; Gritli-Linde, A.; Ahn, S.; Harfe, B.D. Shh pathway activation is present and required within the vertebrate limb bud apical ectodermal ridge for normal auropod patterning. *Proc. Natl. Acad. Sci. USA* **2010**, *107*, 5489–5494. [CrossRef] [PubMed]

188. Scherz, P.J.; McGlinn, E.; Nissim, S.; Tabin, C.J. Extended exposure to Sonic hedgehog is required for patterning the posterior digits of the vertebral limb. *Dev. Biol.* **2007**, *308*, 343–354. [CrossRef]

189. Zhang, R.; Lee, C.; Lawson, L.Y.; Svete, L.J.; McIntyre, L.M.; Harfe, B.D. SHH protein variance in the limb bud is constrained by feedback regulation and correlates with altered digit patterning. *Genes Genomes Genet.* **2017**, *7*, 851–858. [CrossRef]

190. Zhou, J.; Kochlar, D.M. Cellular anomalies underlying retinoic-acid-induced phocomelia. *Reprod. Toxicol.* **2004**, *19*, 103–110. [CrossRef]

International Journal of
Molecular Sciences

Article

Discrete Hedgehog Factor Expression and Action in the Developing Phallus

Gerard A. Tarulli *, Andrew J. Pask and Marilyn B. Renfree

School of BioSciences, The University of Melbourne, Parkville, VIC 3010, Australia;
a.pask@unimelb.edu.au (A.J.P.); m.renfree@unimelb.edu.au (M.B.R.)
* Correspondence: gerard.tarulli@unimelb.edu.au

Received: 8 January 2020; Accepted: 10 February 2020; Published: 12 February 2020

Abstract: Hypospadias is a failure of urethral closure within the penis occurring in 1 in 125 boys at birth and is increasing in frequency. While paracrine hedgehog signalling is implicated in the process of urethral closure, how these factors act on a tissue level to execute closure itself is unknown. This study aimed to understand the role of different hedgehog signalling members in urethral closure. The tammar wallaby (*Macropus eugenii*) provides a unique system to understand urethral closure as it allows direct treatment of developing offspring because mothers give birth to young before urethral closure begins. Wallaby pouch young were treated with vehicle or oestradiol (known to induce hypospadias in males) and samples subjected to RNAseq for differential expression and gene ontology analyses. Localisation of Sonic Hedgehog (SHH) and Indian Hedgehog (IHH), as well as the transcription factor SOX9, were assessed in normal phallus tissue using immunofluorescence. Normal tissue culture explants were treated with SHH or IHH and analysed for *AR*, *ESR1*, PTCH1, GLI2, *SOX9*, *IHH* and *SHH* expression by qPCR. Gene ontology analysis showed enrichment for bone differentiation terms in male samples compared with either female samples or males treated with oestradiol. Expression of SHH and IHH localised to specific tissue areas during development, akin to their compartmentalised expression in developing bone. Treatment of phallus explants with SHH or IHH induced factor-specific expression of genes associated with bone differentiation. This reveals a potential developmental interaction involved in urethral closure that mimics bone differentiation and incorporates discrete hedgehog activity within the developing phallus and phallic urethra.

Keywords: hypospadias; urethra; penis; bone; hedgehog signalling

1. Introduction

Disorders in the development of reproductive organs are the most common birth defects worldwide. Among them, hypospadias effects 1 in 250 births [1,2] and is a defect in urethral closure so that the opening is positioned within the posterior penile meatus rather than the tip of the glans penis. Alarming increases in incidence of up to 2% a year have been reported [1,2], that highlights the current burden of hypospadias and the value in understanding its underlying causes.

While it is known that gene regulatory networks critical for bone/limb development play important roles in regulating phallus development and urethral closure [3–6], it is yet to be determined how these factors are coordinated on a tissue level to execute the process of urethral closure itself. One factor important in bone and limb development is the secreted signalling factor sonic hedgehog (SHH). SHH plays important roles in the development of the genital tubercle from which the mouse phallus forms [7–10] as well as later masculinisation events and the formation of the penile urethra [8,11]. Published evidence indicates that expression of SHH in urethral epithelial cells signals via underlying mesenchyme to promote proliferation and growth of the genital tubercle and urorectal septum mesenchyme [8]. The importance of hedgehog signalling is conserved in humans, as evidenced by polymorphisms in hedgehog pathway genes being associated with an increased risk of delivering a baby with hypospadias [12]. A second hedgehog family member implicated in urethral closure and phallus development is Indian hedgehog (IHH). Far less is known about the role of IHH in this process, though its expression is positively regulated by androgens, and knockout of IHH results in a failure of proper phallus masculinisation in the mouse [13]. However, whether these hedgehog factors have distinct functions in urethral closure has never been investigated.

The importance of androgens in phallus development and urethral closure has a long evolutionary history [3] and is well described in literature [14–19]. The process of urethral closure is highly sensitive to androgen disruption during a critical window of developmental programming that precedes the initiation of closure [20–22]. Therefore factors that can interfere with normal androgen signalling during this critical window, such as oestrogenic chemicals, are associated with hypospadias [14,23,24]. Making direct assessments of the gene regulatory changes that occur in response to hypospadias-inducing chemicals during this programming window are key to capturing the fundamental genetic networks underpinning urethral closure that go awry to contribute to hypospadias. In most species this programming window occurs during fetal development and therefore most animal models of chemical-induced hypospadias use indirect assessments via treating the mothers during pregnancy. This introduces confounding factors such as maternal metabolism and placental transfer that complicate experimental analysis.

One way to overcome these limitations and perform direct assessments in vivo is to use a marsupial model such as the tammar wallaby (*Macropus eugenii*). Marsupials give birth to altricial young, and in the tammar wallaby this occurs before androgen programming and urethral closure begin. This allows for offspring to be treated directly while in the pouch, for the entire duration of androgen programming. The tammar wallaby model was used to identify the gene regulatory networks most enriched when wallaby offspring are treated with ethinylestradiol (E2), a known hypospadias-inducing chemical in this species, during the critical androgen programming window. Comparisons between male and female phallus tissue were also made. Findings of this study identified regulation of ossification and osteoblast differentiation as being the most highly enriched gene ontologies altered by E2, and when comparing normal male and female phallus tissue. Further analysis showed, for the first time, the tissue region-specific expression of different hedgehog factors that we propose is akin to the discrete spatial functions of specific hedgehog factors in developing bone.

2. Results

2.1. Enrichment of Bone Differentiation Ontologies between Male and Female Phallus Tissue, or Males Treated with Oestradiol

Ranked lists of differentially expressed genes from outputs of DESeq2 analysis were assessed by gene ontology using GOrilla [25]. Due to the non-conventional animal model employed and the incomplete annotation of the wallaby genome available in these studies, only 40% of gene IDs were recognised in GOrilla analysis. Comparing male and female phallus tissue, all ontologies with significant enrichment by a false-discovery *q*-value <0.01 are listed in Figure 1a. The most highly enriched ontology was regulation of osteoblast differentiation (3.4-fold) and regulation of ossification (3.2-fold). Additional ontologies of significance include those relating to gene transcription, as well as cell proliferation, migration and adhesion. Gene ontology analysis for differentially expressed genes between male phallus tissue and male phallus tissue from pouch young treated with E2 are listed in Figure 1b. Muscle system process was the most significantly enriched ontology in this comparison, followed by other ontologies associated with muscle and cytoskeletal processes. Regulation of ossification was also enriched, though with a lower significance compared with enrichments in Figure 1a.

Examples of known bone differentiation-associated genes and expression changes from RNAseq data are shown in Figure 1c. Significant changes in a fundamental bone development factor and receptor, bone-morphogenic protein 5 (BMP5) and BMP receptor (BMPR2) were observed when comparing males with males treated with E2. Additionally, significant changes in chondrocyte differentiation markers (COL2A1 and COL9A2) and functional mediators of bone development (aggrecan (ACAN) and fibromodulin (FMOD)) were also observed.

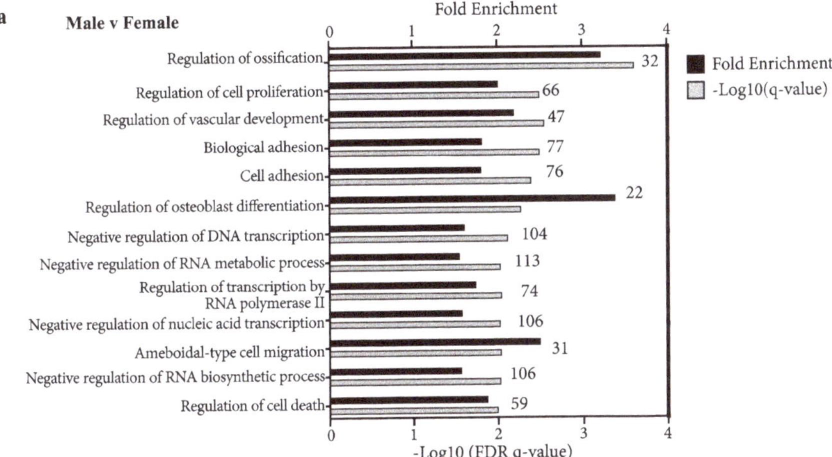

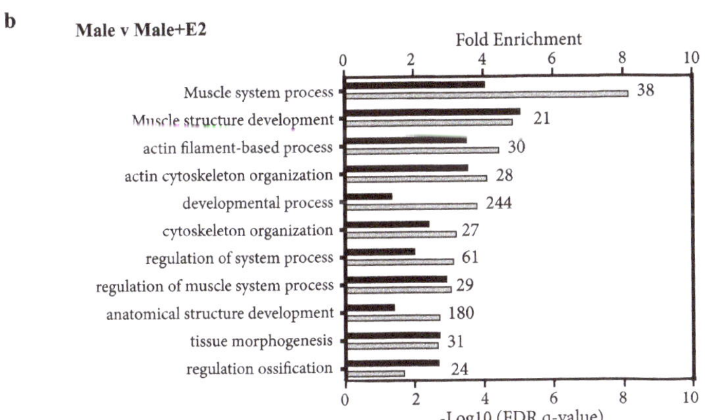

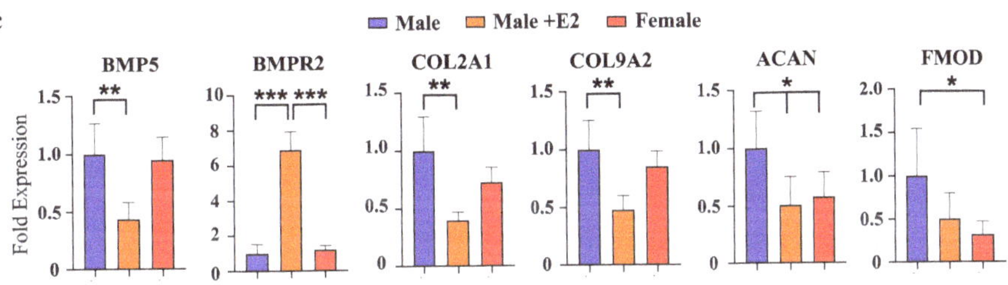

Figure 1. Enrichment of bone development ontologies in male versus female and males treated with oestradiol (**a**) Fold enrichment (black bars) and FDR q-values (grey bars) of process gene ontologies from differentially expressed genes between male and female phallus, or (**b**) male phallus with and without oestradiol treatment. Numbers above bars on graph indicate the number of enriched genes in that ontology (**c**) Representative bone/chondrocyte development genes and their relative levels in RNAseq data. (*n* = 5 animals/group; mean +/− s.d, * = <0.05, ** = <0.01, *** = <0.001).

2.2. Expression of Specific Hedgehog Factors Was Spatially Restricted in the Developing Phallus

As bone differentiation ontologies were most highly enriched when comparing male versus female phallus tissue, to determine whether SHH and IHH may have spatially-restricted functions in the developing phallus and urethra—as they do in bone development and repair—confocal immunofluorescence was performed to localise their expression through phallus development. Normal wallaby phallus tissue was immunolabelled for either SHH or IHH (Figure 2, red) in combination with the mesenchymal marker vimentin (Figure 2, green). In the distal phallus (Figure 2a–f), expression of SHH was observed throughout the urethral epithelium at all stages tested (Figure 2a–a′ (d20), 2c–c′ (d60), 2e–e′ (d90), asterisks). In contrast, IHH expression was either very low or undetectable in distal urethral plate epithelial cells (Figure 2b–b′ (d20), 2d–d′ (d60), 2f–f′ (d90)). However, IHH is known to be highly expressed in the gastrointestinal tract [26,27]. In Figure 2c–c′, the plane of sectioning captured the rectum. As an internal positive control, strong IHH expression is observed in the most apical cells of the rectal epithelium. Notably, the anogenital epithelium with the lowest SHH expression (Figure 2c′, arrow) coincides with that expressing high levels of IHH (Figure 2d′, arrow). In addition, under high z-axis resolution it was possible to observe an ordered arrangement of urethral plate epithelial cells, with cytoplasmic SHH directed towards regions at a critical juncture between mesenchymal cells and urethral plate epithelium (Figure 2e′, arrows pointing to a "focal point" marked by an asterisk).

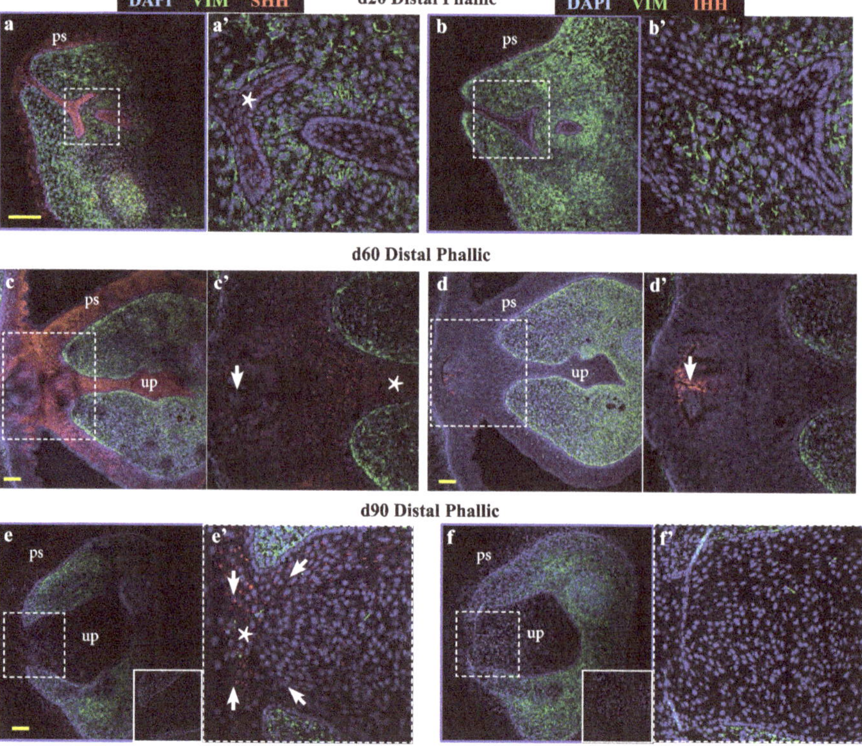

Figure 2. *Cont.*

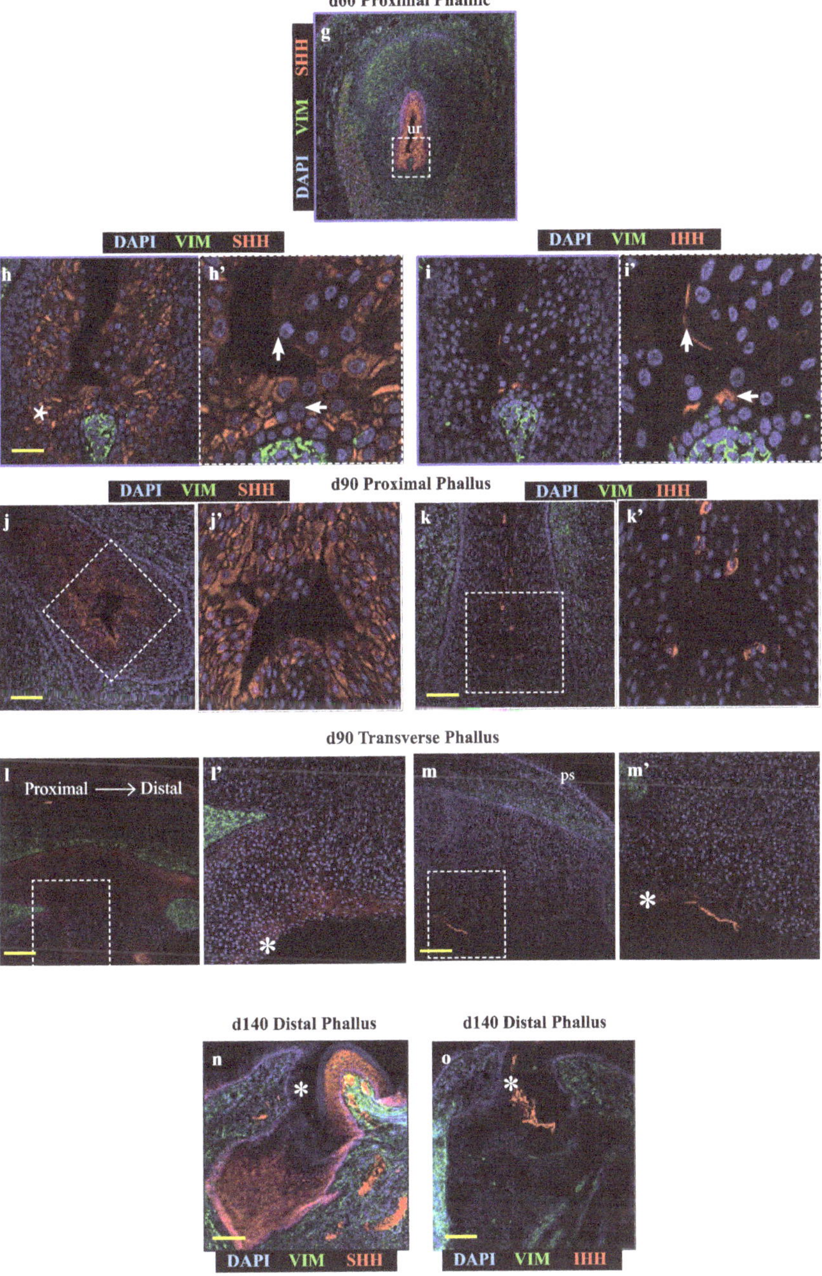

Figure 2. Sonic hedgehog (SHH) and Indian hedgehog (IHH) localize to discrete regions of the developing phallic epithelium. Day 20 male phallus immunolabelled for SHH (red—a, c, e, g, h, j) or IHH (red—b, d, f, i, k) and vimentin (green—all images). Dotted box denotes magnified portion in adjacent image. The expression of SHH in the distal phallic epithelium at day 20 (**a**), day 60 (**c**) and day

90 (**e**) is extensive in urethral plate epithelium (up) and phallic skin (ps). Cell direction can be inferred from cytoplasmic SHH expression (arrows in **e′**) that con- verge on a focal point between mesenchymal tissue (asterisk in **e′**). IHH expression is very low (**b**) or absent (**d,f**) from distal phallic urethral epithelium, though expression is observed in rectal epithelium (**d′**, arrow). (**g**) Day 60 proximal phallus section as reference for images of sequential sections in h–i. (**h**) SHH expression predominates in cells in an intermediate position within the epithelium (asterisk), while (**i**) IHH expression is restricted to cells closer to the luminal edge and in cells with low SHH expression (arrows in 2**h′**–**i′**). (**j**–**k**) Day 90 proximal phallus sections with a similar pattern of expression as at day 60. (**l**–**m**) Longitudinal sections of day 90 phallus showing SHH expression throughout the epithelium but low in the outermost cell populations. (**k**) IHH expression observed in regions with low SHH (asterisks in l–**m**). (**n**–**o**) Distal phallus sections from Day 150 phallus treated with estradiol showing extensive SHH expression (**n**) but absence of expression in regions dominated by IHH expression (**o**). Small image inserts in e and f show sections labeled using isotype control antibodies. (ur = urethra; ps = phallic skin; scale = 100 μm).

In contrast to the distal phallic epithelium, expression of IHH was observed in a small proportion of proximal phallic urethral epithelial cells. To determine whether IHH and SHH were expressed in the same population of cells, sequential sections were assessed at high magnification and z-axis resolution (low magnification image in Figure 2g, dotted box denotes area analysed at high magnification in Figure 2h,i). This revealed that SHH and IHH were expressed in separate areas, with SHH expression highest in cells with an intermediate position within the epithelium (Figure 2h, asterisk), while IHH expression was concentrated to cells nearer to the lumen edge (Figure 2i′, arrows). This segregated expression pattern was most clearly observed in the perineum (Figure S1).

This pattern was also observed in the proximal phallus of d90 animals (Figure 2j,k). Longitudinal sections of the phallus at d90 confirmed that SHH protein was present throughout the urethral epithelium and urethral plate epithelium at this developmental stage, though with differing intensities in different regions (Figure 2l). In contrast, IHH expression was restricted to the outermost layers of cells in the urethral plate and phallic skin (Figure 2m). This pattern of protein localisation was also observed at d140 (Figure 2n,o, asterisks). Overall, SHH and IHH were expressed in different tissue regions within the wallaby phallus epithelium, where SHH was associated with intermediate regions within the epithelium and IHH expression restricted to more luminal cell layers.

2.3. Specific Hedgehog Proteins Elicited Discrete Gene Expression Changes in Developing Phallus

To test whether the segregated expression pattern of SHH and IHH reflects separate molecular functions of these factors in the developing phallus, normal day 60 phallus tissue was cultured as explants and treated with either SHH or IHH for 24 h as outlined in Figure 3a. This developmental stage was selected as it is outside of the androgen programming window and at the very beginning of sexual dimorphic phallus development and urethral closure in the wallaby. After explant treatment, tissue was collected for RNA extraction and mRNA quantification by qPCR. Next it was investigated whether SHH and IHH can act to modulate expression of known regulators of urethral development, such as androgen and oestrogen receptor, as well as a known direct target of SHH in developing bone, the transcription factor SOX9 that itself plays important roles in sexual differentiation.

Whole organ or histological staining of phallus tissue cultured for 72 h showed no obvious tissue necrosis or cell death (Figures 3b and 3c, respectively). Treatment with either SHH or IHH failed to induce changes in oestrogen receptor (*ESR1*) or androgen receptor (*AR*) mRNA when compared to vehicle treated tissue (Figure 3d). In contrast, while SHH treatment induced a significant induction of *GLI2* transcription—a result confirming its in vitro activity—no increase in *GLI2* transcription was observed after treatment with IHH. However, IHH treatment induced a clear increase in *PTCH1* transcription, another SHH/IHH target gene. SHH treatment caused increases in *PTCH1* transcription that were highly variable and did not reach statistical significance. Upon replication with an additional primer set similar results were observed that reflects a likely biological or sampling source for this variation.

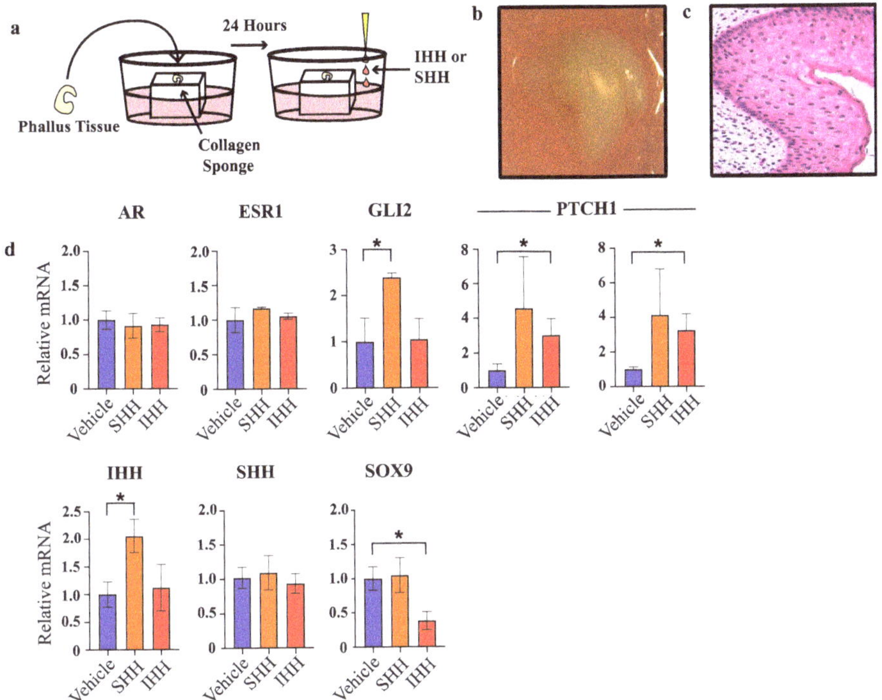

Figure 3. SHH and IHH elicit different transcriptional responses in phallus explant culture. (**a**) Explant culture protocol. (**b**) Gross appearance of phallus tissue after 72 h in culture. (**c**) Haematoxylin and esoin stain of PFA-fixed tissue section of phallus tissue in b. (**d**) Relative mRNA expression in phallus explants as determined by qPCR ($n = 3$, mean +/− s.d, * = <0.05).

Treatment with SHH induced a significant increase in *IHH* transcription (2.1-fold, $n = 3$, $p < 0.04$) but no change in *SHH* or *SOX9*. In contrast, IHH treatment induced a significant decrease in *SOX9* transcription (2.6-fold, $n = 3$, $p < 0.02$) but did not alter the levels of *IHH* or *SHH* transcription. These findings show that at this early stage of urethral closure in the tammar wallaby, SHH induced *IHH* transcription while IHH suppressed *SOX9* transcription.

2.4. Positive Association between SOX9 and SHH Protein Expression, and Negative Association between SOX9 and IHH Expression, in Normal Phallus Tissue

Findings in Figure 3 indicate that SHH and IHH have different actions on transcription in the developing phallus. To determine whether these findings are relevant on a tissue and protein level, immunolocalisation for IHH, SHH or SOX9 was performed on sequential oblique sections from normal d60 phallus tissue (Figure 4), to capture both a closed portion of the urethra and a component of the urethral plate epithelium.

Consistent with Figure 2, SHH localised along the entire urethral epithelium and urethral plate epithelium but was absent from two distinct regions of epithelial cells: the luminal-most cells in the closed urethra as well as regions in the outermost cell layers of the external penile skin (Figure 4a', asterisks). The expression pattern of epithelial IHH was broadly inverse to the expression pattern of SHH, with IHH expression identified only in the most luminal cells of the closed urethra and the outermost cell layer of phallus skin (Figure 4b', asterisks). Consistent with findings from explant cultures of reduced SOX9 transcription after treatment with IHH, regions expressing high IHH exhibit very low SOX9 expression (Figure 4c', asterisks). In these regions with low SOX9 there was significantly lower levels of the proliferation marker phosphorylated histone H3 (P-HH3: Figure 4c′, asterisks). Therefore, the ability of purified IHH to induce a reduction in *SOX9* transcription in phallus explant culture is supported by their protein localisation patterns at single cell level in vivo. Furthermore, expression of high epithelial IHH is associated with cells that do not express a marker of proliferation.

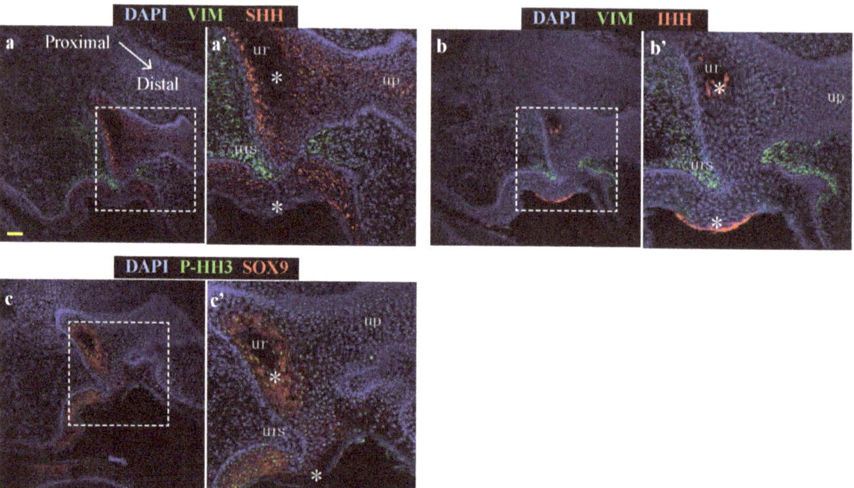

Figure 4. SHH and SOX9 expression is low in urethral epithelial and phallic skin cells expressing IHH (**a**) Expression of SHH (red) and vimentin (green), (**b**) Expression of IHH (red) and vimentin (green) or (**c**) Expression of SOX9 (red) and proliferation marker, P-HH3 (green) in oblique sections of day 60 wallaby phallus tissue. (ur = urethra; up = urethral plate epithelium) Scale = 100 µm.

3. Discussion

This study is the first in any species to identify that specific hedgehog proteins have distinct patterns of localisation and effects on gene expression in the developing phallus. Such region-specific expression and action of SHH and IHH led us to formulate a new model of urethral closure whereby the expression of IHH antagonises the actions of SHH, partly by inhibiting SOX9 expression. In this model, the coordinated timing of SHH and IHH expression at critical regions in the phallus is an important element of appropriate urethral development and closure. This conclusion is supported by the lack of proliferation in cells of the developing phallus that express IHH, while widespread proliferation was observed in SHH-expressing epithelial cells.

Results herein support other studies identifying gene regulatory networks present in bone/limb formation as being important in phallus development in a variety of species [3–6]. Furthermore, a comparative evolutionary study determined that a key bone structure of the phallus in many species, the baculum or os penis, has evolved a minimum of nine times and lost a minimum of 10 times in evolutionary history [28]. While both humans and wallabies lack an os penis, the fact such a structure has re-emerged many times provides a strong indication that the signalling networks involved in chondrogenesis and bone formation persist in the developing phallus of the wallaby. It should not, therefore, be surprising that genes involved in early bone formation were identified as being differentially expressed between male and female, as well as male wallabies treated with E2. However, results of the present study take this perspective further by demonstrating that, far from being surrogates, SHH and IHH likely play critically distinct roles during urethral closure that may mimic signalling networks present in developing bone. This is supported by their expression in discrete tissue compartments, and their ability to elicit different transcriptional responses in a key regulator of cell/tissue development, SOX9. A critical role is played by SOX9 during gonad as well as cartilage and bone development. Removal of SOX9 from SHH-expressing cells failed to induce morphological changes in genital development [29], though SOX9 expression in the mouse GT shows a restricted pattern within a subset of urethral and preputial epithelium that may be akin to those observed in the study presented herein. Therefore, it is possible that the role of SOX9 is critical in the IHH-expressing population that will have escaped SHH-driven knockout of SOX9.

These findings break new ground in our mechanistic understanding of phallus development and urethral closure. In particular these findings indicate that contrary to current models that are limited to epithelial-mesenchymal tissue interactions in hedgehog signalling, the mechanisms underpinning urethral closure may also involves inter/intra-epithelial signalling. While a novel finding in the phallus, distinct expression patterns and functions of SHH and IHH are found in other tissues. For example during craniofacial suture morphogenesis [30,31], as well as during tendon development, attachment and migration [32], where SHH and IHH are believed to occupy distinct tissue regions and play distinct temporal functions during development.

The restricted localisation of IHH protein to the luminal most cells in the proximal penile urethral, perineum and rectum/gastrointestinal tract may reflect an endodermal origin of these cells, that is consistent with a model of urethral closure whereby the urorectal septum divides the urogenital and anorectal tracts [33]. In that model, endodermal and ectodermal tissue fuse along the length of the phallus during urethral closure with the ectodermal cells contributing to the outermost layers of a subset of developing urethral epithelium and penile skin. It is possible that the interactions of IHH and SHH during phallus development described through results of the study presented herein, serve as the signalling network regulating the interactions of these two tissues during urethral closure. Such a model adds an extra dimension to our understanding of the cell and tissue interactions regulating urethral closure.

Overall this study demonstrated evidence of a novel mechanism to explain the process of urethral closure during phallus development, providing important new direction for researchers aiming to understand the tissue signalling networks underpinning phallus development and urethral closure.

4. Materials and Methods

4.1. Animal Treatments and Tissue Collection

Tammar wallabies were collected from wild populations originating on Kangaroo Island (South Australia) and were held in a breeding colony in Melbourne (Victoria) for breeding and experimentation. Females were monitored from 25–30 days after removal of a pouch young that initiates reactivation of their diapause blastocysts, to check for newborns [34]. The sex of newborns was determined by the presence of scrotal or mammary primordia, as previously described. For young of unknown day of birth, this was estimated from measurements of head length and weight from established growth curves [35]. Pouch young were treated from days 20–40 after birth, and treatments were performed as previously described [36].

Phallus tissues were collected from tammar wallaby pouch young after anaesthesia with Zoletil® 100 (Lyppard Australia Cat# ZOLED (Tiletamine HCl 50 mg/mL, Zolazepam HCl 50 mg/mL, used at 1 mL/kg)). Samples were snap-frozen and stored at −80 °C for RNA extraction or immersion-fixed in 4% paraformaldehyde for histological analysis. All animals were treated and tissues collected under appropriate permits, and experiments approved by the University of Melbourne Animal Experimentation Ethics Committees in accordance with the National Health and Medical Research Council of Australia animal ethics guidelines.

4.2. Explant Culture and Treatment

Collagen dental sponges (10 cm cubes—(Dental Solutions Israel, Israel, Cat# DSP-32)) were immersed in DMEM plus 10% fetal calf serum (Thermo-Fisher, Waltham MA, USA, Cat#s 11995115 and 10100147, respectively) until saturated. Dissected phallus tissue from d60 male pouch young were cultured atop saturated sponges for 24 h at 37 °C for tissue acclimatisation. After this, media was removed and replaced with DMEM plus 10% fetal calf serum, with the addition of either vehicle (ethanol—0.25%), SHH (0.25 ug/mL—Abcam, Cambridge, UK, #Ab123773) or IHH (0.25 ug/mL—Abcam, Cambridge, UK, #Ab243268) and media allowed to equilibrate for 5 min at 37 °C. This was subsequently replaced with fresh media supplemented with SHH or IHH and culture for an additional 24 h. Phallus tissue was subsequently snap-frozen for RNA extraction.

4.3. RNA Extraction and Next Generation Sequencing

RNA material was extracted and prepared for sequencing as previously described [37]. Briefly, RNA was subjected to multiplex indexed-RNAseq analyses using TruSeq kit (Illumina, San Diego, CA, USA) and a HiSeq2500 analyzer (Illumina). Roughly $10–14 \times 10^6$ reads (single end 100 bp) were obtained from each sample after Q.V.>30 filtering. RNA sequencing data were analysed with FastQC followed by CutAdapt to remove bases sequenced with high uncertainty. Mapping was performed With RNA Star [38] and the annotated tammar wallaby genome 3.0 (Heider et al., personal communications). The number of reads for each annotated gene was determined using FeatureCounts. Differential gene expression analysis was executed using DESeq2 [39].

4.4. RNA Extraction for cDNA Generation and qPCR Analysis

Frozen phallus tissues were digested in TriZOL® reagent (Thermo Fisher Scientific, Waltham MA, USA, Cat# 15596026) using an IKA T10® basic homogeniser (Lab Gear, Australia, Cat# IKAW3737000), followed by chloroform extraction as per manufacturer's instructions using a PureLink RNA mini kit (Thermo Fisher Scientific, Waltham MA, USA, Cat# 12183018A). Contaminating DNA was removed using DNase (Ambion, Austin, TX, USA Cat# 1906) and the manufacturer's protocol. For cDNA synthesis, 500ng of RNA was used as template employing the SuperScript IV® cDNA synthesis kit (Thermo Fisher Scientific, Waltham MA, USA, Cat# 18091050). For quantification of cDNA, SsoAdvanced® Universal SYBR Green Supermix was employed (BioRad, Hercules, CA, USA, Cat# 1725272) utilising primers for AR, ESR1, SHH, IHH, SOX9 and the housekeeping genes GAPDH and 18S (see Table 1). A 384-well QuantStudio® 5 thermal light cycler (Thermo Fisher Scientific, Waltham MA, USA) was employed for quantification of SYBR intensities. The standard curve method was employed for relative quantification of cDNA.

Table 1. Sequences of qPCR primers used in Figure 4.

Target Gene	Sequences
ACTB	F – TTGCTGACAGGATGCAGAAG
	R – AAAGCCATGCCAATCTCATC
AR	F – CACATTGAAGGCTATGCGTG
	R – CCCATCCAGGAGTACTGAAT
ESR1	F – TGATCAACTGGGCAAAGAGGG
	R – GATGTAGCCAGCAACATGTCA
GAPDH	F – TCCCAATGTATCTGTTGTGGATCTG
	R – AACCATACTCATTGTCATACCAAGAAAT
GLI2	F – GTTCACAGCTAGTGGCTCC
	R – ACTGCTGCCTCACTGCTTTG
IHH	F – CTTCCTGGCCTTCTTGGACC
	R – CTTCCTGGCCTTCTTGGACC
PTCH1 #1	F – AATGAAGACAAGGCAGCAG
	R – TAGCAACTCGGATAACACT
PTCH1 #2	F – AATGACTCCCAAGCAAATGTA
	R – TAGACAGGCATAGGCGAGCAT
SHH	F – CTTCCTGGCCTTCTTGGACC
	R – CTTCCTGGCCTTCTTGGACC
SOX9	F – TGCGAGTCAATGGCTCTAGCAA
	R – CTCCTCCGAGGTTGGTATTTGT

4.5. Immunofluorescence Microscopy

Wallaby young were removed from the pouch and phallus tissue dissected and placed in 4% paraformaldehyde overnight at 4 °C on a carousel. Tissue was washed for 2 × 30 min in PBS (pH7.4) at room temperature followed by transfer to 70% ethanol for tissue processing. Tissue was processed and embedded in paraffin wax, sectioned at 5um thickness before mounting on Superfrost® ultra PLUS slides (Thermo-Fisher Scientific, Waltham MA, USA). Tissue dewaxing, rehydration, antigen retrieval, primary and secondary antibody staining protocols were performed as previously described [40], with an additional blocking step being added between antigen retrieval and primary antibody incubation (see Table 2 for antibody incubation conditions). This blocking step involved incubation of slides with 0.3% Sudan black (ProSciTech, Qld, Australia) in 70% ethanol for 10 min. This was followed by brief immersion in fresh 70% ethanol before incubation for 5 min in PBS and primary antibody incubation. Images were acquired on a Nikon (Minato City, Japan) A1R® spectral confocal microscope. Specificity of the primary antibody was determined by staining positive and negative control tissue, as well as substitution of the target-specific primary antibodies with non-specific IgG control antibodies. All 40x and 63x images were acquired at minimum airy unit values to maximise z-axis resolution.

Table 2. Primary antibodies and incubation conditions.

Target Protein	Supplier and Catalogue Number	Dilution
IHH	LS Bio, Seattle, WA, USA, # LS-40514	1:400
SHH	Abcam, Cambridge, UK, # Ab19897	1:200
SOX9	Merck, Kenilworth, NJ, USA # Ab5535	1:800
Vimentin	Abcam, Cambridge, UK, # Ab8069	1:400

4.6. Haematoxylin and Eosin Staining of Explant-Cultured Phallus Tissue

Explant tissue was processed into paraffin blocks, sectioned, dewaxed and rehydrated as per immunofluorescence protocol. Sections were then immersed in Mayer's haematoxylin (Sigma-Aldrich, St Louis MS, USA) for 5 min, washed in tap water for 1 min, followed by incubation in 0.3% acid ethanol for 1 min, washing in tap water for 1 min, immersion in 1% eosin (Sigma,-Aldrich, St Louis MS, USA) in 75% ethanol and a final 1 min wash in running tap water. This was followed by dehydration in graded ethanol, clearing in histolene and mounting with UltraMount No. 4 (Thermo-Fisher Scientific, Waltham MA, USA).

Supplementary Materials: Supplementary materials can be found at http://www.mdpi.com/1422-0067/21/4/1237/s1.

Author Contributions: Conceptualisation, G.A.T.; methodology, G.A.T.; formal analysis, G.A.T.; original draft writing, G.A.T.; review and editing, G.A.T., A.J.P. and M.B.R.; project administration, G.A.T. and M.B.R.; funding acquisition, M.B.R. and A.J.P. All authors have read and agreed to the published version of the manuscript.

Funding: This research was funded by a research grant from the National Health & Medical Research Council of Australia.

Acknowledgments: The authors wish to thank Geoff Shaw and Jean Wilson for contributions to funding acquisition, Asao Fujiyama for providing RNA sequencing services and Yu Chen for assistance with sample preparation and animal treatments.

Conflicts of Interest: The authors declare no conflict of interest.

Abbreviations

ACAN	Aggrecan
AR	Androgen Receptor
BMP5	Bone morphogenic protein 5
E2	Oestradiol
ESR1	Oestrogen Receptor Alpha
FMOD	Fibromodulin
IHH	Indian Hedgehog
SHH	Sonic Hedgehog

References

1. Nassar, N.; Bower, C.; Barker, A. Increasing prevalence of hypospadias in Western Australia, 1980–2000. *Arch. Dis. Child.* **2007**, *92*, 580–584. [CrossRef] [PubMed]
2. Springer, A.; Heijkant, M.V.D.; Baumann, S.; Information, P.E.K.F.C. Worldwide prevalence of hypospadias. *J. Pediatr. Urol.* **2016**, *12*, 152.e1–152.e7. [CrossRef] [PubMed]
3. O'Shaughnessy, K.L.; Dahn, R.D.; Cohn, M.J. Molecular development of chondrichthyan claspers and the evolution of copulatory organs. *Nat. Commun.* **2015**, *6*, 6698. [CrossRef]
4. Tschopp, P.; Sherratt, E.; Sanger, T.J.; Groner, A.C.; Aspiras, A.C.; Hu, J.K.; Pourquie, O.; Gros, J.; Tabin, C.J. A relative shift in cloacal location repositions external genitalia in amniote evolution. *Nature* **2014**, *516*, 391–394. [CrossRef] [PubMed]
5. Herrera, A.M.; Cohn, M.J. Embryonic origin and compartmental organization of the external genitalia. *Sci. Rep.* **2014**, *4*, 6896. [CrossRef] [PubMed]
6. Infante, C.R.; Mihala, A.G.; Park, S.; Wang, J.S.; Johnson, K.K.; Lauderdale, J.D.; Menke, D.B. Shared Enhancer Activity in the Limbs and Phallus and Functional Divergence of a Limb-Genital cis-Regulatory Element in Snakes. *Dev. Cell* **2015**, *35*, 107–119. [CrossRef]
7. Perriton, C.L.; Powles, N.; Chiang, C.; Maconochie, M.K.; Cohn, M.J. Sonic hedgehog Signaling from the Urethral Epithelium Controls External Genital Development. *Dev. Boil.* **2002**, *247*, 26–46. [CrossRef]
8. Seifert, A.W.; Bouldin, C.M.; Choi, K.-S.; Harfe, B.D.; Cohn, M.J. Multiphasic and tissue-specific roles of sonic hedgehog in cloacal septation and external genitalia development. *Development* **2009**, *136*, 3949–3957. [CrossRef]
9. Seifert, A.W.; Zheng, Z.; Ormerod, B.K.; Cohn, M.J. Sonic hedgehog controls growth of external genitalia by regulating cell cycle kinetics. *Nat. Commun.* **2010**, *1*, 1–9. [CrossRef]
10. Haraguchi, R.; Suzuki, K.; Murakami, R.; Sakai, M.; Kamikawa, M.; Kengaku, M.; Sekine, K.; Kawano, H.; Kato, S.; Ueno, N.; et al. Molecular analysis of external genitalia formation: The role of fibroblast growth factor (Fgf) genes during genital tubercle formation. *Development* **2000**, *127*, 2471–2479.
11. Miyagawa, S.; Matsumaru, D.; Murashima, A.; Omori, A.; Satoh, Y.; Haraguchi, R.; Motoyama, J.; Iguchi, T.; Nakagata, N.; Hui, C.-C.; et al. The role of sonic hedgehog-Gli2 pathway in the masculinization of external genitalia. *Endocrinology* **2011**, *152*, 2894–2903. [CrossRef] [PubMed]
12. Carmichael, S.L.; Ma, C.; Choudhry, S.; Lammer, E.J.; Witte, J.S.; Shaw, G.M. Hypospadias and genes related to genital tubercle and early urethral development. *J. Urol.* **2013**, *190*, 1884–1892. [CrossRef] [PubMed]
13. Zheng, Z.; Armfield, B.A.; Cohn, M.J. Timing of androgen receptor disruption and estrogen exposure underlies a spectrum of congenital penile anomalies. *Proc. Natl. Acad. Sci. USA* **2015**, *112*, E7194–E7203. [CrossRef] [PubMed]
14. Bouty, A.; Ayers, K.L.; Pask, A.; Héloury, Y.; Sinclair, A.H. The Genetic and Environmental Factors Underlying Hypospadias. *Sex. Dev.* **2015**, *9*, 239–259. [CrossRef] [PubMed]
15. Dean, A.; Smith, L.B.; MacPherson, S.; Sharpe, R.M. The effect of dihydrotestosterone exposure during or prior to the masculinization programming window on reproductive development in male and female rats. *Int. J. Androl.* **2012**, *35*, 330–339. [CrossRef]
16. Ittiwut, C.; Pratuangdejkul, J.; Supornsilchai, V.; Muensri, S.; Hiranras, Y.; Sahakitrungruang, T.; Watcharasindhu, S.; Suphapeetiporn, K.; Shotelersuk, V. Novel mutations of the SRD5A2 and AR genes in Thai patients with 46, XY disorders of sex development. *J. Pediatr. Endocrinol. Metab.* **2017**, *30*, 19–26. [CrossRef]

17. Matsushita, S.; Suzuki, K.; Murashima, A.; Kajioka, D.; Acebedo, A.R.; Miyagawa, S.; Haraguchi, R.; Ogino, Y.; Yamada, G. Regulation of masculinization: Androgen signalling for external genitalia development. *Nat. Rev. Urol.* **2018**, *15*, 358–368. [CrossRef]

18. Murakami, R. A histological study of the development of the penis of wild-type and androgen-insensitive mice. *J. Anat.* **1987**, *153*, 223–231.

19. Welsh, M.; MacLeod, D.J.; Walker, M.; Smith, L.B.; Sharpe, R.M. Critical androgen-sensitive periods of rat penis and clitoris development. *Int. J. Androl.* **2010**, *33*, e144–e152. [CrossRef]

20. MacLeod, D.J.; Sharpe, R.M.; Welsh, M.; Fisken, M.; Scott, H.M.; Hutchison, G.R.; Drake, A.J.; Driesche, S.V.D. Androgen action in the masculinization programming window and development of male reproductive organs. *Int. J. Androl.* **2010**, *33*, 279–287. [CrossRef]

21. Sinclair, A.W.; Cao, M.; Pask, A.; Baskin, L.; Cunha, G.R. Flutamide-induced hypospadias in rats: A critical assessment. *Differ.* **2017**, *94*, 37–57. [CrossRef] [PubMed]

22. Driesche, S.V.D.; Kilcoyne, K.R.; Wagner, I.; Rebourcet, D.; Boyle, A.; Mitchell, R.; McKinnell, C.; MacPherson, S.; Donat, R.; Shukla, C.J.; et al. Experimentally induced testicular dysgenesis syndrome originates in the masculinization programming window. *JCI Insight* **2017**, *2*, e91204. [PubMed]

23. Shao, M.; Ghosh, A.; Cooke, V.G.; Naik, U.P.; Martin-DeLeon, P.A. JAM-A is present in mammalian spermatozoa where it is essential for normal motility. *Dev. Biol.* **2008**, *313*, 246–255. [CrossRef] [PubMed]

24. Sinclair, A.W.; Cao, M.; Baskin, L.; Cunha, G.R. Diethylstilbestrol-induced mouse hypospadias: "window of susceptibility". *Differentation* **2016**, *91*, 1–18. [CrossRef] [PubMed]

25. Eden, E.; Navon, R.; Steinfeld, I.; Lipson, D.; Yakhini, Z. GOrilla: A tool for discovery and visualization of enriched GO terms in ranked gene lists. *BMC Bioinform.* **2009**, *10*, 48. [CrossRef]

26. Brink, G.R.V.D. Hedgehog Signaling in Development and Homeostasis of the Gastrointestinal Tract. *Physiol. Rev.* **2007**, *87*, 1343–1375. [CrossRef]

27. Brink, G.R.V.D.; Bleuming, S.A.; Hardwick, J.C.H.; Schepman, B.L.; Offerhaus, G.J.; Keller, J.J.; Nielsen, C.; Gaffield, W.; Van Deventer, S.J.H.; Roberts, U.J.; et al. Indian Hedgehog is an antagonist of Wnt signaling in colonic epithelial cell differentiation. *Nat. Genet.* **2004**, *36*, 277–282. [CrossRef]

28. Schultz, N.G.; Lough-Stevens, M.; Abreu, E.; Orr, T.; Dean, M.D. The Baculum was Gained and Lost Multiple Times during Mammalian Evolution. *Integr. Comp. Boil.* **2016**, *56*, 644–656. [CrossRef]

29. Sreenivasan, R.; Gordon, C.T.; Benko, S.; De Iongh, R.; Bagheri-Fam, S.; Lyonnet, S.; Harley, V. Altered SOX9 genital tubercle enhancer region in hypospadias. *J. Steroid Biochem. Mol. Boil.* **2017**, *170*, 28–38. [CrossRef]

30. Lenton, K.; James, A.W.; Manu, A.; Brugmann, S.A.; Birker, D.; Nelson, E.R.; Leucht, P.; Helms, J.A.; Longaker, M.T. Indian hedgehog positively regulates calvarial ossification and modulates bone morphogenetic protein signaling. *Genesis* **2011**, *49*, 784–796. [CrossRef]

31. Santagati, F.; Rijli, F.M. Cranial neural crest and the building of the vertebrate head. *Nat. Rev. Neurosci.* **2003**, *4*, 806–818. [CrossRef] [PubMed]

32. Felsenthal, N.; Rubin, S.; Stern, T.; Krief, S.; Pal, D.; Pryce, B.A.; Schweitzer, R.; Zelzer, E. Development of migrating tendon-bone attachments involves replacement of progenitor populations. *Development* **2018**, *145*, dev165381. [CrossRef] [PubMed]

33. Seifert, A.W.; Harfe, B.D.; Cohn, M.J. Cell lineage analysis demonstrates an endodermal origin of the distal urethra and perineum. *Dev. Boil.* **2008**, *318*, 143–152. [CrossRef] [PubMed]

34. Wai-Sum, O.; Short, R.V.; Renfree, M.B.; Shaw, G. Primary genetic control of somatic sexual differentiation in a mammal. *Nature* **1988**, *331*, 716–717. [CrossRef]

35. Poole, W.E.; CSIRO. *Division of Wildlife and Ecology, Tables for Age Determination of the Kangaroo Island Wallaby (Tammar), Macropus Eugenii, from Body Measurements*; CSIRO Division of Wildlife and Ecology: Canberra, Austrialia, 1991; p. 37.

36. Coveney, D.; Shaw, G.; Renfree, M.B. Estrogen-induced gonadal sex reversal in the tammar wallaby. *Boil. Reprod.* **2001**, *65*, 613–621. [CrossRef]

37. Chen, Y.; Yu, H.; Pask, A.J.; Fujiyama, A.; Suzuki, Y.; Sugano, S.; Shaw, G.; Renfree, M.B. Hormone-responsive genes in the SHH and WNT/beta-catenin signaling pathways influence urethral closure and phallus growth. *Biol. Reprod.* **2018**, *99*, 806–816.

38. Dobin, A.; Davis, C.A.; Schlesinger, F.; Drenkow, J.; Zaleski, C.; Jha, S.; Batut, P.; Chaisson, M.; Gingeras, T.R. STAR: Ultrafast universal RNA-seq aligner. *Bioinformatics* **2013**, *29*, 15–21. [CrossRef]

39. Love, M.I.; Huber, W.; Anders, S. Moderated estimation of fold change and dispersion for RNA-seq data with DESeq2. *Genome Biol.* **2014**, *15*, 002832. [CrossRef]

40. Tarulli, G.A.; Stanton, P.G.; Lerchl, A.; Meachem, S.J. Adult Sertoli Cells Are Not Terminally Differentiated in the Djungarian Hamster: Effect of FSH on Proliferation and Junction Protein Organization1. *Boil. Reprod.* **2006**, *74*, 798–806. [CrossRef]

International Journal of
Molecular Sciences

Review

The Roles of Indian Hedgehog Signaling in TMJ Formation

Till E. Bechtold [1,2], Naito Kurio [3,†], Hyun-Duck Nah [3], Cheri Saunders [1], Paul C. Billings [1] and Eiki Koyama [1,*]

[1] Division of Orthopaedic Surgery, Department of Surgery, The Children's Hospital of Philadelphia, Philadelphia, PA 19104, USA; tilledward@hotmail.com (T.E.B.); SaundersC1@email.chop.edu (C.S.); BillingsP@email.chop.edu (P.C.B.)

[2] Department of Orthodontics, Dentofacial Orthopedics and Pedodontics, Center for Dental and Craniofacial Sciences (CC3), Charité–University Hospital Berlin, D-14197 Berlin, Germany

[3] Division of Plastic and Reconstructive Surgery, Department of Surgery, The Children's Hospital of Philadelphia, Philadelphia, PA 19104, USA; kurio.naito@tokushima-u.ac.jp (N.K.); NAH@email.chop.edu (H.-D.N.)

* Correspondence: koyamae@email.chop.edu; Tel.: +267-425-2073; Fax: +267-426-7814

† Current address: Department of Oral Surgery, Institute of Biomedical Sciences, Tokushima University Graduate School, Tokushima 770-8504, Japan.

Received: 20 November 2019; Accepted: 10 December 2019; Published: 13 December 2019

Abstract: The temporomandibular joint (TMJ) is an intricate structure composed of the mandibular condyle, articular disc, and glenoid fossa in the temporal bone. Apical condylar cartilage is classified as a secondary cartilage, is fibrocartilaginous in nature, and is structurally distinct from growth plate and articular cartilage in long bones. Condylar cartilage is organized in distinct cellular layers that include a superficial layer that produces lubricants, a polymorphic/progenitor layer that contains stem/progenitor cells, and underlying layers of flattened and hypertrophic chondrocytes. Uniquely, progenitor cells reside near the articular surface, proliferate, undergo chondrogenesis, and mature into hypertrophic chondrocytes. During the past decades, there has been a growing interest in the molecular mechanisms by which the TMJ develops and acquires its unique structural and functional features. Indian hedgehog (Ihh), which regulates skeletal development including synovial joint formation, also plays pivotal roles in TMJ development and postnatal maintenance. This review provides a description of the many important recent advances in Hedgehog (Hh) signaling in TMJ biology. These include studies that used conventional approaches and those that analyzed the phenotype of tissue-specific mouse mutants lacking Ihh or associated molecules. The recent advances in understanding the molecular mechanism regulating TMJ development are impressive and these findings will have major implications for future translational medicine tools to repair and regenerate TMJ congenital anomalies and acquired diseases, such as degenerative damage in TMJ osteoarthritic conditions.

Keywords: TMJ; synovial joint; articular disc; *Ihh*; *PTHrP*; osteoarthritis

---

1. Introduction

The temporomandibular joint (TMJ), like joints in the shoulder, hip, and knee, is a highly specialized synovial joint and plays a pivotal role in the functioning of the mammalian jaw [1–5]. The TMJ consists of the glenoid fossa in the temporal bone, a condylar head of the mandible, and a fibrocartilaginous articular disc intervening between the fossa and condyle (Figure 1A). Condylar cartilage, unlike the cartilage present in developing limbs, is classified as secondary cartilage, undergoes endochondral ossification, and displays characteristic developmental and growth processes [6,7]. In mammalian embryos, the first overt sign of mandibular condylar development is the appearance of a neural crest-derived cell condensation at the supra-lateral site of the jaw anlagen. The condensation is likely of the periosteal origin within the jaw anlagen [8] or may derive from a separate distinct condensation [9,10]. At this early stage, there is no obvious sign of an intervening articular disc primordium. The condylar condensation differentiates into cartilage and forms a growth plate-like structure, which displays the characteristic zonal organization, consisting of fibroblasts, chondroprogenitor cells, and chondrocytes, along its main axis. These layers are characterized by differences in cell shape and properties and are designated (from the surface): (1) superficial cell layer, (2) fibrous/polymorphic progenitor cell layer, (3) zone of flattened chondrocytes, and (4) zone of hypertrophic chondrocytes (Figure 1B) [11–13].

In embryos, the condyle undergoes rapid growth and elongation toward the differentiating temporal bone. Interestingly, the longitudinal growth of the condyle during embryonic and postnatal life primarily results from appositional growth at its apical end, where chondro-progenitor cells residing in the polymorphic cell layer proliferate and differentiate into chondrocytes that in turn become incorporated into the underlying condylar cartilage. Hence, condylar cartilage functions as a growth site of the developing mandibular bone. Therefore, condyle elongation differs from that taking place in other developing skeletal elements, such as long bones or cranial base synchondroses, in which elongation is contributed to by mitotic activity of chondrocytes within the growth plates. With time, the newly differentiated condylar chondrocytes undergo maturation and hypertrophy and are eventually replaced by endochondral bone connecting to the condylar process [14–17]. A recent study indicated that a small number of chondrocytes may directly differentiate into osteoblasts and form the underlying subarticular bone of the condyle [18].

The development of the articular disc initiates with the formation of a separate flat-shaped ecto-mesenchymal cell condensation located between the developing condylar apex and the glenoid fossa of the temporal bone [19]. With time, the articular disc primordium becomes apparent by the creation of upper and lower articular cavities filled with synovial fluid. The disc subsequently develops into a fibrocartilage structure displaying (1) a biconcave shape with thicker peripheral portions (designated as anterior and posterior bands, respectively) and attaching to the TMJ capsules or the lateral pterygoid muscle and (2) a relatively thin central portion-intermediate zone [19,20]. The TMJ disc and joint cavities enable the condyle to rotate and translate along the glenoid fossa and eminence of the temporal bone during TMJ function.

Although the general development of condyles and articular discs in the TMJ is well understood, comparatively little is known regarding the molecular mechanisms controlling glenoid fossa formation. The glenoid fossa of the temporal bone derives from cranial neural crest cells [4,21,22]. Compared to the articular surface of the mandibular condyle, the articular surface of the glenoid fossa is quite distinct: sporadically distributed chondrocyte progenitors display less proliferative activity and hypertrophic chondrocyte synthesize very little, if any, cartilage matrix [23]. As taking place in developing condylar cartilage, chondrocytes differentiate amongst type I collagen (*Col I*)-expressing mesenchymal cells in a presumptive articulating layer covering the temporal bone where the condyle articulates, exhibiting features of secondary cartilage-like condylar cartilage [24,25]. With time, these chondrocytes undergo endochondral ossification, become entrapped in the intramembranous bony matrix of the temporal bone, and form chondroid bone. Absence or dislocation of the condyle results in arrested glenoid fossa development, suggesting that proper signals and/or mechanical stimulation by the condyle are required to sustain proper glenoid fossa development [26].

Indian hedgehog (Ihh), a member of the Hh family of signaling molecules, is widely recognized as a critical regulator of skeletal development [27–30]. Ihh is expressed in prehypertrophic and early hypertrophic chondrocytes of the developing growth plate [31,32] and regulates a number of processes including (1) intramembranous bone collar formation [32–34], (2) chondrocyte proliferation and maturation rate [35], (3) expression of parathyroid hormone-related protein (PTHrP) in periarticular tissue [36], and (4) endochondral ossification [37,38]. Binding of Ihh to Patched1 (Ptch1), its 12-pass transmembrane receptor, leads to the displacement of Ptch1 from primary cilium, an organelle that bulges from the cell surface. This allows Smoothened (Smo), a 7-pass transmembrane receptor, to be phosphorylated and activate glioma-associated oncogene (Gli) proteins, a family of zinc-finger transcription factors that include Gli1, Gli2, and Gli3. In the absence of Hh ligands, Ptch1 localizes at the base of the primary cilia, preventing Smo from activating the cilium [39–45]. Under these conditions, Gli2 and Gli3 are subjected to proteolytic cleavage to generate C-terminal truncated forms that repress the transcription of Hh target genes [46–48], whereas GLi1, due to a lack of the protein kinase A recognition site necessary for phosphorylation and subsequent cleavage, is thought to function exclusively as an activator [49–51]. Studies utilizing *Ihh*-null mouse embryos have provided not only conclusive evidence that Ihh plays multiple roles in long bone development, but also regulates synovial joint formation [1,37,52,53]. The digits of *Ihh*-mutant embryos remain uninterrupted, while heterozygous or wild-type littermates displayed obvious joints. Despite the remarkable nature of these observations and their potentially fundamental implications for other joints in the body, it has remained largely obscure, until quite recently, how Ihh regulates TMJ joint formation, growth, and maintenance [4,54].

In this review, we discuss the important findings on the involvement of Hh signaling in TMJ development during embryonic and early postnatal stages as well as in TMJ establishment and maintenance at postnatal life. We also discuss the possible involvement of Hh pathways in osteoarthritic conditions.

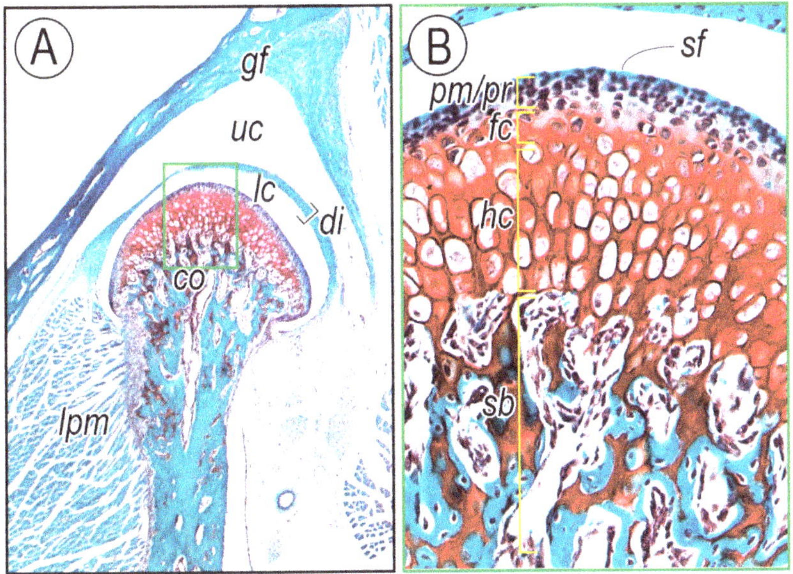

Figure 1. Histology of the TMJ. (**A**) TMJs from 3-month-old wild-type mice were sectioned along their longitudinal axis and sections were stained with safranin O/fast green. (**B**) High-magnification picture of the green boxed area in (**A**), showing the characteristic cellular organization of the condylar cartilage with superficial layer (*sf*), polymorphic/progenitor layer (*pm/pr*), flattened chondrocyte zone (*fc*), hypertrophic chondrocyte zone (*hc*), and subchondral bone (*sb*). *gf*, glenoid fossa; *uc*, upper joint cavity; *di*, articular disc; *lc*, lower joint cavity; *cd*, condyle; *lpm*, lateral pterygoid muscle.

2. Recent Experimental Findings

2.1. Abnormal TMJ Development in Ihh-Null Mice at Embryonic and Early Postnatal Life

It is well established that Indian hedgehog (Ihh) signaling is essential for early axial and appendicular skeletal development [35–37,55]. Thus, initial analyses were carried out in skulls from embryonic and newborn (P0) wild-type mice and their corresponding *Ihh*$^{-/-}$ littermates. Wild-type mandibles exhibited their typical elongated morphologies and much of the mandibular body was ossified and stained with alizarin red (Figure 2A). The condylar process (*co*) was prominent and its most apical region contained a typical cartilaginous condyle (Figure 2B). The outer surface of the central and basal regions of the condylar process was surrounded by newly differentiated intramembranous bone (Figure 2B). The angular process (*ap*), where secondary cartilage develops at the apical end, was prominent as well (Figure 2A). In corresponding *Ihh*$^{-/-}$ littermates, the overall length of the mandibular body was reduced as much as 30% and other components, including the condylar process, condyle cartilage, and angular process (*ap*), were all affected (Figure 2C,D).

Detailed histological examination revealed additional structural defects and cellular derangement in *Ihh* mutant TMJs. E15.5 wild-type condyle anlagen contained chondrocytes in their central portions that were circumscribed by a distinct mesenchymal condensation corresponding to disc primordium. By E18.5 to newborn, a complete disc along with upper and lower cavities had formed (Figure 2E, single and double arrows in the right side panel), while condylar chondrocytes displayed typical growth plate-like zonal organization, including a superficial (*sf*) layer, a polymorphic (*pm*)/chondro-progenitor layer, a flattened chondrocyte (*fc*) layer, and hypertrophic chondrocyte (*hc*) layer (Figure 2E). In *Ihh*$^{-/-}$ embryos, condylar chondrocytes were also present by E15.5, but, strikingly, the disc primordium was absent or not discernable. The absence of disc and joint cavities was evident at E18.5 (Figure 2F), such that the condyle directly opposed the glenoid fossa (*gf*). In addition, most of the mutant chondrocytes had undergone hypertrophy by E18.5 with a concurrent reduction in thickness of both the flattened chondrocyte layer (*fc*) and polymorphic (*pm*) layer (Figure 2F). Interestingly, while some phenotypic defects caused by Ihh deficiency are rescued by the concurrent absence of *Gli3* in developing limbs [56], the disc phenotype of *Ihh*$^{-/-}$ mutants was not rescued in double *Ihh*$^{-/-}$;*Gli3*$^{-/-}$ mutants, suggesting unique functions of Ihh in the TMJ [11]. Abnormal formation of the mandibular condyle, articular disc, and joint cavity are also reported in mice carrying mutations in genes (1) directly involved in hedgehog signaling (*Smo*, *Glis*) [57,58], (2) that interact with the hedgehog signaling pathway (*Trps1*) [59], or (3) that reduce or eliminate *Ihh* expression (*Shox2*, *Sox9*) [22,26]. Yang et al. reported that augmented Ihh signaling in cranial neural crest cells caused severe craniofacial abnormalities, including TMJs, where the glenoid fossa was completely absent [60]. Notably, human patients carrying mutations in *Gli2* exhibit a range of facial defects, including mandibular hypoplasia [4,61]. Thus, these observations provide strong evidence that Ihh signaling dictates the cellular organization of the condyle and regulates disc formation and subsequent joint cavitation.

Several lines of evidence indicate that Ihh and PTHrP interact in a negative feedback loop and regulate the onset of chondrocyte hypertrophy in developing long bones [35,36,55]. In the current model, Ihh expressed in prehypertrophic/early hypertrophic chondrocytes signals to the periarticular region and early proliferative chondrocytes at the top of growth plate cartilage to induce PTHrP expression. PTHrP in turn acts on PTHrP receptor-expressing chondrocytes to maintain them in a proliferating and less differentiated state. In developing condylar cartilage, PTHrP is expressed in the superficial and fibrous/chondroprogenitor cells at the apical region of wild-type condylar cartilage by E17.5 (Figure 2G). Importantly, PTHrP expression was drastically reduced or absent in corresponding cell populations in condylar cartilage in *Ihh*$^{-/-}$ embryos (Figure 2H). Given the fact that the number of proliferating chondroprogenitor cells was drastically decreased (ca. 50%) and chondrocytes underwent accelerated hypertrophy in *Ihh*$^{-/-}$ condyles, it is likely that PTHrP induced by Ihh signaling may (1) regulate the proliferation of chondro-progenitor cells and (2) maintain newly differentiated chondrocytes in a less differentiated stage.

Recent studies have suggested that Ihh also acts on chondrocytes to increase rates of proliferation and hypertrophy in a PTHrP-independent manner [62,63]. Expression of PTHrP significantly decreases in the apical region of early postnatal wild-type condyles and is nearly undetectable in juvenile condyles, while chondro-progenitors are still proliferating. Thus, the Ihh-PTHrP feedback loop appears to function primarily during embryogenesis and early postnatal life, while Ihh signaling in juvenile and early adult mice may govern proliferation of chondroprogenitor cells and chondrocyte maturation in a PTHrP-independent manner. Since $Ihh^{-/-}$ mice die during embryogenesis or soon after birth, the role(s) that Ihh plays in TMJ growth and maintenance should be investigated using alternative approaches, such as conditional gene knockout techniques employing appropriate inducible Cre mouse lines. Conditional *PTHrP* and compound *PTHrP/Ihh* mutant mice may provide new insights into this important and intriguing area of research. Taken together, studies in embryonic and early postnatal *Ihh*-mutant mice suggest that Ihh is essential for the coordination of (1) intramembranous bone collar formation, (2) progenitor cell proliferation, (3) expression of PTHrP in periarticular tissues, (4) endochondral ossification, and (5) disc and synovial cavity formation.

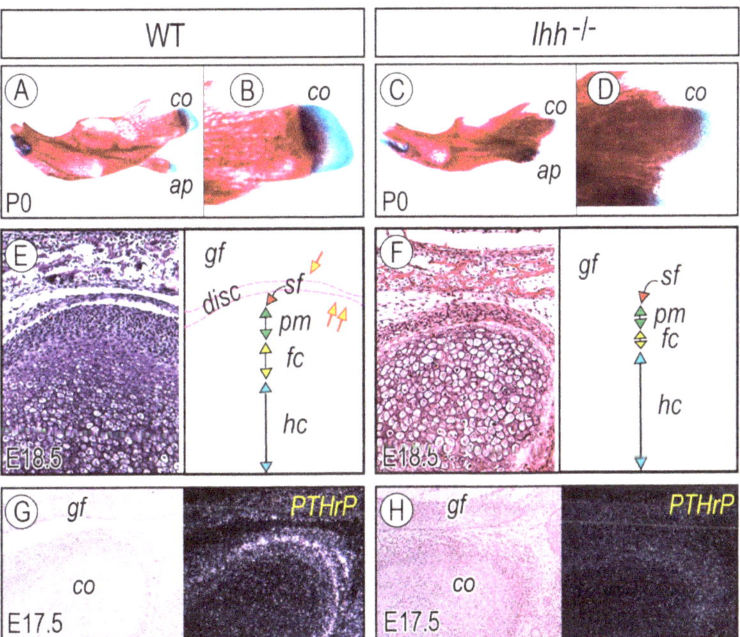

Figure 2. Mandible and TMJ abnormalities in $Ihh^{-/-}$ embryos and newborn mice. Mandibles from postnatal day 0 (P0) (**A–D**) of (**A,B**) wild-type (WT) and (**C,D**) $Ihh^{-/-}$ skulls were stained with alizarin red and alcian blue. Histological analysis of condylar cartilage from embryonic day 18.5 (E18.5) of (**E**) wild-type and (**F**) $Ihh^{-/-}$. Red, green, yellow, and blue vertical lines point to a superficial layer, a polymorphic layer, a flattened chondrocyte layer, and a hypertrophic chondrocyte layer, respectively. Note the absence of the articular disc tissue (disc), the upper joint cavity (arrow), and the lower joint cavity (double arrow). TMJ parasagittal serial sections from E17.5 of (**G**) wild-type (WT) and (**H**) $Ihh^{-/-}$ were processed for in situ hybridization with isotope-labeled riboprobe for *PTHrP*. *co*, condyle; *co*, coronoid process; *ap*, angular process; *gf*, glenoid fossa. Figure modified from Shibukawa et al. [11].

2.2. Role of Ihh in TMJ Growth and Maintenance during Postnatal Stages

2.2.1. Cellular Organization of Condylar Cartilage in Postnatal Stages

The apical layer in developing and adult condyles contains superficial cells producing Proteoglycan 4 (Prg4) and polymorphic cells that display stem cell-like characteristics [64]. Polymorphic cells give rise to chondrocytes for condylar growth and play a role in homeostasis and/or remodeling of condylar cartilage in response to mechanical stress [65,66]. Condylar cartilage length along its longitudinal axis in mice was ca. 470 μm at newborn stages and decreased to ca. 120 μm by 1 month, a thickness maintained through adulthood (Figure 3A,E). Condylar head width along the mediolateral axis was about 150 µm at newborn stages, increased to about 500 µm by 1 month, and remained so thereafter (Figure 3A,E). Subchondral bone plate (*sb*) was fully formed by 3 months of age, supporting articular cartilage (Figure 3E, bracket). Histomorphometric and in situ hybridization analyses revealed that the superficial/polymorphic (*sf/pm*) layers positive for fast green staining and less so for Safranin-O were characterized by the lack of type II collagen (*Col-II*) expression (Figure 3C,D). The thickness of superficial/polymorphic layers was ca. 50 µm at the newborn stage, became almost 3 times thinner (ca. 15 µm) by 3 months, and remained so thereafter (Figure 3F,G). Clearly, development, growth, and homeostasis of condylar cartilage during postnatal stages involve a dynamic structural organization of the apical layer (superficial and polymorphic-progenitor layers), whereby chondro-progenitor cells and their progeny cells provide newly differentiated chondrocytes to condylar articular cartilage.

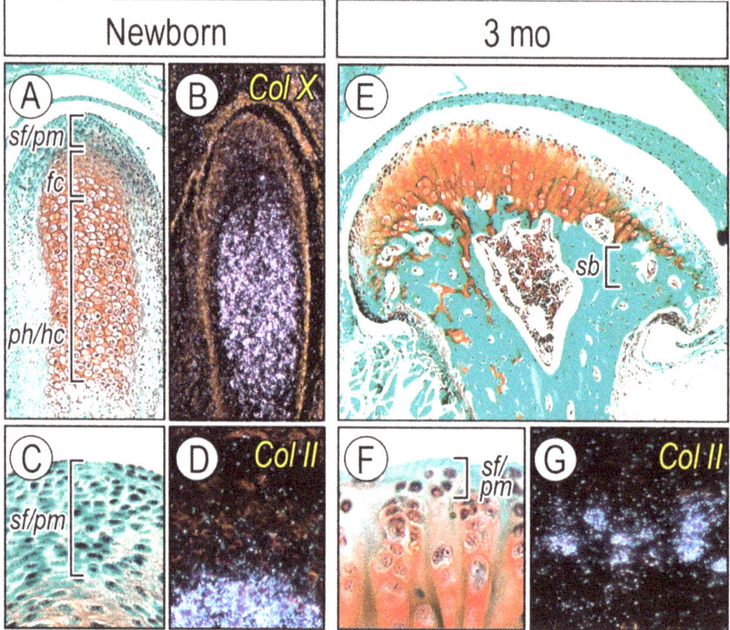

Figure 3. Condylar articular cartilage development and structural organization of a superficial (*sf*) and a polymorphic/progenitor (*pm/pr*) layer and chondrocytes with age. Frontal sections from (**A–D**) newborn and (**E–G**) 3-month-old (3 mo) wild-type mice. (**B**) *Col-X* and (**D,G**) *Col-II* gene expression. *sf*, superficial layer; *fc*, flattened chondrocyte layer; *ph/hc*, prehypertrophic and hypertrophic chondrocyte layer; *sb*, subchondral bone. Figure modified from Kurio et al. [67].

2.2.2. Topography of Hedgehog Signaling

Expression of Hh target genes in the condylar cartilage has been investigated to determine whether Hh signaling acts directly or indirectly on joint formation and maintenance in postnatal mice. *Ihh* transcripts were restricted to the prehypertrophic and early hypertrophic chondrocytes (Figure 4A,B). Interestingly, expression of *Ptch1*, a hedgehog receptor and transcriptional target, exhibited a gradient of expression, with relatively low expression levels in the central chondrocyte area of the condyle and higher levels toward the flattened chondrocyte (*fc*), polymorphic (*pm*), and superficial (*sf*) layers and articular disc (*di*) (Figure 4A,C). To determine the actual range of Ihh bioactivity, heterozygous *Gli1-nLacZ* embryos, widely used as a functional readout of hedgehog signaling activity, were investigated [67,68]. β-galactosidase activity was detectable over much of the growing condylar cartilage (*co*), the coronoid process (*cp*), and the angular process (*ap*) (Figure 4D) in postnatal day 1 (P1) condyles after processing for whole mount β-galactosidase staining, but was stronger over the entire apical layer of condyles and glenoid fossa (Figure 4F,G) in 8-week-old mice. β-galactosidase activity was also detected in cells lining the disc, with a tendency of β-galactosidase-positive lining cells being more abundant in those facing the lower, rather than upper, joint cavity (Figure 4F). The significance of hedgehog signaling maintained in the postnatal disc cells and the glenoid fossa apical cells needs to be elucidated. These studies indicate that Ihh signaling is active in condylar chondro-progenitors, superficial cells, and disc cells even postnatally, and is likely to influence those cells through life.

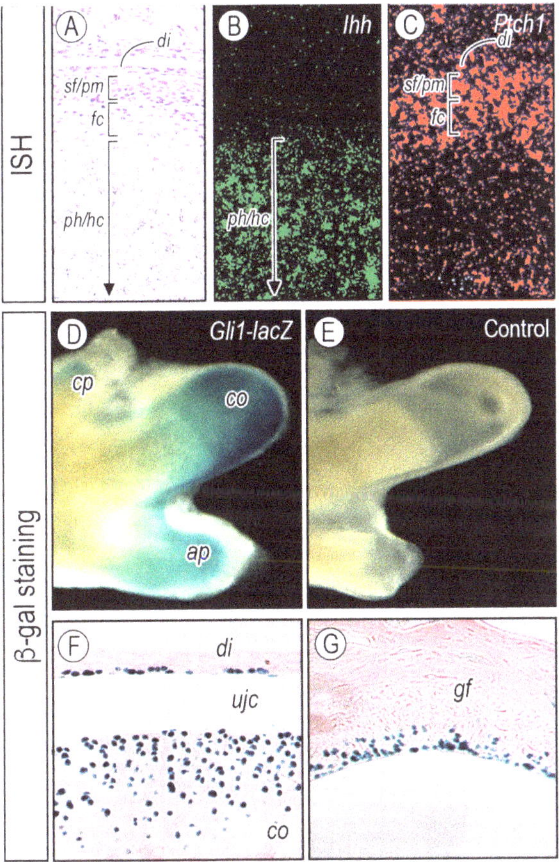

Figure 4. *Ihh*-expressing chondrocytes and its target cells depicted by *Ptch1* expression and *LacZ*-positive cells in postnatal *Gli1-nLacZ*-reporter mice. Parasagittal sections from newborn (P1) mice (**A–C**) were processed for in situ hybridization with radioisotope-labeled RNA probes of (**B**) *Ihh* and (**C**) *Ptch1*. Whole mount *LacZ* staining of mandible of (**D**) *Gli1-nLacZ*-reporter and (**E**) wild-type mice. Histological analyses of (**F**) *lacZ*-stained condyle and (**G**) glenoid fossa of 8-week-old *Gli1-nLacZ*-reporter mice. *gf*, glenoid fossa; *di*, articular disc; *ujc*, upper joint cavity; *sf*, superficial layer; *pr*, progenitor layer; *fc*, flattened chondrocyte layer; *ph/hc*, perhypertrophic/hypertrophic chondrocyte layer; *cp*, coronoid process; *co*, condyle; *ap*, angular process. Figure modified from Ochiai et al. [69].

2.2.3. Effect of Conditional Ihh Signaling Ablation in Postnatal Stages

As noted above, expression of Hh target genes and β-galactosidase activity in hedgehog reporter mice indicated that an Ihh signaling gradient across the condylar cartilage may contribute to cell function in the progenitor layer and zonal organization in postnatal condylar cartilage. Genetic studies in mice have provided experimental evidence for the significance of Ihh signaling in postnatal TMJ maintenance. To ablate *Ihh* expression in condylar cartilage in the postnatal period, an Aggrecan (*Agc*) *CreER* mouse line was employed [70] (Figure 5A). Compound *Ihh^{fl/fl};Agc-CreER;Gli1-nLacZ*, and control (*Ihh^{fl/fl};Gli1-nLacZ*) mice received tamoxifen injections at P14, P21 and P28, and Cre-mediated recombination and subsequent inactivation of Ihh signaling were confirmed by a significant decrease of *Gli1-nLacZ*-positive cells in the condylar cartilage. Mutant condylar cartilage displayed decreased numbers of superficial cells and proliferating chondro-progenitor cells and ectopic chondrocyte hypertrophy observed near the articular surface by 3 months old (Figure 5D, arrowhead and double arrowhead, respectively). By 5 months old, µCT analysis revealed that mutant subchondral bone became porous (Figure 5E), leading to decreased bone volume fraction and increased trabecular spacing compared to age-matched controls (Figure 5C). It is likely that decreased Hh signaling is associated with age-related TMJ degenerative changes [66,71]. In senescence-accelerated-prone 8 (SAMP8) mice, which develop early osteoarthritis-like changes in synovial joints at a high frequency [72], condylar cartilage in young SAMP8 mice displayed early-onset degenerative changes, concomitant with reductions in superficial/chondro-progenitor cells, proteoglycan/collagen content, and *Ihh*-expressing chondrocytes [66]. These data clearly demonstrate that Ihh signaling is essential for condylar superficial/progenitor cell layer development and function in postnatal condylar cartilage of TMJs, and its ablation and/or decreased expression in juvenile mice leads to degenerative changes in TMJ condyles, manifesting abnormal chondrocyte maturation and subchondral bone formation in the condyle.

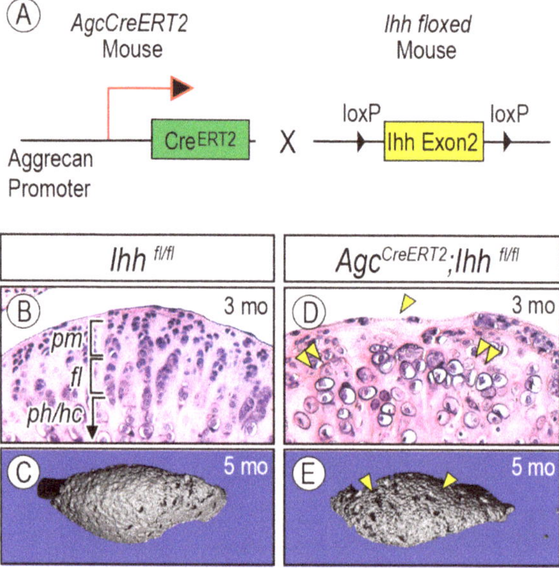

Figure 5. Condylar articular cartilage zonal organization and cellularity are abnormal in *AgcCreER;Ihh^{fl/fl}* mice over time. (**A**) Schematic showing the inducible Cre-Lox system where the floxed-Ihh gene is removed from chondrocytes that express Cre recombinase (arrow). Mice received multiple tamoxifen injections at P14, P21, and P28. TMJs from (**B,D**) 3-month-old and (**C,E**) 5-month-old of (**B,C**) control (*Ihh^{fl/fl}*) mice and (**D,E**) *AgcCreER;Ihh^{fl/fl}* mice were analyzed by (**B,D**) hematoxylin and eosin staining and (**C,E**) µCT. Note the decreased superficial cell number (arrowhead) and the presence of ectopic hypertrophic chondrocytes closer to the condylar surface (double arrowhead) in (**D**). Note that subchondral bone is irregular and porous (arrowheads) in (**E**). *pr*, polymorphic/progenitor layer; *fl*, flattened chondrocyte layer; *hl*, hypertrophic chondrocyte layer. Figure modified from Kurio et al. [67].

2.2.4. Hh Signaling in Degenerative TMJs

Osteoarthritis (OA) is characterized by the chronic degeneration of various hard and soft tissues around the affected joints. Stress bearing joints of the body, such as the hip and knee, are most commonly affected, but the TMJ is affected as well. TMJ osteoarthritis alters the condylar and glenoid fossa cartilage, subchondral bone, articular disc, and the synovial membrane that, in turn, cause pain and dysfunctional jaw movement [12,73–76]. There are several contributing factors to TMJ OA inception and progression, including parafunction, occlusion, psychosocial aspects, trauma, and genetics.

Recent studies indicate that decreased lubrication is also associated with the initiation and progression of OA in patients as well as in rodent models after anterior cruciate ligament injury [77–80]. Lubricin, a mucinous glycoprotein encoded by the proteoglycan 4 (*Prg4*) gene and a major component of synovial fluid, functions as both boundary lubrication and a chondro-protective agent in synovial joints [81,82]. Patients with camptodactyly-arthropathy-coxa vara-pericarditis (CACP) fail to express *PRG4* and subsequently develop polyarthropathy [83,84]. *Prg4*-mutant mice develop OA-like phenotypes in synovial joints, implying that Prg4 may have important roles in joint maintenance [85,86]. While TMJs in *Prg4*$^{-/-}$ mice developed normally, mutant mice developed degenerative changes. *Prg4*-mutant mice exhibited hyperplasia in the glenoid fossa articular cartilage, articular disc, and synovial membrane as early as 2 weeks of age and osteoarthritic changes in articular cartilage of the glenoid fossa and condyle by 6 months, in which loss of proteoglycans, an increase in osteoclast activity, and subchondral bone loss were observed [24,87]. Interestingly, these degenerative changes occurred earlier and were more severe than those in knee and hip joints, indicating that TMJs are more vulnerable to the loss of lubricin than other joints [87]. It has been reported that compound mutants of biglycan and fibromodulin, members of the small leucine-rich repeat proteoglycan family, display OA-like phenotypes in the knee joints much earlier than in TMJ [88–90]. Thus, these results suggest that synovial fluid plays an important role(s) in TMJ function and maintenance.

Osteophyte, a fibrocartilage-capped bonny outgrowth, is a hallmark radiographic feature of degenerative TMJ joint disease [91]. Joint instability likely contributes to osteophyte formation in the articular surface of the condyle and glenoid fossa. Compared to the development of OA in synovial joints of appendicular skeletal elements, the prevalence of osteophyte formation in TMJ OA is relatively rare. However, once developed, it causes various clinical symptoms and subsequently compromises joint function [92]. While Transforming growth factor β (TGFβ), Bone morphogenetic proteins (BMPs), Fibroblast growth factors (FGFs), or insulin-like growth factor-1 (IGF-1) have been detected in the developing osteophyte [93–97], what causes osteophytes in TMJs remains obscure. Interestingly, *Prg4*$^{-/-}$ mice exhibit increased osteophyte formation in the condylar cartilage and glenoid fossa with age (Figure 6A, arrowhead) [25,87,98]. This study showed that expression levels of *Ihh, Gli-1, Sox9*, and *Aggrecan* (*Agc*) (the latter 2 genes are markers of chondro-progenitors and chondrocytes, respectively) increased in osteophytes developing in the affected glenoid fossa. Immunohistochemistry revealed that IHH was preferentially distributed in the peripheral cells of osteophytes and underlying chondrocytes (Figure 6B). *Gli-1* transcripts were expressed in cells residing at the apical region of developing osteophytes (Figure 6C), indicative of Hh signaling activation as well as chondrogenesis taking place at this site. Expression of *PTHrP* and its receptor *Pth1r* was increased in *Prg4*$^{-/-}$ glenoid fossa. In glenoid fossa cells in culture, Hh signaling stimulated chondrocyte differentiation and maturation, evaluated by increased chondrocyte proteoglycan synthesis and alkaline phosphatase activity, respectively, while treatment with hedgehog inhibitor, Hh Antag, prevented such maturation process [25]. In line with these results, data with *Col2-CreER;Pth1r*$^{fl/fl}$*;Smo*$^{fl/fl}$ mice suggest that inhibition of Ihh signaling in osteoarthritis-like TMJs prevents chondrocyte terminal differentiation through a Pth1r-dependent mechanism [99]. Further studies are warranted to determine the pathophysiology underlying activation of Ihh and PTHrP signaling in osteoarthritic TMJs.

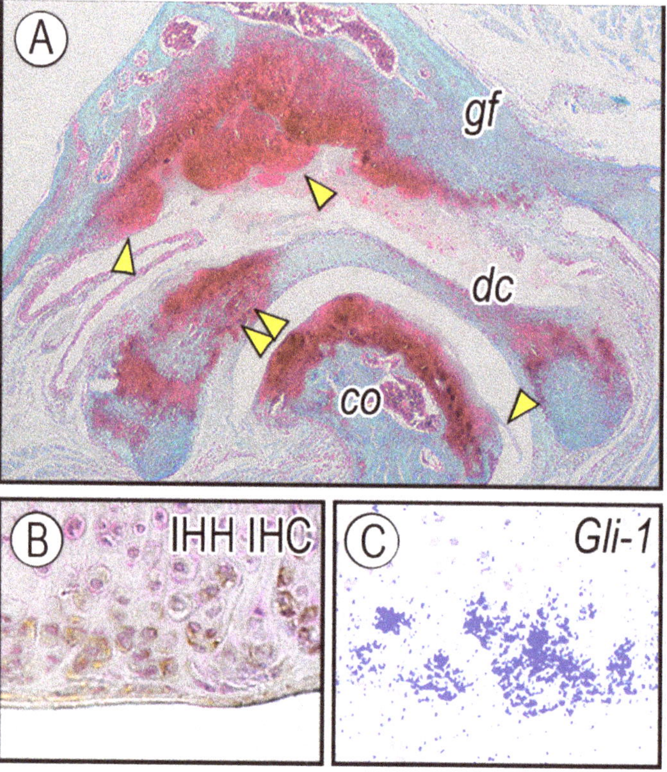

Figure 6. Ectopic expression of *Ihh* in osteophyte-developing glenoid fossa cartilage in *Prg4*-null mice. TMJs from (**A**) 15-month-old *Prg4$^{-/-}$* mice were analyzed by Safranin O/fast green staining. Note that osteophytes are developing from the glenoid fossa, along with condylar cartilage (arrowhead) and ectopic cartilage formation in disc (double arrowhead). (**B**) Immunohistochemistry (IHC) with IHH antibody and (**C**) in situ hybridization of *Gli-1* mRNA in the developing osteophytes. Figure modified from Bechtold et al. [25].

3. Perspectives

While a number of studies have addressed the importance of the Hh signaling pathway in TMJ biology, there are many questions that remain unanswered.

First, data summarized in this review show long-range signaling of Ihh proteins during embryonic development and postnatal growth. However, the underlying molecular mechanisms regulating Ihh protein release from the cell surface need to be further clarified. Multiple studies indicate that such long-range signaling of hedgehog requires lipid modifications that promote the formation of multimeric complexes, the formation of which depends on the palmitoylation and addition of cholesterol to the N-terminal hedgehog fragments [100–102]. Heparan sulfate proteoglycans (HSPGs) with which hedgehog proteins interact through their Cardin–Weintraub motif, could allow formation of hedgehog multimers, facilitating Hh protein oligomerization [103–107]. Following oligomerization, Hh proteins bind the membrane protein Dispatched (Disp) in a cholesterol-dependent manner and the combined action of Disp and Scube2, a secreted protein, release oligomerized Hh proteins from the cells [108–111]. Studies suggest that the Golgi-associated *N*-sulfotransferase 1(Ndst-1), which catalyzes the sulfation of HSPG glycosaminoglycan chains, is critical for organogenesis [112–114], including mandibular condyles and TMJ development, and allows HSPGs to exert their roles via regulation of Ihh signaling topography and action [112–114].

Second, there is a critical need for in vivo and in vitro studies to further define interactions between Ihh and other signaling pathways that regulate postnatal morphogenesis and growth of TMJs. In mouse embryos, Ihh signaling promotes expression of PTHrP at the apical end of the presumptive condylar cartilage, which leads to increased numbers of presumptive chondro-progenitors [11,35–37]. Notably, the size of the condylar cartilage in young adult mice, the length along anteroposterior and mediolateral axes, are about 3 times larger compared with prenatal mice [67]. As indicated above, the expression of PTHrP is high during embryogenesis and early postnatal life, but declines in juvenile mice. Thus, further studies are required to define the role of Ihh signaling during the growth and development of the TMJ and associated tissues in the presence and absence of PTHrP.

Third, the role(s) that altered Ihh signaling and associated pathways, such as primary cilia components, play in the degenerative changes that accompany osteoarthritis are not fully understood in the TMJ or synovial joints [25,69,98,99,115,116]. For example, it has been observed that activation of Hh signaling leads to the induction of ectopic chondrocyte hypertrophy in degenerative articular cartilage [28,115,117]. Further, numerous studies have demonstrated that several signaling pathways, including TGFβ, BMP, IGF-1, and FGF, are up-regulated during osteophyte formation in synovial joints [93–97]. Interestingly, altered Hh and *PTHrP* signaling has been detected in osteophytes developing at the surface of the glenoid fossa and condylar cartilage [25].

While there has been much progress defining the roles that different signaling pathways play during normal TMJ growth and development and the pathological changes giving rise to osteoarthritis, much remains to explored. Future studies need to define the interactions between multiple signaling pathways and determine how this 'crosstalk' directs TMJ morphogenesis and more broadly, bone and cartilage differentiation. The results from these studies will provide a solid basis leading to the development of new and novel approaches for repairing TMJ congenital anomalies and acquired degenerative damage resulting from osteoarthritic conditions.

Author Contributions: Conception and design: E.K.; Analysis and Interpretation of the data: T.E.B., N.K., C.S., T.E.B., H.-D.N.; Drafting the Article: E.K., P.C.B.; Critical revision of the article for important intellectual content: E.K.; Final approval of the article: T.E.B., N.K., C.S., H.-D.N., P.C.B., E.K.

Funding: This research was funded by NIDCR grant (RO1DE023841 to E.K. and H.-D.N.).

Acknowledgments: We express our gratitude to the Penn Center for Musculoskeletal Disorders (PCMD) supported by NIHP30 grant AR069619 where the mouse μCT scans were performed, analyzed, and quantified. We also thank Matt Warman, Boston Children's Hospital, for providing us with *Prg4*-mutant mice.

References

1. Rux, D.; Decker, R.S.; Koyama, E.; Pacifici, M. Joints in the appendicular skeleton: Developmental mechanisms and evolutionary influences. *Curr. Top. Dev. Biol.* **2019**, *133*, 119–151. [CrossRef] [PubMed]

2. Pacifici, M.; Decker, R.S.; Koyama, E. Limb Synovial Joint Development From the Hips Down: Implications for Articular Cartilage Repair and Regeneration. In *Developmental Biology and Musculoskeletal Tissue Engineering*; Academic Press: Cambridge, MA, USA, 2018; pp. 67–101. Available online: https://www.sciencedirect.com/science/article/pii/B9780128114674000048 (accessed on 12 December 2019).

3. Koyama, E.; Yasuda, T.; Minugh-Purvis, N.; Kinumatsu, T.; Yallowitz, A.R.; Wellik, D.M.; Pacifici, M. Hox11 genes establish synovial joint organization and phylogenetic characteristics in developing mouse zeugopod skeletal elements. *Development* **2010**, *137*, 3795–3800. [CrossRef] [PubMed]

4. Hinton, R.J. Genes that regulate morphogenesis and growth of the temporomandibular joint: A review. *Dev. Dyn.* **2014**, *243*, 864–874. [CrossRef] [PubMed]

5. Suzuki, A.; Iwata, J. Mouse genetic models for temporomandibular joint development and disorders. *Oral Dis.* **2016**, *22*, 33–38. [CrossRef] [PubMed]

6.	Symons, N.B. A histochemical study of the secondary cartilage of the mandibular condyle in the rat. *Arch. Oral Biol.* **1965**, *10*, 579–584. [CrossRef]

7.	Koyama, E.; Shibukawa, Y.; Nagayama, M.; Sugito, H.; Young, B.; Yuasa, T.; Okabe, T.; Ochiai, T.; Kamiya, N.; Rountree, R.B.; et al. A distinct cohort of progenitor cells participates in synovial joint and articular cartilage formation during mouse limb skeletogenesis. *Dev. Biol.* **2008**, *316*, 62–73. [CrossRef]

8.	Shibata, S.; Fukada, K.; Suzuki, S.; Ogawa, T.; Yamashita, Y. In situ hybridization and immunohistochemistry of bone sialoprotein and secreted phosphoprotein 1 (osteopontin) in the developing mouse mandibular condylar cartilage compared with limb bud cartilage. *J. Anat.* **2002**, *200*, 309–320. [CrossRef]

9.	Vinkka, H. Secondary cartilages in the facial skeleton of the rat. *Proc. Finn. Dent. Soc.* **1982**, *78* (Suppl. 7), 1–137.

10.	Radlanski, R.J.; Renz, H.; Klarkowski, M.C. Prenatal development of the human mandible. 3D reconstructions, morphometry and bone remodelling pattern, sizes 12-117 mm CRL. *Anat. Embryol.* **2003**, *207*, 221–232. [CrossRef]

11.	Shibukawa, Y.; Young, B.; Wu, C.; Yamada, S.; Long, F.; Pacifici, M.; Koyama, E. Temporomandibular joint formation and condyle growth require Indian hedgehog signaling. *Dev. Dyn.* **2007**, *236*, 426–434. [CrossRef]

12.	Wadhwa, S.; Kapila, S. TMJ disorders: Future innovations in diagnostics and therapeutics. *J. Dent. Educ.* **2008**, *72*, 930–947. [PubMed]

13.	Luder, H.U.; Leblond, C.P.; von der Mark, K. Cellular stages in cartilage formation as revealed by morphometry, radioautography and type II collagen immunostaining of the mandibular condyle from weanling rats. *Am. J. Anat* **1988**, *182*, 197–214. [CrossRef] [PubMed]

14.	Silbermann, M.; Frommer, J. The nature of endochondral ossification in the mandibular condyle of the mouse. *Anat. Rec.* **1972**, *172*, 659–667. [CrossRef] [PubMed]

15.	Sarnat, B.G. Developmental facial abnormalities and the temporomandibular joint. *Dent. Clin. N. Am.* **1966**, *79*, 587–600. [CrossRef]

16.	Petrovic, A.G. Mechanisms and regulation of mandibular condylar growth. *Acta Morphol. Neerl. Scand.* **1972**, *10*, 25–34. [PubMed]

17.	Kantomaa, T.; Tuominen, M.; Pirttiniemi, P. Effect of mechanical forces on chondrocyte maturation and differentiation in the mandibular condyle of the rat. *J. Dent. Res.* **1994**, *73*, 1150–1156. [CrossRef]

18.	Jing, Y.; Zhou, X.; Han, X.; Jing, J.; von der Mark, K.; Wang, J.; de Crombrugghe, B.; Hinton, R.J.; Feng, J.Q. Chondrocytes Directly Transform into Bone Cells in Mandibular Condyle Growth. *J. Dent. Res.* **2015**, *94*, 1668–1675. [CrossRef]

19.	Frommer, J. Prenatal Development of the Mandibular Joint in Mice. *Anat. Rec.* **1964**, *150*, 449–461. [CrossRef]

20.	Bhaskar, S.N. Growth pattern of the rat mandible from 13 days insemination age to 30 days after birth. *Am. J. Anat* **1953**, *92*, 1–53. [CrossRef]

21.	Mori-Akiyama, Y.; Akiyama, H.; Rowitch, D.H.; de Crombrugghe, B. Sox9 is required for determination of the chondrogenic cell lineage in the cranial neural crest. *Proc. Natl. Acad. Sci. USA* **2003**, *100*, 9360–9365. [CrossRef]

22.	Gu, S.; Wei, N.; Yu, L.; Fei, J.; Chen, Y. Shox2-deficiency leads to dysplasia and ankylosis of the temporomandibular joint in mice. *Mech. Dev.* **2008**, *125*, 729–742. [CrossRef] [PubMed]

23.	Yasuda, T.; Nah, H.D.; Laurita, J.; Kinumatsu, T.; Shibukawa, Y.; Shibutani, T.; Minugh-Purvis, N.; Pacifici, M.; Koyama, E. Muenke syndrome mutation, FgfR3P(2)(4)(4)R, causes TMJ defects. *J. Dent. Res.* **2012**, *91*, 683–689. [CrossRef] [PubMed]

24.	Koyama, E.; Saunders, C.; Salhab, I.; Decker, R.S.; Chen, I.; Um, H.; Pacifici, M.; Nah, H.D. Lubricin is Required for the Structural Integrity and Post-natal Maintenance of TMJ. *J. Dent. Res.* **2014**, *93*, 663–670. [CrossRef] [PubMed]

25.	Bechtold, T.E.; Saunders, C.; Decker, R.S.; Um, H.B.; Cottingham, N.; Salhab, I.; Kurio, N.; Billings, P.C.; Pacifici, M.; Nah, H.D.; et al. Osteophyte formation and matrix mineralization in a TMJ osteoarthritis mouse model are associated with ectopic hedgehog signaling. *Matrix Biol.* **2016**, *52-54*, 339–354. [CrossRef] [PubMed]

26.	Wang, Y.; Liu, C.; Rohr, J.; Liu, H.; He, F.; Yu, J.; Sun, C.; Li, L.; Gu, S.; Chen, Y. Tissue interaction is required for glenoid fossa development during temporomandibular joint formation. *Dev. Dyn.* **2011**, *240*, 2466–2473. [CrossRef]

27.	Lee, R.T.; Zhao, Z.; Ingham, P.W. Hedgehog signalling. *Development* **2016**, *143*, 367–372. [CrossRef]

28. Alman, B.A. The role of hedgehog signalling in skeletal health and disease. *Nat. Rev. Rheumatol.* **2015**, *11*, 552–560. [CrossRef]

29. Yao, E.; Chuang, P.T. Hedgehog signaling: From basic research to clinical applications. *J. Formos. Med. Assoc.* **2015**, *114*, 569–576. [CrossRef]

30. Abramyan, J. Hedgehog Signaling and Embryonic Craniofacial Disorders. *J. Dev. Biol.* **2019**, *7*. [CrossRef]

31. Bitgood, M.J.; McMahon, A.P. Hedgehog and Bmp genes are coexpressed at many diverse sites of cell-cell interaction in the mouse embryo. *Dev. Biol.* **1995**, *172*, 126–138. [CrossRef]

32. Koyama, E.; Leatherman, J.L.; Noji, S.; Pacifici, M. Early chick limb cartilaginous elements possess polarizing activity and express hedgehog-related morphogenetic factors. *Dev. Dyn.* **1996**, *207*, 344–354. [CrossRef]

33. Nakamura, T.; Aikawa, T.; Iwamoto-Enomoto, M.; Iwamoto, M.; Higuchi, Y.; Pacifici, M.; Kinto, N.; Yamaguchi, A.; Noji, S.; Kurisu, K.; et al. Induction of osteogenic differentiation by hedgehog proteins. *Biochem. Biophys. Res. Commun.* **1997**, *237*, 465–469. [CrossRef] [PubMed]

34. Long, F.; Chung, U.I.; Ohba, S.; McMahon, J.; Kronenberg, H.M.; McMahon, A.P. Ihh signaling is directly required for the osteoblast lineage in the endochondral skeleton. *Development* **2004**, *131*, 1309–1318. [CrossRef] [PubMed]

35. Vortkamp, A.; Lee, K.; Lanske, B.; Segre, G.V.; Kronenberg, H.M.; Tabin, C.J. Regulation of rate of cartilage differentiation by Indian hedgehog and PTH-related protein. *Science* **1996**, *273*, 613–622. [CrossRef] [PubMed]

36. Lanske, B.; Karaplis, A.C.; Lee, K.; Luz, A.; Vortkamp, A.; Pirro, A.; Karperien, M.; Defize, L.H.; Ho, C.; Mulligan, R.C.; et al. PTH/PTHrP receptor in early development and Indian hedgehog-regulated bone growth. *Science* **1996**, *273*, 663–666. [CrossRef] [PubMed]

37. St-Jacques, B.; Hammerschmidt, M.; McMahon, A.P. Indian hedgehog signaling regulates proliferation and differentiation of chondrocytes and is essential for bone formation. *Genes Dev.* **1999**, *13*, 2072–2086. [CrossRef]

38. Chung, U.I.; Schipani, E.; McMahon, A.P.; Kronenberg, H.M. Indian hedgehog couples chondrogenesis to osteogenesis in endochondral bone development. *J. Clin. Investig.* **2001**, *107*, 295–304. [CrossRef] [PubMed]

39. Alcedo, J.; Ayzenzon, M.; Von Ohlen, T.; Noll, M.; Hooper, J.E. The Drosophila smoothened gene encodes a seven-pass membrane protein, a putative receptor for the hedgehog signal. *Cell* **1996**, *86*, 221–232. [CrossRef]

40. van den Heuvel, M.; Ingham, P.W. smoothened encodes a receptor-like serpentine protein required for hedgehog signalling. *Nature* **1996**, *382*, 547–551. [CrossRef]

41. Chen, W.; Burgess, S.; Hopkins, N. Analysis of the zebrafish smoothened mutant reveals conserved and divergent functions of hedgehog activity. *Development* **2001**, *128*, 2385–2396.

42. Goodrich, L.V.; Milenkovic, L.; Higgins, K.M.; Scott, M.P. Altered neural cell fates and medulloblastoma in mouse patched mutants. *Science* **1997**, *277*, 1109–1113. [CrossRef] [PubMed]

43. Haycraft, C.J.; Serra, R. Cilia involvement in patterning and maintenance of the skeleton. *Curr. Top. Dev. Biol.* **2008**, *85*, 303–332. [CrossRef] [PubMed]

44. Sreekumar, V.; Norris, D.P. Cilia and development. *Curr. Opin. Genet. Dev.* **2019**, *56*, 15–21. [CrossRef] [PubMed]

45. Nachury, M.V.; Mick, D.U. Establishing and regulating the composition of cilia for signal transduction. *Nat. Rev. Mol. Cell Biol.* **2019**, *20*, 389–405. [CrossRef]

46. Ruiz i Altaba, A. Gli proteins encode context-dependent positive and negative functions: Implications for development and disease. *Development* **1999**, *126*, 3205–3216.

47. Sasaki, H.; Nishizaki, Y.; Hui, C.; Nakafuku, M.; Kondoh, H. Regulation of Gli2 and Gli3 activities by an amino-terminal repression domain: Implication of Gli2 and Gli3 as primary mediators of Shh signaling. *Development* **1999**, *126*, 3915–3924.

48. Wong, S.Y.; Reiter, J.F. The primary cilium at the crossroads of mammalian hedgehog signaling. *Curr. Top. Dev. Biol.* **2008**, *85*, 225–260. [CrossRef]

49. Hynes, M.; Stone, D.M.; Dowd, M.; Pitts-Meek, S.; Goddard, A.; Gurney, A.; Rosenthal, A. Control of cell pattern in the neural tube by the zinc finger transcription factor and oncogene Gli-1. *Neuron* **1997**, *19*, 15–26. [CrossRef]

50. Karlstrom, R.O.; Tyurina, O.V.; Kawakami, A.; Nishioka, N.; Talbot, W.S.; Sasaki, H.; Schier, A.F. Genetic analysis of zebrafish gli1 and gli2 reveals divergent requirements for gli genes in vertebrate development. *Development* **2003**, *130*, 1549–1564. [CrossRef]

51. Lee, J.; Platt, K.A.; Censullo, P.; Ruiz i Altaba, A. Gli1 is a target of Sonic hedgehog that induces ventral neural tube development. *Development* **1997**, *124*, 2537–2552.

52. Koyama, E.; Ochiai, T.; Rountree, R.B.; Kingsley, D.M.; Enomoto-Iwamoto, M.; Iwamoto, M.; Pacifici, M. Synovial joint formation during mouse limb skeletogenesis: Roles of Indian hedgehog signaling. *Ann. N. Y. Acad. Sci.* **2007**, *1116*, 100–112. [CrossRef] [PubMed]

53. Decker, R.S.; Koyama, E.; Pacifici, M. Genesis and morphogenesis of limb synovial joints and articular cartilage. *Matrix Biol.* **2014**, *39*, 5–10. [CrossRef] [PubMed]

54. Kubiak, M.; Ditzel, M. A Joint Less Ordinary: Intriguing Roles for Hedgehog Signalling in the Development of the Temporomandibular Synovial Joint. *J. Dev. Biol.* **2016**, *4*. [CrossRef] [PubMed]

55. Maeda, Y.; Nakamura, E.; Nguyen, M.T.; Suva, L.J.; Swain, F.L.; Razzaque, M.S.; Mackem, S.; Lanske, B. Indian Hedgehog produced by postnatal chondrocytes is essential for maintaining a growth plate and trabecular bone. *Proc. Natl. Acad. Sci. USA* **2007**, *104*, 6382–6387. [CrossRef] [PubMed]

56. Hilton, M.J.; Tu, X.; Cook, J.; Hu, H.; Long, F. Ihh controls cartilage development by antagonizing Gli3, but requires additional effectors to regulate osteoblast and vascular development. *Development* **2005**, *132*, 4339–4351. [CrossRef]

57. Purcell, P.; Joo, B.W.; Hu, J.K.; Tran, P.V.; Calicchio, M.L.; O'Connell, D.J.; Maas, R.L.; Tabin, C.J. Temporomandibular joint formation requires two distinct hedgehog-dependent steps. *Proc. Natl. Acad. Sci. USA* **2009**, *106*, 18297–18302. [CrossRef] [PubMed]

58. Mo, R.; Freer, A.M.; Zinyk, D.L.; Crackower, M.A.; Michaud, J.; Heng, H.H.; Chik, K.W.; Shi, X.M.; Tsui, L.C.; Cheng, S.H.; et al. Specific and redundant functions of Gli2 and Gli3 zinc finger genes in skeletal patterning and development. *Development* **1997**, *124*, 113–123.

59. Michikami, I.; Fukushi, T.; Honma, S.; Yoshioka, S.; Itoh, S.; Muragaki, Y.; Kurisu, K.; Ooshima, T.; Wakisaka, S.; Abe, M. Trps1 is necessary for normal temporomandibular joint development. *Cell Tissue Res.* **2012**, *348*, 131–140. [CrossRef]

60. Yang, L.; Gu, S.; Ye, W.; Song, Y.; Chen, Y. Augmented Indian hedgehog signaling in cranial neural crest cells leads to craniofacial abnormalities and dysplastic temporomandibular joint in mice. *Cell Tissue Res.* **2016**, *364*, 105–115. [CrossRef]

61. Bertolacini, C.D.; Ribeiro-Bicudo, L.A.; Petrin, A.; Richieri-Costa, A.; Murray, J.C. Clinical findings in patients with GLI2 mutations—Phenotypic variability. *Clin. Genet.* **2012**, *81*, 70–75. [CrossRef]

62. Mak, K.K.; Kronenberg, H.M.; Chuang, P.T.; Mackem, S.; Yang, Y. Indian hedgehog signals independently of PTHrP to promote chondrocyte hypertrophy. *Development* **2008**, *135*, 1947–1956. [CrossRef] [PubMed]

63. Karp, S.J.; Schipani, E.; St-Jacques, B.; Hunzelman, J.; Kronenberg, H.; McMahon, A.P. Indian hedgehog coordinates endochondral bone growth and morphogenesis via parathyroid hormone related-protein-dependent and -independent pathways. *Development* **2000**, *127*, 543–548. [PubMed]

64. Embree, M.C.; Chen, M.; Pylawka, S.; Kong, D.; Iwaoka, G.M.; Kalajzic, I.; Yao, H.; Shi, C.; Sun, D.; Sheu, T.J.; et al. Exploiting endogenous fibrocartilage stem cells to regenerate cartilage and repair joint injury. *Nat. Commun.* **2016**, *7*, 13073. [CrossRef] [PubMed]

65. Kaul, R.; O'Brien, M.H.; Dutra, E.; Lima, A.; Utreja, A.; Yadav, S. The Effect of Altered Loading on Mandibular Condylar Cartilage. *PLoS ONE* **2016**, *11*, e0160121. [CrossRef] [PubMed]

66. Ishizuka, Y.; Shibukawa, Y.; Nagayama, M.; Decker, R.; Kinumatsu, T.; Saito, A.; Pacifici, M.; Koyama, E. TMJ degeneration in SAMP8 mice is accompanied by deranged Ihh signaling. *J. Dent. Res.* **2014**, *93*, 281–287. [CrossRef] [PubMed]

67. Kurio, N.; Saunders, C.; Bechtold, T.E.; Salhab, I.; Nah, H.D.; Sinha, S.; Billings, P.C.; Pacifici, M.; Koyama, E. Roles of Ihh signaling in chondroprogenitor function in postnatal condylar cartilage. *Matrix Biol.* **2018**, *67*, 15–31. [CrossRef]

68. Bai, C.B.; Auerbach, W.; Lee, J.S.; Stephen, D.; Joyner, A.L. Gli2, but not Gli1, is required for initial Shh signaling and ectopic activation of the Shh pathway. *Development* **2002**, *129*, 4753–4761.

69. Ochiai, T.; Shibukawa, Y.; Nagayama, M.; Mundy, C.; Yasuda, T.; Okabe, T.; Shimono, K.; Kanyama, M.; Hasegawa, H.; Maeda, Y.; et al. Indian hedgehog roles in post-natal TMJ development and organization. *J. Dent. Res.* **2010**, *89*, 349–354. [CrossRef]

70. Henry, S.P.; Jang, C.W.; Deng, J.M.; Zhang, Z.; Behringer, R.R.; de Crombrugghe, B. Generation of aggrecan-CreERT2 knockin mice for inducible Cre activity in adult cartilage. *Genesis* **2009**, *47*, 805–814. [CrossRef]

71. Chen, W.H.; Hosokawa, M.; Tsuboyama, T.; Ono, T.; Iizuka, T.; Takeda, T. Age-related changes in the temporomandibular joint of the senescence accelerated mouse. SAM-P/3 as a new murine model of degenerative joint disease. *Am. J. Pathol.* **1989**, *135*, 379–385.

72. Hosokawa, M.; Kasai, R.; Higuchi, K.; Takeshita, S.; Shimizu, K.; Hamamoto, H.; Honma, A.; Irino, M.; Toda, K.; Matsumura, A.; et al. Grading score system: A method for evaluation of the degree of senescence in senescence accelerated mouse (SAM). *Mech. Ageing Dev.* **1984**, *26*, 91–102. [CrossRef]

73. Scrivani, S.J.; Keith, D.A.; Kaban, L.B. Temporomandibular disorders. *N. Engl. J. Med.* **2008**, *359*, 2693–2705. [CrossRef] [PubMed]

74. Dimitroulis, G. Management of temporomandibular joint disorders: A surgeon's perspective. *Aust. Dent. J.* **2018**, *63* (Suppl. 1), S79–S90. [CrossRef]

75. Mercuri, L.G. Osteoarthritis, osteoarthrosis, and idiopathic condylar resorption. *Oral Maxillofac. Surg. Clin. N. Am.* **2008**, *20*, 169–183. [CrossRef] [PubMed]

76. Schiffman, E.; Ohrbach, R.; Truelove, E.; Look, J.; Anderson, G.; Goulet, J.P.; List, T.; Svensson, P.; Gonzalez, Y.; Lobbezoo, F.; et al. Diagnostic Criteria for Temporomandibular Disorders (DC/TMD) for Clinical and Research Applications: Recommendations of the International RDC/TMD Consortium Network* and Orofacial Pain Special Interest Groupdagger. *J. Oral Fac. Pain Headache* **2014**, *28*, 6–27. [CrossRef] [PubMed]

77. Elsaid, K.A.; Machan, J.T.; Waller, K.; Fleming, B.C.; Jay, G.D. The impact of anterior cruciate ligament injury on lubricin metabolism and the effect of inhibiting tumor necrosis factor alpha on chondroprotection in an animal model. *Arthritis Rheum.* **2009**, *60*, 2997–3006. [CrossRef]

78. Elsaid, K.A.; Fleming, B.C.; Oksendahl, H.L.; Machan, J.T.; Fadale, P.D.; Hulstyn, M.J.; Shalvoy, R.; Jay, G.D. Decreased lubricin concentrations and markers of joint inflammation in the synovial fluid of patients with anterior cruciate ligament injury. *Arthritis Rheum.* **2008**, *58*, 1707–1715. [CrossRef]

79. Teeple, E.; Elsaid, K.A.; Fleming, B.C.; Jay, G.D.; Aslani, K.; Crisco, J.J.; Mechrefe, A.P. Coefficients of friction, lubricin, and cartilage damage in the anterior cruciate ligament-deficient guinea pig knee. *J. Orthop. Res.* **2008**, *26*, 231–237. [CrossRef]

80. Kosinska, M.K.; Ludwig, T.E.; Liebisch, G.; Zhang, R.; Siebert, H.C.; Wilhelm, J.; Kaesser, U.; Dettmeyer, R.B.; Klein, H.; Ishaque, B.; et al. Articular Joint Lubricants during Osteoarthritis and Rheumatoid Arthritis Display Altered Levels and Molecular Species. *PLoS ONE* **2015**, *10*, e0125192. [CrossRef]

81. Jay, G.D.; Waller, K.A. The biology of lubricin: Near frictionless joint motion. *Matrix Biol.* **2014**, *39*, 17–24. [CrossRef]

82. Das, N.; Schmidt, T.A.; Krawetz, R.J.; Dufour, A. Proteoglycan 4: From Mere Lubricant to Regulator of Tissue Homeostasis and Inflammation: Does proteoglycan 4 have the ability to buffer the inflammatory response? *Bioessays* **2019**, *41*, e1800166. [CrossRef] [PubMed]

83. Bahabri, S.A.; Suwairi, W.M.; Laxer, R.M.; Polinkovsky, A.; Dalaan, A.A.; Warman, M.L. The camptodactyly-arthropathy-coxa vara-pericarditis syndrome: Clinical features and genetic mapping to human chromosome 1. *Arthritis Rheum.* **1998**, *41*, 730–735. [CrossRef]

84. Marcelino, J.; Carpten, J.D.; Suwairi, W.M.; Gutierrez, O.M.; Schwartz, S.; Robbins, C.; Sood, R.; Makalowska, I.; Baxevanis, A.; Johnstone, B.; et al. CACP, encoding a secreted proteoglycan, is mutated in camptodactyly-arthropathy-coxa vara-pericarditis syndrome. *Nat. Genet.* **1999**, *23*, 319–322. [CrossRef] [PubMed]

85. Hill, A.; Waller, K.A.; Cui, Y.; Allen, J.M.; Smits, P.; Zhang, L.X.; Ayturk, U.M.; Hann, S.; Lessard, S.G.; Zurakowski, D.; et al. Lubricin restoration in a mouse model of congenital deficiency. *Arthritis Rheumatol.* **2015**, *67*, 3070–3081. [CrossRef] [PubMed]

86. Rhee, D.K.; Marcelino, J.; Baker, M.; Gong, Y.; Smits, P.; Lefebvre, V.; Jay, G.D.; Stewart, M.; Wang, H.; Warman, M.L.; et al. The secreted glycoprotein lubricin protects cartilage surfaces and inhibits synovial cell overgrowth. *J. Clin. Investig.* **2005**, *115*, 622–631. [CrossRef]

87. Hill, A.; Duran, J.; Purcell, P. Lubricin protects the temporomandibular joint surfaces from degeneration. *PLoS ONE* **2014**, *9*, e106497. [CrossRef]

88. Chen, J.; Gupta, T.; Barasz, J.A.; Kalajzic, Z.; Yeh, W.C.; Drissi, H.; Hand, A.R.; Wadhwa, S. Analysis of microarchitectural changes in a mouse temporomandibular joint osteoarthritis model. *Arch. Oral Biol.* **2009**, *54*, 1091–1098. [CrossRef]

89. Wadhwa, S.; Embree, M.; Ameye, L.; Young, M.F. Mice deficient in biglycan and fibromodulin as a model for temporomandibular joint osteoarthritis. *Cells Tissues Organs* **2005**, *181*, 136–143. [CrossRef]

90. Embree, M.C.; Kilts, T.M.; Ono, M.; Inkson, C.A.; Syed-Picard, F.; Karsdal, M.A.; Oldberg, A.; Bi, Y.; Young, M.F. Biglycan and fibromodulin have essential roles in regulating chondrogenesis and extracellular matrix turnover in temporomandibular joint osteoarthritis. *Am. J. Pathol.* **2010**, *176*, 812–826. [CrossRef]

91. Larheim, T.A.; Abrahamsson, A.K.; Kristensen, M.; Arvidsson, L.Z. Temporomandibular joint diagnostics using CBCT. *Dentomaxillofac. Radiol.* **2015**, *44*, 20140235. [CrossRef]

92. Rehan, O.M.; Saleh, H.A.K.; Raffat, H.A.; Abu-Taleb, N.S. Osseous changes in the temporomandibular joint in rheumatoid arthritis: A cone-beam computed tomography study. *Imaging Sci. Dent.* **2018**, *48*, 1–9. [CrossRef] [PubMed]

93. Murata, K.; Kokubun, T.; Onitsuka, K.; Oka, Y.; Kano, T.; Morishita, Y.; Ozone, K.; Kuwabara, N.; Nishimoto, J.; Isho, T.; et al. Controlling joint instability after anterior cruciate ligament transection inhibits transforming growth factor-beta-mediated osteophyte formation. *Osteoarthr. Cartil.* **2019**, *27*, 1185–1196. [CrossRef] [PubMed]

94. van der Kraan, P.M.; van den Berg, W.B. Osteophytes: Relevance and biology. *Osteoarthr. Cartil.* **2007**, *15*, 237–244. [CrossRef] [PubMed]

95. Blaney Davidson, E.N.; Vitters, E.L.; van der Kraan, P.M.; van den Berg, W.B. Expression of transforming growth factor-beta (TGFbeta) and the TGFbeta signalling molecule SMAD-2P in spontaneous and instability-induced osteoarthritis: Role in cartilage degradation, chondrogenesis and osteophyte formation. *Ann. Rheum. Dis.* **2006**, *65*, 1414–1421. [CrossRef]

96. Jingushi, S.; Shida, J.; Iwamoto, Y.; Kinoshita, T.; Hiyama, Y.; Tamura, M.; Izumi, T. Transient exposure of fibroblast growth factor-2 induced proliferative but not destructive changes in mouse knee joints. *Connect. Tissue Res.* **2006**, *47*, 242–248. [CrossRef]

97. Okazaki, K.; Jingushi, S.; Ikenoue, T.; Urabe, K.; Sakai, H.; Ohtsuru, A.; Akino, K.; Yamashita, S.; Nomura, S.; Iwamoto, Y. Expression of insulin-like growth factor I messenger ribonucleic acid in developing osteophytes in murine experimental osteoarthritis and in rats inoculated with growth hormone-secreting tumor. *Endocrinology* **1999**, *140*, 4821–4830. [CrossRef]

98. Bechtold, T.E.; Saunders, C.; Mundy, C.; Um, H.; Decker, R.S.; Salhab, I.; Kurio, N.; Billings, P.C.; Pacifici, M.; Nah, H.D.; et al. Excess BMP Signaling in Heterotopic Cartilage Forming in Prg4-null TMJ Discs. *J. Dent. Res.* **2016**, *95*, 292–301. [CrossRef]

99. Yang, H.; Zhang, M.; Liu, Q.; Zhang, H.; Zhang, J.; Lu, L.; Xie, M.; Chen, D.; Wang, M. Inhibition of Ihh Reverses Temporomandibular Joint Osteoarthritis via a PTH1R Signaling Dependent Mechanism. *Int. J. Mol. Sci.* **2019**, *20*. [CrossRef]

100. Cohen, M.M., Jr. The hedgehog signaling network. *Am. J. Med. Genet. A* **2003**, *123A*, 5–28. [CrossRef]

101. Pepinsky, R.B.; Zeng, C.; Wen, D.; Rayhorn, P.; Baker, D.P.; Williams, K.P.; Bixler, S.A.; Ambrose, C.M.; Garber, E.A.; Miatkowski, K.; et al. Identification of a palmitic acid-modified form of human Sonic hedgehog. *J. Biol. Chem.* **1998**, *273*, 14037–14045. [CrossRef]

102. Porter, J.A.; Ekker, S.C.; Park, W.J.; von Kessler, D.P.; Young, K.E.; Chen, C.H.; Ma, Y.; Woods, A.S.; Cotter, R.J.; Koonin, E.V.; et al. Hedgehog patterning activity: Role of a lipophilic modification mediated by the carboxy-terminal autoprocessing domain. *Cell* **1996**, *86*, 21–34. [CrossRef]

103. Chen, M.H.; Li, Y.J.; Kawakami, T.; Xu, S.M.; Chuang, P.T. Palmitoylation is required for the production of a soluble multimeric Hedgehog protein complex and long-range signaling in vertebrates. *Genes Dev.* **2004**, *18*, 641–659. [CrossRef] [PubMed]

104. Zeng, X.; Goetz, J.A.; Suber, L.M.; Scott, W.J., Jr.; Schreiner, C.M.; Robbins, D.J. A freely diffusible form of Sonic hedgehog mediates long-range signalling. *Nature* **2001**, *411*, 716–720. [CrossRef] [PubMed]

105. Gallet, A.; Ruel, L.; Staccini-Lavenant, L.; Therond, P.P. Cholesterol modification is necessary for controlled planar long-range activity of Hedgehog in Drosophila epithelia. *Development* **2006**, *133*, 407–418. [CrossRef] [PubMed]

106. Goetz, J.A.; Singh, S.; Suber, L.M.; Kull, F.J.; Robbins, D.J. A highly conserved amino-terminal region of sonic hedgehog is required for the formation of its freely diffusible multimeric form. *J. Biol. Chem.* **2006**, *281*, 4087–4093. [CrossRef] [PubMed]

107. Billings, P.C.; Pacifici, M. Interactions of signaling proteins, growth factors and other proteins with heparan sulfate: Mechanisms and mysteries. *Connect. Tissue Res.* **2015**, *56*, 272–280. [CrossRef]

108. Radhakrishnan, A.; Sun, L.P.; Kwon, H.J.; Brown, M.S.; Goldstein, J.L. Direct binding of cholesterol to the purified membrane region of SCAP: Mechanism for a sterol-sensing domain. *Mol. Cell* **2004**, *15*, 259–268. [CrossRef]

109. Tukachinsky, H.; Kuzmickas, R.P.; Jao, C.Y.; Liu, J.; Salic, A. Dispatched and scube mediate the efficient secretion of the cholesterol-modified hedgehog ligand. *Cell Rep.* **2012**, *2*, 308–320. [CrossRef]

110. Kawakami, T.; Kawcak, T.; Li, Y.J.; Zhang, W.; Hu, Y.; Chuang, P.T. Mouse dispatched mutants fail to distribute hedgehog proteins and are defective in hedgehog signaling. *Development* **2002**, *129*, 5753–5765. [CrossRef]

111. Jakobs, P.; Exner, S.; Schurmann, S.; Pickhinke, U.; Bandari, S.; Ortmann, C.; Kupich, S.; Schulz, P.; Hansen, U.; Seidler, D.G.; et al. Scube2 enhances proteolytic Shh processing from the surface of Shh-producing cells. *J. Cell Sci.* **2014**, *127*, 1726–1737. [CrossRef]

112. Yasuda, T.; Mundy, C.; Kinumatsu, T.; Shibukawa, Y.; Shibutani, T.; Grobe, K.; Minugh-Purvis, N.; Pacifici, M.; Koyama, E. Sulfotransferase Ndst1 is needed for mandibular and TMJ development. *J. Dent. Res.* **2010**, *89*, 1111–1116. [CrossRef] [PubMed]

113. Crawford, B.E.; Garner, O.B.; Bishop, J.R.; Zhang, D.Y.; Bush, K.T.; Nigam, S.K.; Esko, J.D. Loss of the heparan sulfate sulfotransferase, Ndst1, in mammary epithelial cells selectively blocks lobuloalveolar development in mice. *PLoS ONE* **2010**, *5*, e10691. [CrossRef] [PubMed]

114. Grobe, K.; Inatani, M.; Pallerla, S.R.; Castagnola, J.; Yamaguchi, Y.; Esko, J.D. Cerebral hypoplasia and craniofacial defects in mice lacking heparan sulfate Ndst1 gene function. *Development* **2005**, *132*, 3777–3786. [CrossRef] [PubMed]

115. Long, H.Q.; Tian, P.F.; Guan, Y.X.; Liu, L.X.; Wu, X.P.; Li, B. Expression of Ihh signaling pathway in condylar cartilage after bite-raising in adult rats. *J. Mol. Histol.* **2019**, *50*, 459–470. [CrossRef]

116. Kinumatsu, T.; Shibukawa, Y.; Yasuda, T.; Nagayama, M.; Yamada, S.; Serra, R.; Pacifici, M.; Koyama, E. TMJ development and growth require primary cilia function. *J. Dent. Res.* **2011**, *90*, 988–994. [CrossRef]

117. Buckland, J. Osteoarthritis: Blocking hedgehog signaling might have therapeutic potential in OA. *Nat. Rev. Rheumatol.* **2010**, *6*, 61. [CrossRef]

International Journal of
Molecular Sciences

Review

Recent Insights into Long Bone Development: Central Role of Hedgehog Signaling Pathway in Regulating Growth Plate

Ryuma Haraguchi [1,*], Riko Kitazawa [1,2], Yukihiro Kohara [1], Aoi Ikedo [3], Yuuki Imai [3] and Sohei Kitazawa [1]

[1] Department of Molecular Pathology, Ehime University Graduate School of Medicine, Shitsukawa, Toon City 791-0295, Japan; riko@m.ehime-u.ac.jp (R.K.); kohara.yukihiro.yu@ehime-u.ac.jp (Y.K.); kitazawa@m.ehime-u.ac.jp (S.K.)
[2] Department of Diagnostic Pathology, Ehime University Hospital, Shitsukawa, Toon City 791-0295, Japan
[3] Division of Integrative Pathophysiology, Proteo-Science Center, Ehime University Graduate School of Medicine, Shitsukawa, Toon City 791-0295, Japan; aikedo@m.ehime-u.ac.jp (A.I.); y-imai@m.ehime-u.ac.jp (Y.I.)
* Correspondence: ryumaha@m.ehime-u.ac.jp; Tel.: +81-89-960-5265

Received: 11 October 2019; Accepted: 18 November 2019; Published: 20 November 2019

Abstract: The longitudinal growth of long bone, regulated by an epiphyseal cartilaginous component known as the "growth plate", is generated by epiphyseal chondrocytes. The growth plate provides a continuous supply of chondrocytes for endochondral ossification, a sequential bone replacement of cartilaginous tissue, and any failure in this process causes a wide range of skeletal disorders. Therefore, the cellular and molecular characteristics of the growth plate are of interest to many researchers. Hedgehog (Hh), well known as a mitogen and morphogen during development, is one of the best known regulatory signals in the developmental regulation of the growth plate. Numerous animal studies have revealed that signaling through the Hh pathway plays multiple roles in regulating the proliferation, differentiation, and maintenance of growth plate chondrocytes throughout the skeletal growth period. Furthermore, over the past few years, a growing body of evidence has emerged demonstrating that a limited number of growth plate chondrocytes transdifferentiate directly into the full osteogenic and multiple mesenchymal lineages during postnatal bone development and reside in the bone marrow until late adulthood. Current studies with the genetic fate mapping approach have shown that the commitment of growth plate chondrocytes into the skeletal lineage occurs under the influence of epiphyseal chondrocyte-derived Hh signals during endochondral bone formation. Here, we discuss the valuable observations on the role of the Hh signaling pathway in the growth plate based on mouse genetic studies, with some emphasis on recent advances.

Keywords: hedgehog; growth plate; endochondral ossification; chondrocyte; osteoblast; bone disease

---

1. Introduction

The growth plate is a layer of cartilage in developing long bones between the epiphysis and the metaphysis. The elongation of long bones occurs at the growth plate, where cartilage is formed and then replaced by bone tissue (Figure 1). In mammals, the growth plate is composed of three types of highly organized and specialized cartilage: resting, proliferative, and hypertrophic zone, which originate from embryonic cartilage primordia by the condensation of undifferentiated limb bud mesenchymal cells. Resting zone chondrocytes supply stem-like cells that give rise to clones of proliferative zone chondrocytes, and determine the spatial orientation of adjacent proliferative columns parallel to the long axis of the bone [1]. The proliferative zone is the region of active cell replication [2]. When a proliferative zone chondrocyte divides, its derivatives, proliferating rapidly, line up along the long axis of the bone. As a result, clones of chondrocytes are arranged in columns parallel to this axis, and this orientation determines longitudinal bone growth in a specific direction. Proliferative chondrocytes gradually stop dividing and expand to become hypertrophic chondrocytes [3,4]. Hypertrophic zone chondrocytes, terminally differentiated chondrocytes, produce a kind of scaffold, mineralized by their extracellular matrix, that supports bone formation by osteoblastic cells before they undergo apoptosis. Hypertrophic chondrocytes also promote vascular invasion at the chondro-osseous junction (COJ), the junction between calcified and non-calcified cartilage matrices, which is a critical process for recruiting osteoblast and osteoclast progenitors [5,6]. The overall process described above, commonly referred to as "endochondral ossification", has been studied widely because it regulates the longitudinal growth of bone [7]. Also, a limited number of chondrocytes within the growth plate are themselves generated from stem-cell-like progenitor cells called chondro-progenitors, and directly transform in considerable numbers into the osteogenic lineage in developing bone [8].

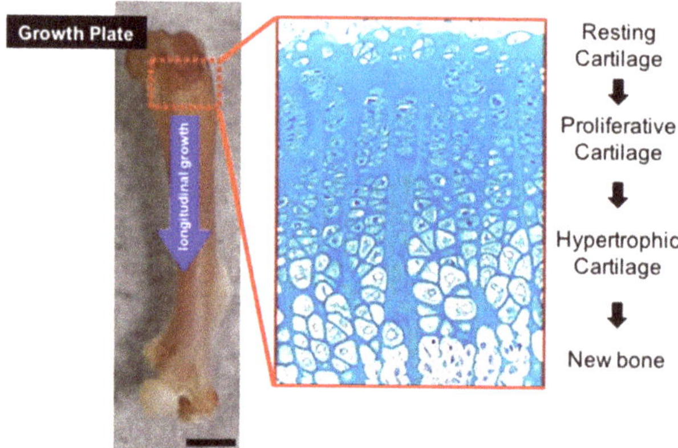

Figure 1. The longitudinal growth of long bone by the growth plate. The growth plate is composed of highly organized and specialized three types of cartilage: the resting, proliferative, and hypertrophic zone. Resting zone chondrocytes supply stem-like cells that give rise to clones of proliferative zone chondrocytes, and determine the spatial orientation of adjacent proliferative columns parallel to the long axis of the bone. The proliferative zone is the region of active cell replication. Hypertrophic zone chondrocytes provide a cartilaginous template, mineralized by their extracellular matrix, supporting the new bone formation by osteoblastic cells. Scale bars indicate 1.25 mm.

Hedgehog (Hh) signaling is known to be among the most important regulators in many aspects of insect and vertebrate development [9–15]. In mammals, the three Hh proteins, Sonic hedgehog (Shh), Indian hedgehog (Ihh), and Desert hedgehog (Dhh), undergo several steps of post-translational modification, including proteolytic cleavage, glycosylation, and lipid modification, after which they are released by Hh-secreting cells with the help of Dispatched (Disp), a membrane transporter protein [10,14,15]. Hedgehog ligands are equally essential during vertebrate embryonic development. Shh is expressed at various tissues, including the brain, skeleton, tooth, skin, gastrointestinal tract, urogenital tract, and lung, Ihh in gastrointestinal tract and cartilage, Dhh in the peripheral nerves and testicular cells [16]. Recently, it has been discovered that Shh also controls the behavior of cells with stem cell properties in the maintenance and regeneration of adult tissues [17]. After post-translational modification Hh proteins transmit signals through a receptor complex that includes the G-protein-coupled receptor, the twelve-pass transmembrane receptor Patched-1 (Ptc-1) and the seven-pass transmembrane protein smoothened (Smo), to control gene expression by modulating the activity of Gli transcription factors (Figure 2) [9–11,13,15]. In the absence of Hh ligands, Ptc-1 negatively regulates Hh pathway activation through the constitutive repression of positive Hh effector smoothened and is also transcriptionally controlled. Once the Hh ligand binds to Ptc-1, the repressive action to Smo is released, and Gli-mediated transcription leading to the regulation of downstream target Hh is activated. In mammals, among Gli transcription factors (Gli1/2/3) that collectively mediate all Hh signaling, Gli-2 and Gli-3 are the initial responders to Hh signaling [9,14]. Gli-1 is a positive transcriptional mediator and one of the direct downstream target genes in the Hh pathway. Gli-2 is considered to function predominantly as a transcriptional activator, whereas Gli-3 functions mainly as a repressor (for detailed review, see [9,14,18]).

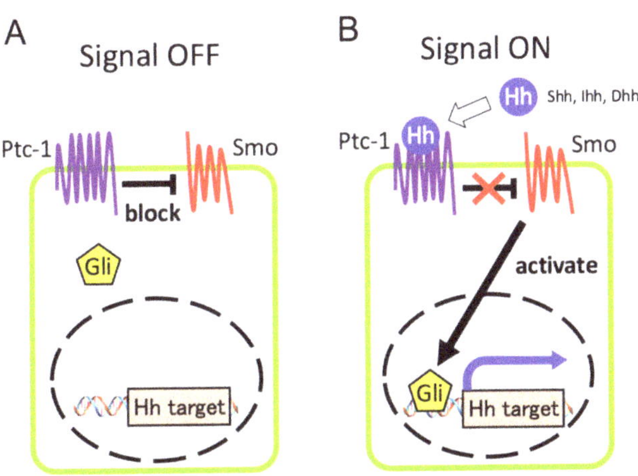

Figure 2. Overview of hedgehog signaling pathway. (**A**) In the absence of Hh ligands, Ptc-1 blocks Hh pathway activation through the repression of Smo. (**B**) Once the Hh ligand binds to Ptc-1, the repressive action to Smo is released, and Gli-mediated transcription leading to the regulation of downstream target Hh is activated.

In general, Ihh participates in the process of endochondral bone formation and is expressed in the hypertrophic zone chondrocytes of the growth plate [19]. Based mainly on in vivo studies with the use of genetically modified mice, remarkable progress has been made in understanding how hedgehog signaling from growth plate chondrocytes regulates skeletal development and interacts with other signaling factors. The multiple functions of Hh signaling in developing long bone have previously been summarized by several research groups [9,15,20]. The present review is aimed at summarizing current findings that would assist in understanding this area of research, with an emphasis on the multiple roles of Hh during early-chondrogenesis and endochondral ossification processes, the coupling function of Hh and cholesterol biosynthesis during chondrocyte differentiation, the potential role of Hh in the cell trans-differentiation of chondrocytes into bone cells, and the involvement of Hh in skeletal disorders.

2. Hedgehog Signal Is a Critical Regulator of Early Chondrogenesis

Chondrogenesis is the earliest phase of skeletogenesis that results in the formation of the growth plate and leads to endochondral ossification during the growth of the long bone [21–23]. The onset of chondrogenesis is marked by the condensation of dividing undifferentiated mesenchymal cells in the limb bud, and condensed cells subsequently differentiate into clusters of cartilage cells known as chondrocytes that continue to proliferate until their hypertrophic differentiation. This cartilaginous tissue eventually becomes vascularized, initiating the formation of growth plates. Before chondrocyte hypertrophy, the onset of the chondrocyte maturation process, enlargement of the cartilage template by the proliferation of condensed pre-cartilage or cartilage cells, is vital for a well-organized growth plate.

Among the Hh protein family in mammals, Ihh is known to function as a principal source of the Hh ligand that activates the Hh signaling pathway on skeletogenesis, which is primarily expressed in proliferating limb bud mesodermal cells that eventually differentiate into skeletal chondrocytes immediately after mesenchymal condensation [19]. Studies on mice with mutations in the *Ihh* gene have provided in vivo evidence that Hh signaling is requisite for adequate cell proliferation of the condensed pre-cartilage mesenchyme responsible for forming a framework for endochondral ossification [19,24]. Global Ihh knockout mice show a remarkable reduction in longitudinal growth, and most of Ihh-null mutants died at birth, due to respiratory failure [19]. The long bones of Ihh-null mutants are only about one-third the length of those in wild-types. These defects are not directly affected by the chondrocyte maturation process, as is apparent from as early as the mid-embryonic stage prior to cartilage hypertrophy. Moreover, Ptc-1, some of the transmembrane receptor complex for Hh ligands and the direct downstream target of the Hh signaling pathway are expressed at the dividing condensed mesenchyme adjacent to Ihh-expressing cells, and their expression in the Ihh mutant limb is markedly decreased with a significant reduction in the proliferation of cartilaginous tissue [19]. These observations suggest a direct role of the Hh pathway in the cartilaginous growth of limb skeletal elements. In contrast, whereas the Hh pathway is required for limb bud chondrocyte proliferation, its aberrant activation also leads to the dysregulation of chondrogenic skeletal formation. Ligand-independent activation of the Hh pathway has an inhibitory effect on early chondrogenesis [24]. The authors have demonstrated that conditional deletion of the *Ptc-1* gene in the undifferentiated limb mesenchyme with the use of Prx1-Cre, causes cell-autonomously activated Hh signaling cascade, resulting in marked disorganization of skeletal tissues that are severely truncated cartilage elements with a negative Alcian blue staining. Furthermore, an in vitro micro-mass culture system has revealed that activation of ligand-independent Hh signaling prevents early chondrogenesis. Micro-mass cultures derived from Prx1-Cre:Ptc-1$^{c/c}$ limbs show a significant decrease in cartilage cluster formation. Moreover, a decrease in the expression of the *Col2a1* gene, an early chondrogenic marker reflecting the onset of chondrocyte differentiation, is detected in mutant cultures with the upregulation of universal downstream Hh target genes. Under the same experimental conditions, despite an increase in the level of Hh targets, no difference is observed in the expression level of Sox-9, the earliest master regulator of chondrogenesis, or of N-Cadherin, a marker for mesenchymal condensation, in Prx1-Cre:Ptc-1$^{c/c}$ versus the control. These findings of in vitro experiments using the limb micro-mass culture system support the concept that the inhibitory effect of cell-autonomously activated cells of the Hh pathway on early-chondrogenesis underlie below mesenchymal cell condensation and above chondrocyte differentiation. In contrast to the Prx1-Cre:Ptc-1$^{c/c}$ model, exogenous Hh ligand treatment of micro-mass cultures, which is an activation of the ligand-dependent Hh pathway, causes continuous increases in the expression of chondrogenic markers involved in the formation of mature cartilage clusters [24]. Results of Hh ligand treatment on micro-mass cultures is consistent with global Ihh knockout early-stage limb phenotypes. Thus, Hh signaling is most likely related to the rapid enlargement of cartilage tissues during early-chondrogenesis, and this developmental process requires the balancing of positive and negative input involved in the control of the activation level of the Hh pathway.

In addition to the fundamental effector molecules such as *Ptc-1*, *Smo*, and *Gli*, functional genes that directly control Hh signal transduction have been identified by using differential screening and phenotypic analyses of mutant animal models [10,12,13,15]. These genes, including *Kif7*, *Sufu*, *Hhip*, *Cdo*, *Boc*, and *Gas1*, are capable of modulating the activation level of the Hh pathway through direct interaction with Hh ligands and their cytoplasmic components. Functional mutation in these genes exhibits various chondrogenic defects with an alteration of Hh signaling activity during early embryogenesis [25–27]. These observations also support the importance of fine-tuned Hh signal activity in early cartilage development, and the mechanisms underlying developmental defects caused by the dysfunction of Hh modifiers need to be elucidated further.

3. Central Role of Hedgehog Signaling in Regulation of Growth Plate

In the metaphysis at both ends of a long bone, the growth plate is orchestrated in the limb skeletal cartilage through multistage processes: vascular invasion, the formation of primary/secondary ossification centers and osteoblast/osteoclast recruitment under the influence of regulatory molecules (for detailed review, see [28]). Growth plate chondrocytes undergo a tightly regulated developmental program of proliferation, pre-hypertrophy, hypertrophy, and apoptosis in the specialized cartilage layers and are eventually replaced by osteoblasts at the distal edge of the growth plate (also termed COJ: chondro-osseous junction). The precise regulation of growth plate chondrocytes aligned according to their defined differentiation phase, absolutely crucial for longitudinal growth of endochondral bones, is achieved under the adequately controlled activity of the Hedgehog (Hh) signaling pathway.

3.1. Hh Pathway and Growth Plate Formation

In general, among Hh ligands, Indian hedgehog (Ihh) acts in the process of growth plate development [20,28]. As noted above, the *Ihh* gene is initially expressed in condensed limb mesenchymal cells or in chondrocytes of the cartilaginous skeletal elements. During growth plate development, Ihh expression becomes gradually restricted to postmitotic pre-hypertrophic chondrocytes adjacent to proliferative zone chondrocytes.

In vivo studies using Ihh mutant mouse models and our data have revealed that Ihh is indispensable for the process of growth plate organization (Figure 3) [19,29–32]. These models show abnormal endochondral bone formation with a complete absence of the growth plate and the superiority of mature chondrocytes. Mice carrying null mutations of the *Ihh* gene show a severely disrupted growth plate with abnormal chondrocyte proliferation and maturation at embryonic stages [19]. The conditional ablation of Ihh in the full skeletal lineages of the limb by using *Prx1* promoter markedly inhibits skeletal development in the absence of the normal growth plate and the secondary ossification center in the postnatal period [29,30]. Newborn Prx1-Cre·Ihh$^{c/c}$ growth plate cartilage lacks a zone of aligned columnar chondrocytes and both pre-hypertrophic and hypertrophic chondrocytes are barely formed. Before postnatal day 10, mutant humerus bone revealed a total absence of a growth plate and no secondary ossification center. Ablation of Ihh by using *Col2a1* promoter/enhancer also reveals severe skeletal deformities with loss of a normal growth plate exhibiting the characteristic zones of chondrocyte differentiation (resting, proliferating, pre-hypertrophic and hypertrophic) [32]. The Col2a1-Cre:Ihh$^{c/c}$ disorganized growth plate shows approximately half the number of BrdU-labeling cells and abnormal location (in the central region of the long bone) of hypertrophic cells expressing Type X collagen at late embryonic stages. The mutant growth plate also showed a delay in chondrocyte differentiation, as indicated by the delayed expression of Type X collagen and osteopontin. Moreover, ablation of the *Ihh* gene from postnatal chondrocytes by using tamoxifen-inducible Col2a1-CreER transgenic mouse lines has been shown to cause premature closure of the growth plate: disrupted columnar structure of chondrocytes, and the appearance of abnormal maturation of hypertrophic chondrocytes near the articular surface [31]. The growth plate of neonatal-tamoxifen-injected Col2a1-CreER:Ihh$^{c/c}$ mice has shown complete loss of the columnar structure of proliferating chondrocytes. The mutant growth plate is composed mainly of hypertrophic chondrocytes that express Type X collagen but not Type II collagen, showing an incorrect progressive maturation of cartilaginous cells [31]. This abnormal process that starts at postnatal day seven eventually leads to a total loss of the growth plate in mutant tibial bones at postnatal day 14. Thus, the actions of Ihh in the skeletal cartilage adequately regulate chondrocyte proliferation and maturation required for the organization of the normal growth plate during embryonic and postnatal periods.

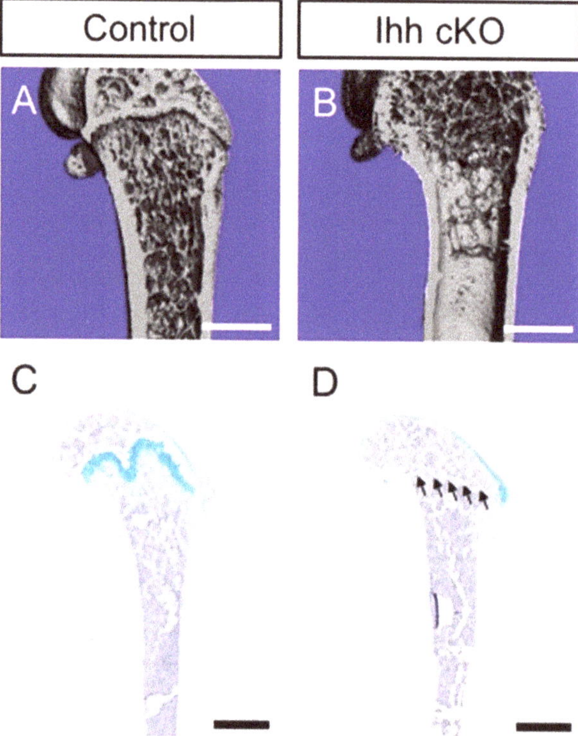

Figure 3. Chondrocyte-derived Ihh is required for the maintenance of a normal growth plate. (**A,B**) Longitudinal view of μCT images in distal femur from control and Gli1-CreER; Ihh$^{c/c}$ (Ihh cKO). Control and Ihh cKO littermate mice were treated with tamoxifen at four weeks of age and analyzed after eight weeks to inactivate the *Ihh* gene. Note decreased trabecular mass and completely lacked growth plate in Ihh cKO mice (**B**). (**C,D**) Representative images of femur stained with hematoxylin and alcian blue. Alcian blue positive cartilage matrix in the distal femur is absent in Ihh cKO mice (**D**, arrowheads show). Scale bars indicate 1 mm (**A,B**) and 1.25 mm (**C,D**).

3.2. Hh Pathway Controls Regulation of Growth Plate Differentiation through Interaction with PTH-PTHrP Signaling

The Hh pathway through Ihh signaling is critical for not only the initial morphogenesis but also the subsequent onset and advancement of chondrocyte differentiation in the growth plate. In vivo genetic studies have verified that the activated Hh pathway controls these processes through interaction with PTH-PTHrP signaling [7,19,20,28,33,34]. Parathyroid hormone-related peptide (PTHrP), which is similar to parathyroid hormone (PTH), plays a crucial role in chondrocyte proliferation and hypertrophy of the growth plate [35,36]. In the growth plate, PTHrP is expressed at high levels in periarticular resting cells and at low levels in proliferating chondrocytes adjacent to the pre-hypertrophic zone, while its receptor, Parathyroid hormone 1 receptor (PTH1R), is produced at low levels by proliferating chondrocytes and at a high level in pre-hypertrophic cells [19,34]. In studies on mice, loss-of- and gain-of-function of PTHrP and PTH1R have indicated that the PTH-PTHrP signal maintains chondrocytes proliferating in the growth plate and suppresses their excessive hypertrophy, resulting in premature mineralization of growth plate chondrocytes [37,38]. Basic studies on animal models have demonstrated that PTHrP production in the growth plate is controlled by the Hh pathway through Ihh signaling. As noted above, Ihh is expressed and secreted by pre-hypertrophic and hypertrophic chondrocytes in the growth plate, while PTHrP is expressed in periarticular resting cells and proliferating chondrocytes adjacent to the Ihh expressed pre-hypertrophic zone. A study on a chicken embryo model has demonstrated that overexpression of Ihh increases PTHrP expression in periarticular chondrocytes in the growth plate [34]. Hh pathway may not directly control the promoter activity of PTHrP [39,40]. PTHrP expression is absent from the growth plate in Ihh-null mice that exhibit a skeletal phenotype (leading to accelerated hypertrophy of chondrocytes) similar to that caused by the *PTHrP* gene deletion [19]. Studies with the use of compound mutant mice have demonstrated that constitutive activation of the PTHrP signal in the Ihh-null growth plate partially rescues its abnormality. Double mutant growth plates do not accelerate chondrocyte hypertrophy, suggesting that the Hh pathway (through Ihh) controls growth plate development by a PTHrP-dependent pathway [41,42]. PTHrP regulated by Ihh probably plays a critical role in fine-tuning between chondrocyte proliferation and maturation.

Ihh promotes chondrocyte differentiation hypertrophic chondrocytes; at the same time, PTHrP expression induced by Ihh maintains the proliferating state of chondrocytes and blocks their hypertrophic differentiation [7,20,28]. Maintenance of the proliferation state in the growth plate eventually delays Ihh production by hypertrophic chondrocytes. Thus, Ihh and PTHrP form a negative feedback loop that both synchronizes and controls the ratio of chondrocyte proliferation and differentiation in the growth plate.

3.3. Crosstalk between Hh and Other Signaling Pathways as a Basis for Regulatory Mechanisms of Growth Plate Development and Function

In vivo studies based on genetic manipulation of mice strongly suggest the possibility of signaling crosstalk underlying strict regulation of growth plate development and function by Hh pathways and other signaling pathways [20,28,43,44].

3.3.1. Hh and Wnt/β-Catenin Signaling

Wnt/β-catenin signaling that regulates osteoblast maturation could be affected by the functional deletion of Ihh from postnatal chondrocytes. It has been identified as playing fundamental roles in growth plate formation and in the terminal differentiation of osteoblasts from their progenitors adjacent to the growth plate [31].

In neonatal-tamoxifen-injected Col2a1-CreER:Ihh$^{c/c}$ mice, efficient deletion of the *Ihh* gene from postnatal growth plate chondrocytes has shown a significant reduction in β-catenin expression in the bone collar and primary trabeculae of mutants. Moreover, a remarkable reduction in the expression of Dickkopf1 (Dkk1) and osteoprotegerin (OPG), the downstream target gene of the Wnt/β-catenin signaling pathway, has been evident in the mutants [31]. Furthermore, compound mutant analysis has shown that the Wnt/β-catenin signaling pathway is a critical downstream target of Hh signaling from chondrocytes for the regulation of osteoblast differentiation during endochondral ossification. Also, Hh signaling is activated and Wnt/β-catenin is inactivated by a chondrocyte-specific deletion of Ptc-1 and β-catenin in mice treated with *Col2a1* promoter/enhancer [45]. By expression analyses of Hh signaling target genes, Hhip and Gli-1 marked activation has been found in both Ptc-1 mutant and double mutant mice in terms of Ptc-1 and β-catenin, which indicates that Wnt/β-catenin is requisite for bone formation and acts downstream of the Hh pathway. Thus, these observations support the view that growth plate chondrocyte-derived Ihh is critical for skeletal formation through the activation of the Wnt/β-catenin pathway and for regulating its action.

3.3.2. Hh and FGF Signaling

Fibroblast growth factor (FGF) signaling has been identified as playing fundamental roles in the proliferation and differentiation process of growth plate chondrocytes [46]. The FGF family comprises at least 22 ligands that bind to at least four receptors, among which FGF receptor-3 (Fgfr-3) critically regulates endochondral bone formation in the growth plate [47,48]. Mice carrying null mutations of Fgfr-3 display accelerated long bone elongation, a high rate of chondrocyte proliferation and enlargement of chondrocyte columns in the hypertrophic zone [49,50]. Conversely, the gain-of-function mutation of Fgfr-3 reduces chondrocyte proliferation and results in a markedly shortened long bone with disorganized chondrocyte columns [51–55]. Minina et al. have shown that the inhibition of growth plate chondrocyte proliferation, by upregulated FGF signaling through Fgfr-3 activation, is caused partly through the inactivation of the Hh pathway through Ihh signaling [56]. They have observed that dominantly activated Fgfr-3 reduces Ihh expression in hypertrophic chondrocytes of the growth plate. In vitro studies using the limb culture system have shown similar results, indicating an antagonistic action of FGF signaling in the control of chondrocyte proliferation and in Ihh expression [56]. Also, the Hh signaling pathway is dysregulated in the Fgfr-3 mutant growth plate [57]. As noted above, normally Ihh is expressed in growth plate chondrocytes of pre-hypertrophic and hypertrophic zones. By contrast, Fgfr-3-deficient mice show a markedly upregulated Ihh expression in the disorganized mutant growth plate, revealing an increase in proliferating chondrocytes and expansion of the hypertrophic zone. Moreover, systemic inhibition of the Hh pathway by smoothened inhibitor treatment partially prevents growth plate defects of Fgfr-3 mutants [57]. Thus, FGF signaling through Fgfr-3 activation controls the balance between proliferation and maturation of growth plate chondrocytes by fine-tuning the Hh pathway and suppressing Ihh expression.

3.3.3. Hh and BMP Signaling

Bone morphogenetic protein (BMP) signaling plays a vital role in endochondral bone development [7,28,58–60]. In vitro studies have shown that the addition of BMPs to limb explant culture systems enhances chondrocyte proliferation, and their effect is blocked by noggin, an inhibitor of several BMPs; also, that Bmp-2 stimulation delays terminal differentiation of hypertrophic chondrocytes [61]. Presently, various in vivo studies on loss-of-function have confirmed these actions of BMP signaling. Bmp-2 conditional deletion mice using the Col2a1-CreER line have demonstrated the failure of chondrocyte proliferation and maturation in the growth plate [62]. Chondrocyte specific deletion mice of BMP signal core receptors, BMPR-IA expressed throughout the growth plate, have demonstrated a disorganized epiphysis and absence of the growth plate [59,63]. Also, conditional deletion of Smad proteins (Smad1 and Smad5), main signal transducers for BMPR, leads to severe malformed growth plates and impaired chondrocyte survival [64]. Previous reports have described the interaction of Hh and BMP signaling in endochondral bone development [61,64]. Studies with the use of the limb explant culture system have demonstrated that BMP and Ihh signaling interact to coordinate chondrocyte proliferation and differentiation, for example, Bmp-2 treated limb explants show increased Ihh expression by hypertrophic differentiation, and promote both the proliferation of chondrocytes and the elongation of proliferative chondrocyte columns. Moreover, enhanced Ihh signaling by *Col2* promoter transgenic model delays hypertrophic differentiation with the upregulation of *Bmp* genes [61]. Furthermore, the canonical Smad pathway triggered by BMPs actually acts as an upstream regulator of Ihh/PTHrP signaling in the growth plate. Chondrocyte specific Smad1/5 conditional KO mice demonstrate severe chondrodysplasia (as mentioned above). The expression of Ihh and PTHrP receptors is completely lacking in the mutant growth plate [64]. Studies have shown that Ihh is a target of BMP and FGF pathways in chondrocytes and that Smad proteins can bind to the *Ihh* promoter region [65]. Interestingly, Smad1/5 cKO analyses have revealed an imbalance of BMPs and FGF signaling in mutant cartilage: disrupted expression of BMP signaling components and advanced phosphorylation and nuclear entry of STAT1, one of the core mediators in FGF signaling [64]. Furthermore, characterization and functional analysis of the promoter of *Ihh* and *Msx2*, one of the downstream targets in BMP signaling, under the influence of BMP and FGF signaling, has disclosed that BMP signaling controls the Ihh/PTHrP signaling loop by inhibiting the antagonistic effect of FGFs on Ihh signaling (as noted above). From these results, it has been proposed that the enhancement of Ihh expression triggered by BMP signaling might be negatively controlled by FGF activation through BMP canonical Smad phosphorylation [64]. Taken together, these observations suggest that BMP signaling has two established functions by cooperating with Ihh (or FGF) signaling in growth plate development: 1) activation of chondrocytes within the resting zone to enter a proliferative state, and 2) prevention of chondrocyte hypertrophy.

3.3.4. Hh Signaling and Angiogenic Factors

At the chondro-osseous junction (COJ), the onset of the final step of chondrocyte hypertrophy is initiated by loss of hypertrophic marker genes, tightly synchronized with the induction of vascular endothelial growth factors (VEGFs) and metalloproteinases (MMPs) [66]. VEGF signaling plays critical roles in promoting vascular invasion and consequent remodeling of cartilage matrices by recruiting osteoblast and osteoclast progenitors. Systemic inhibition of VEGFs by the administration of its soluble receptor suppresses blood vessel invasion and metaphyseal bone trabeculae formation with an increased width of the hypertrophic zone of the growth plate [5]. Mmp-9 and Mmp-13 are prerequisites to the promotion of vascular invasion into the non-calcified hypertrophic matrix (the lacunae of dying hypertrophs, which do have a mineralized ECM [67]): compound KO mice with deleting both genes show marked enlargement in the hypertrophic zone of the growth plate [68]. Previous studies have demonstrated that Ihh signaling by hypertrophic chondrocytes plays critical roles in orchestrating the above vascular invasion and bone remodeling processes at COJ [31,69]. Ihh-null mutant mice display a significant decrease of VEGF-A, Mmp-9 and Mmp-13 in the disorganized growth plate, with no osteoclast staining positive for a TRAP [69]. Chondrocyte specific Ihh-deficient mice generated by using *Col2a1* promoter/enhancer also show atypical vascular invasion in the central region of the mutant growth plate [31]. It has also been suggested that Ihh indirectly regulates VEGF expression of hypertrophic chondrocytes through Runx2 [70,71]. Runx2, a member of the runt family of transcription factors, plays critical roles in the maturation processes of growth plate chondrocytes under the influence of Ihh signaling [70]. Furthermore, Runx2 plays a role in the induction of VEGF expression in hypertrophic chondrocytes of the growth plate [71]. Thus, these observations strongly suggest the existence of a certain link between Ihh signaling and terminal phase regulation of chondrocyte hypertrophy, which includes vascular invasion and cartilage matrix remodeling.

As shown above, this chapter described crosstalk between the Hh pathway and other signaling factors, Wnt/β-catenin, FGFs, BMPs, and VEGF, which are considered essential for normal growth plate development. In addition to the above, however, numerous other signaling pathways contribute to the regulation of growth plate development, and the relation between the Hh pathway and those pathways has yet to be exhaustively elucidated. We believe that the identification of regulatory signaling interaction with the Hh pathway may reveal additional fundamental molecular mechanisms, like Ihh/PTHrP signaling, that dominate growth plate development.

3.4. Coupling Role of Hh Signaling Pathway and Intracellular Cholesterol Production in Growth Plate Development

Dysregulation of cholesterol synthesis is involved in multiple developmental abnormalities. Numerous human mutational and clinical studies provide the notion that cholesterol is vital for normal skeletal development [72]. Also, previous experimental studies, on animal models treated with cholesterol synthesis inhibitors, have demonstrated severe skeletal malformations, including digit patterning defects, and decreased width of the long bone growth plate [73,74]. In mammals, cholesterol is produced from steroid hormones, bile acids, and vitamins, and intracellular cholesterol biosynthesis is tightly controlled by proteins in the endoplasmic reticulum (ER), including sterol regulatory element-binding proteins (SREBPs) and SREBP cleavage-activating proteins (SCAP) [75]. *SREBP* genes are activated in response to low cellular cholesterol levels by events of protease cleavage and transport into the nucleus. SCAP constitutes a complex with SREBP and acts as a cholesterol sensor. In the case of low cholesterol levels, SCAP recruits SREBP to the Golgi where proteases cleave SREBP, thereby releasing the N-terminal active domain of SREBP into the nucleus. More recently, studies with genetically modified mice have revealed that adequate regulation of cellular cholesterol biosynthesis in the growth plate chondrocyte is requisite for normal endochondral ossification and maintenance of chondrocyte homeostasis [76]. SCAP conditional deletion mice using the Col2a1-Cre line have displayed disordered growth plates and severe dwarfism. The mutant growth plate has displayed abnormal primary ossification, disorganized round cells in the resting zone, disrupted columnar structures in the proliferation zone, and reduction in the hypertrophic zone.

Although it has been strongly suggested that cell-autonomous cellular cholesterol production is critical for the organization of the normal growth plate, it is additionally represented that the Hh pathway and cellular cholesterol biosynthesis regulate each other during growth plate formation [76]. As mentioned above, the Hh pathway is involved in the control of chondrocyte differentiation in growth plate development, as a fundamental regulatory signal. Previous work has shown that Hh signaling regulates genes encoding intracellular cholesterol biosynthesis in chondrocytes [77]. In the *Col2* promoter transgenic model, enhanced Hh signaling by the overexpression of Gli-2 using *Col2* promoter induces higher levels of cholesterol and lipid accumulation in chondrocytes. By contrast, cholesterol is also capable of controlling Hh signaling at multiple phases in its signaling process, from ligand processing to coordination of receptors and intracellular effectors. Cholesterol modification of Hh ligands is needed for the construction of soluble multimeric Hh protein complexes that are freely diffusible, accumulate in a gradient, and are able to directly activate signaling over long distances [78]. Cholesterol also activates membrane protein smoothened by binding to its extracellular domain [79]. Furthermore, a study has indicated that chondrocyte-specific ablation of SCAP leads to the reduction of the type X collagen positive hypertrophic zone with decreasing expression of Ihh and Hh target genes, and exogenous cholesterol treatment slightly reinstates the reduction of Hh target gene expression in Scap-deficient chondrocytes. Also, enhanced activation of the Hh pathway by Gli-2 overexpression partially rescues the truncated limb phenotype of SCAP deficient mice [76]. Thus, these observations suggest that cholesterol biosynthesis is controlled by the Hh pathway, which is, in turn, controlled by intracellular cholesterol levels in chondrocytes. Detailed analyses of this relationship need to be prioritized in future studies on long bone development.

3.5. Hh Pathway and Developmental Contribution of Growth Plate Chondrocytes to Skeletal Bone Formation

Longitudinal bone growth progresses by continuous bone replacement of the growth plate, which is organized into distinct zones of chondrocytes: resting, proliferative, pre-hypertrophic, and hypertrophic. During longitudinal bone growth throughout postnatal and juvenile periods until early adulthood, chondrocytes of the growth plate continue to produce new cartilage matrices that are replaced by bone at the chondro-osseous junction (COJ). Subsequently, chondrocytes at the edge of the developing hypertrophic zone largely disappear by apoptosis as the cartilage matrix is degraded, a process concurrent with the invasion of blood vessels, hematopoietic cells, and progenitors for osteoblasts and marrow adipocytes. Nonetheless, in contrast with the above canonical pathway of endochondral bone formation, there is now a new emerging concept: direct trans-differentiation (chondrocyte-to-osteoblast) of growth plate chondrocytes into bone cells during longitudinal bone growth [8]. This concept is supported by recent genetic lineage tracing studies of growth plate chondrocytes, using constitutively active and inducible Cre-based transgenic mice, such as Acan-Cre-, Col2-Cre-, Col10-Cre- and Sox9-Cre-lines [80–83]. These studies have demonstrated that reporter gene expressing cells derived from growth plate chondrocytes are detected in the osteoblasts and osteocytes of trabecular and cortical bone, and in the bone marrow stroma during longitudinal bone growth. These lineage tracing experiments have also revealed that early-postnatal labeled chondrocytes in the growth plate contribute to multiple skeletal lineages and continue to supply these progeny cells for the long-span, over a year. Thus, growth plate chondrocytes provide opportunities for controlling skeletal formation that occurs rapidly and uniquely in longitudinally growing bone.

More recently, genetic lineage tracing analyses focusing on the Hh pathway have provided evidence that the contribution of growth plate chondrocytes to skeletal lineage formation is regulated under the influence of Hh responsiveness in growing long bone [84,85]. As mentioned above, the Hh pathway through Ihh signaling by hypertrophic chondrocytes has been shown as a critical factor for adequate differentiation of immature growth plate chondrocytes into a hypertrophic state through crosstalk with various signaling pathways. Fate mapping studies by using the Gli1-CreER line, in which the endogenous *Gli-1* gene (one of the Hh pathways downstream of target genes) promoter contains Cre recombinase, have demonstrated that Gli1-CreER genetically labeled cells are observed in hypertrophic chondrocytes and osteoprogenitors at the chondro-osseous junction (COJ). Genetically labeled osteoprogenitors then commit to the osteogenic lineage in the periosteum, trabecular, and cortical bone along the developing longitudinal axis, and continue to supply these progenitor cells for over a year. Our data and studies by others support the concept that correctly regulated Hh-signal responsive cells within the growth plate are functionally crucial for maintaining skeletal bone formation during postnatal life (Figures 3 and 4) [85]. Furthermore, these studies have shown that in aged bone, where longitudinal bone growth ceases, Hh-signal responsiveness and its implication in osteogenic lineage commitment is markedly reduced [84,85]. This observation affirms that age-related regulation of Hh-responsiveness in the growth plate may be one of the key regulatory factors that affect cessation of longitudinal bone growth with age.

The major finding from the above studies (Haraguchi et al. [84] and Shi et al. [85]) is that Hh-responsive cells in the growth plate comprise osteogenic progenitors that can differentiate into osteoblast directly. Congenital and traumatic defects of the growth plate produce a wide range of skeletal disorders including growth retardation, fragmentation, and degeneration with resultant abnormalities of growth [86,87]. The dysregulation of the Hh or other signaling pathways resulting in a permanent anomaly of the growth plate derived osteogenic lineage is one of the causative mechanisms of skeletal dysplasia in humans. Further understanding of the molecular regulatory mechanism of growth plate chondrocytes transitioning to the growing long bone may help to improve the treatment of skeletal growth disorders.

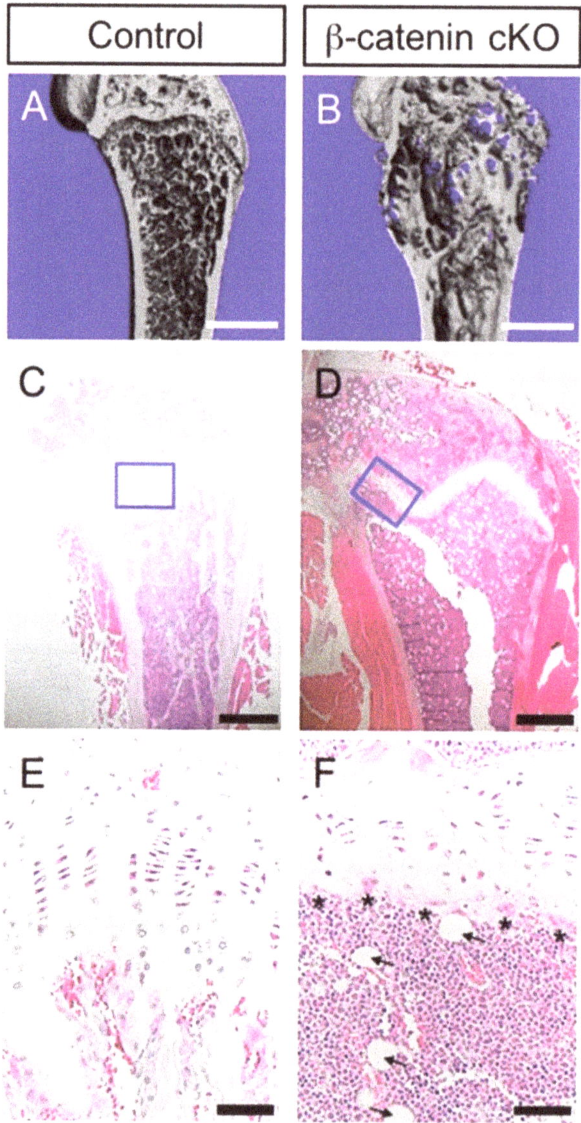

Figure 4. Loss of *β-catenin* gene in growth plate derived Hh-signal responded cells results in osteopenia and fatty bone marrow. (**A,B**) Longitudinal view of μCT images in distal femur from control and Gli1-CreER; β-catenin^c/c (β-catenin cKO). Control and β-catenin cKO littermate mice were treated with tamoxifen at 4 weeks of age and analyzed after 10 weeks to inactivate the *β-catenin* gene. μCT imaging revealed that β-catenin deletion resulted in an abnormal bone formation in the distal femur. (**C–F**) Representative images of femur stained with hematoxylin and eosin. (**E,F**) Higher magnification of blue boxes in (**C,D**). Histology revealed a lack of trabecular bone under the abnormal growth plate of β-catenin cKO mice (**D,F**). This bone phenotype was likely due to increased osteoclastic bone resorption (F, Asterisks mark increased osteoclasts). Histology of the femur also indicated a significant increase in adipocytes at the metaphysis (F, arrowheads show). Scale bars indicate 1 mm (**A,B**), 500 μm (**C,D**) and 50 μm (**E,F**).

4. Aberrant Hedgehog Signaling in Skeletal Disease

The Hh pathway entails a complicated sequence of regulatory events, including the production and spread of the mature Hh from ligand-secreting cells, tuning of Hh-signal responsiveness in ligands receiving cells, and intercellular coordination of Hh signal transduction activity. Abnormalities of the Hh pathway in the above events cause various bone diseases. As shown with evidence from animal studies, several other reports have shown that Hh signaling regulates and is requisite for bone development and growth in humans.

4.1. Hedgehog Signalling and Brachydactyly Syndrome

Brachydactylies are one group of congenital skeletal abnormalities that feature mainly truncated phalanges and/or metacarpals [88–90]. Mutational analyses have indicated that three heterozygous missense mutations in IHH cause brachydactyly type A1 (BDA1; OMIM 112500), which features truncated or lacking phalanges [88]. Analyses of Ihh deficient mice have defined the relation between IHH mutations and BDA1 as disturbed Hh pathway through Ihh signaling leading to truncated limbs [91]. Mutations responsible for BDA1 have been restricted to the N-terminal domain of IHH, and for the most part have altered codon positions 95, 100, and 131 [92–94]. The DBA1 mouse model, generated with the use of Ihh point mutated mice, had one of the mutations, E95K, inserted into the mouse *Ihh* gene locus; the result was that the point mutated mice demonstrated shortened middle phalanges in digits II and V [91]. Thus, the BDA1 mutation (E95K) results in an alteration of the signaling range and binding capacity of the IHH protein in the interaction with Hh co-receptors, such as PTC-1 and antagonist HHIP. Structural analyses have revealed features of the mutations that cause BDA: 1) E95K mutation is involved in the morphogenetic gradient of the IHH protein in vivo, 2) E95K and D100E mutations result in instability of the N-terminal domain of IHH (IHH-N) with enhanced intracellular degradation at the lysosome, 3) E95K and E131EK mutations affect multimeric formation and cholesterol modification of IHH-N, 4) all three mutations affect the binding capacity of IHH-N to the receptor PTC-1 [93]. These observations imply that Hh mutations impair interaction with Hh receptors and strongly implicate changed Hh signaling capacity and range in the pathogenesis of brachydactyly.

4.2. Hedgehog Signaling and Cartilage Tumorigenesis

Cartilaginous tumors are the most frequently occurring benign neoplasms in the skeleton [95–97]. The common lesions, enchondroma and osteochondroma, that form adjacent to growth plates during skeletal development, have the potential for malignant change to chondrosarcoma. Cartilaginous tumors arise as a result of mutations in several genes [98–103]. Patients with enchondromatosis (Ollier disease and Maffucci syndrome, OMIM 166000) are endowed with inactivating mutations in the Parathyroid hormone 1 receptor (PTH1R), while mice with the PTH1R mutation at codon 150 develop multiple enchondroma-like lesions with upregulated Hh signaling [100,104]. Hereditary multiple exostoses syndrome (HME; OMIM 133700) is associated with heterozygous mutations in *EXT* genes (*EXT1* and *EXT2*), which encode glycosyltransferases that catalyze the polymerization of heparan sulphate (HS) chains [101,102]. Ext1 or Ext2 deleted cells do not synthesize sufficient amounts of HS-rich proteoglycan (HSPG), which is vital for the regulation of the binding and diffusion of Hh ligands on the cell surface [105,106]. Although Ext1/2 KO mice develop skeletal lesions similar to osteochondroma in HME with an abnormal extracellular distribution of Hh ligands [107], recent studies have demonstrated that the autosomal dominant disorder metachondromatosis (MC; OMIM 156250), a rare disease characterized by enchodroma and osteochondroma, is found to be involved in heterozygous loss-of-function mutations in tyrosine-protein phosphatase non-receptor type 11 (PTPN11), encoding protein tyrosine phosphatase SHP2 that relays signals from the activated Ras/extracellular signal-regulated kinase (ERK) pathway [98,103]. Analysis of Ptpn11-deficient mice has revealed the association between PTPN11 mutations and MC, just as the inactivated Ptpn11 pathway in KO mice leads to lesions very similar to MC, and mutant chondroprogenitors enhances Ihh expression. Interestingly, in all the above syndromes showing cartilaginous tumors, aberrant activation of the Hh pathway is observed in their cartilaginous lesions [108]. These findings strongly support the view that over-activated Hh signaling at the growth plate is sufficient to cause cartilaginous neoplasms and that some regulatory signaling including PTH/PTHrP, EXTs, and PTPN11 acts as a tumor suppressor in cartilaginous tissues through the inhibition of Hh signaling.

4.3. Hedgehog Signaling and Heterotopic Ossification

Progressive osseous heteroplasia (POH; OMIM 166350) is an autosomal dominant skeletal disorder characterized by widespread heterotopic ossification of skeletal muscle and deep soft connective tissue [109,110]. POH has been described as caused by loss-of-function mutation of GNAS encoding the stimulatory alpha subunit, Gαs, that transduces signals from G protein-coupled receptors (GPCRs) [111,112]. The main phenotypical indication of POH is advanced articular deformation and growth retardation, which are caused by ectopic ossification from embryonic mesenchymal progenitor cells. Analysis of Gnas-deficient mice has revealed the underlying molecular mechanism of POH pathogenesis. Mice carrying tissue-specific mutations of Gnas using Prx1-Cre transgenic driver line display POH-like skeletal anomalies with ectopic expression of osteogenic markers, and aberrant mineralization disclosed by Von Kossa staining [112]. Interestingly, in Gnas-deficient cells, the Hh pathway is activated as indicated by the higher expression of Hh target genes, Ptc-1, Gli-1, and Hhip, and Hh signaling is upregulated in patients with POH. Furthermore, another analysis has also demonstrated that Gnas acts through cAMP and PKA, downstream pathways of Gnas, to suppress Hh signaling and that reducing Hh signaling activity partially improves the phenotypes of POH [112]. These findings have provided strong evidence that Hh signaling is closely associated with Gnas in skeletal development. In soft tissues without ossification, such as muscle and skin, the activity of the Hh pathway may be rigorously regulated by the GPCR pathway through GNAS to prevent ectopic bone formation during early skeletal genesis.

Abundant genetic evidence, that the Hh pathway plays a central role during skeletal formation, has been accumulated over the past two decades, and ongoing studies for the integrated understanding of its dysregulation and development in human skeletal disorders continue to the present day. A great number of researchers and clinicians suggest that the Hh pathway represents a novel drug target with therapeutic potential in diseases, and some pharmacological materials that adjust Hh signaling activity are being utilized annually. Blocking the Hh pathway may help to improve the treatment of heterotopic ossification, or cartilaginous tumors. Conversely, activation of Hh signaling may be effective in the promotion of osteogenesis for tissue repair and recovery from skeletal deformities, or traumas. Thus, maintaining adequate Hh signaling activity can be thought of as a key element for sustaining healthy skeletal homeostasis.

5. Concluding Remarks

In this paper, we have reviewed the multiple roles of the Hh pathway in the regulation of growth plate formation and differentiation. During early chondrogenesis, the Hh pathway promotes cartilaginous growth in condensed limb mesenchymal cells. After organizing the growth plate, the Hh pathway, through Ihh signaling by hypertrophic chondrocytes, regulates chondrocyte differentiation by interacting with PTH-PTHrP signaling, which is termed the PTHrP-Ihh feedback loop system. Other regulatory pathways, such as Wnt/β-catenin, FGFs, BMPs, and VEGF, also interact with the Hh pathway in regulating the growth plate. Moreover, given that the Hh pathway and cellular cholesterol biosynthesis regulate each other during growth plate formation, Hh may be associated with bone diseases related to steroid hormones. Furthermore, recent fate-mapping studies have provided particular evidence showing that epiphyseal hypertrophic chondrocytes under the influence of Hh signaling include osteogenic progenitors that can differentiate into the skeletal lineage for longitudinally growing bone.

The action of the Hh signal in developing long bone is one of the most promising paradigms for understanding the key developmental mechanisms controlled by a growth plate. Future studies are needed to define the precise developmental role of signaling cascades, which is important for understanding skeletal formation (Fgf, Wnt, Bmp, etc.) within Hh-signal-responsive cell lineages originating from the growth plate. Elucidation focused on the regulatory mechanisms of growth plate by Hh pathway would have a positive impact on the full understanding of longitudinal bone development and skeletal disorders.

Funding: This work was supported in part by a Grant-in-Aid for Scientific Research from the Ministry of Education, Culture, Sports, Science and Technology, Japan (18K06832 to R.H.) and from The JSBMR Rising Star Grant for fiscal year 2018.

Acknowledgments: We express our appreciation to Yuki Takaoka, Chie Shiraishi, Yukimi Utsunomiya, Kyoko Shimazu, Miyako Nagao and Hitoshi Iwata for their valuable assistance. We also thank Gen Yamada (Wakayama Medical University) for his helpful discussion.

Conflicts of Interest: The authors declare no conflicts of interest. The funders had no role in study design, data collection and analysis, decision to publish, or preparation of the manuscript.

References

1. Abad, V.; Meyers, J.L.; Weise, M.; Gafni, R.I.; Barnes, K.M.; Nilsson, O.; Bacher, J.D.; Baron, J. The role of the resting zone in growth plate chondrogenesis. *Endocrinology* **2002**, *143*, 1851–1857. [CrossRef]
2. Prein, C.; Warmbold, N.; Farkas, Z.; Schieker, M.; Aszodi, A.; Clausen-Schaumann, H. Structural and mechanical properties of the proliferative zone of the developing murine growth plate cartilage assessed by atomic force microscopy. *Matrix Biol.* **2016**, *50*, 1–15. [CrossRef] [PubMed]
3. Breur, G.J.; VanEnkevort, B.A.; Farnum, C.E.; Wilsman, N.J. Linear relationship between the volume of hypertrophic chondrocytes and the rate of longitudinal bone growth in growth plates. *J. Orthop. Res.* **1991**, *9*, 348–359. [CrossRef] [PubMed]
4. Farnum, C.E.; Lee, R.; O'Hara, K.; Urban, J.P. Volume increase in growth plate chondrocytes during hypertrophy: The contribution of organic osmolytes. *Bone* **2002**, *30*, 574–581. [CrossRef]
5. Gerber, H.P.; Vu, T.H.; Ryan, A.M.; Kowalski, J.; Werb, Z.; Ferrara, N. VEGF couples hypertrophic cartilage remodeling, ossification and angiogenesis during endochondral bone formation. *Nat. Med.* **1999**, *5*, 623–628. [CrossRef]
6. Kojima, T.; Hasegawa, T.; de Freitas, P.H.; Yamamoto, T.; Sasaki, M.; Horiuchi, K.; Hongo, H.; Yamada, T.; Sakagami, N.; Saito, N.; et al. Histochemical aspects of the vascular invasion at the erosion zone of the epiphyseal cartilage in MMP-9-deficient mice. *Biomed. Res.* **2013**, *34*, 119–128. [CrossRef]
7. Kronenberg, H.M. Developmental regulation of the growth plate. *Nature* **2003**, *423*, 332–336. [CrossRef]
8. Aghajanian, P.; Mohan, S. The art of building bone: Emerging role of chondrocyte-to-osteoblast transdifferentiation in endochondral ossification. *Bone Res.* **2018**, *6*, 19. [CrossRef]
9. Alman, B.A. The role of hedgehog signalling in skeletal health and disease. *Nat. Rev. Rheumatol.* **2015**, *11*, 552–560. [CrossRef]
10. Carballo, G.B.; Honorato, J.R.; de Lopes, G.P.F.; Spohr, T. A highlight on Sonic hedgehog pathway. *Cell Commun. Signal.* **2018**, *16*, 11. [CrossRef]
11. Ingham, P.W.; McMahon, A.P. Hedgehog signaling in animal development: Paradigms and principles. *Genes Dev.* **2001**, *15*, 3059–3087. [CrossRef] [PubMed]
12. Lee, R.T.; Zhao, Z.; Ingham, P.W. Hedgehog signalling. *Development* **2016**, *143*, 367–372. [CrossRef] [PubMed]
13. McMahon, A.P. More surprises in the Hedgehog signaling pathway. *Cell* **2000**, *100*, 185–188. [CrossRef]
14. Ryan, K.E.; Chiang, C. Hedgehog secretion and signal transduction in vertebrates. *J. Biol. Chem.* **2012**, *287*, 17905–17913. [CrossRef]
15. Yang, J.; Andre, P.; Ye, L.; Yang, Y.Z. The Hedgehog signalling pathway in bone formation. *Int. J. Oral Sci.* **2015**, *7*, 73–79. [CrossRef]
16. Bitgood, M.J.; McMahon, A.P. Hedgehog and Bmp genes are coexpressed at many diverse sites of cell-cell interaction in the mouse embryo. *Dev. Biol.* **1995**, *172*, 126–138. [CrossRef]
17. Petrova, R.; Joyner, A.L. Roles for Hedgehog signaling in adult organ homeostasis and repair. *Development* **2014**, *141*, 3445–3457. [CrossRef]

18. Varjosalo, M.; Taipale, J. Hedgehog: Functions and mechanisms. *Genes Dev.* **2008**, *22*, 2454–2472. [CrossRef]
19. St-Jacques, B.; Hammerschmidt, M.; McMahon, A.P. Indian hedgehog signaling regulates proliferation and differentiation of chondrocytes and is essential for bone formation. *Genes Dev.* **1999**, *13*, 2072–2086. [CrossRef]
20. Ohba, S. Hedgehog Signaling in Endochondral Ossification. *J. Dev. Biol.* **2016**, *4*, 20. [CrossRef]
21. Goldring, M.B.; Tsuchimochi, K.; Ijiri, K. The control of chondrogenesis. *J. Cell. Biochem.* **2006**, *97*, 33–44. [CrossRef] [PubMed]
22. Long, F.; Ornitz, D.M. Development of the endochondral skeleton. *Cold Spring Harb. Perspect. Biol.* **2013**, *5*, a008334. [CrossRef] [PubMed]
23. Quintana, L.; zur Nieden, N.I.; Semino, C.E. Morphogenetic and regulatory mechanisms during developmental chondrogenesis: New paradigms for cartilage tissue engineering. *Tissue Eng. Part B Rev.* **2009**, *15*, 29–41. [CrossRef] [PubMed]
24. Bruce, S.J.; Butterfield, N.C.; Metzis, V.; Town, L.; McGlinn, E.; Wicking, C. Inactivation of Patched1 in the mouse limb has novel inhibitory effects on the chondrogenic program. *J. Biol. Chem.* **2010**, *285*, 27967–27981. [CrossRef]
25. Chuang, P.T.; McMahon, A.P. Vertebrate Hedgehog signalling modulated by induction of a Hedgehog-binding protein. *Nature* **1999**, *397*, 617–621. [CrossRef]
26. Pan, A.; Chang, L.; Nguyen, A.; James, A.W. A review of hedgehog signaling in cranial bone development. *Front. Physiol.* **2013**, *4*, 61. [CrossRef]
27. Zhulyn, O.; Hui, C.C. Sufu and Kif7 in limb patterning and development. *Dev. Dyn.* **2015**, *244*, 468–478. [CrossRef]
28. Kozhemyakina, E.; Lassar, A.B.; Zelzer, E. A pathway to bone: Signaling molecules and transcription factors involved in chondrocyte development and maturation. *Development* **2015**, *142*, 817–831. [CrossRef]
29. Amano, K.; Densmore, M.; Fan, Y.; Lanske, B. Ihh and PTH1R signaling in limb mesenchyme is required for proper segmentation and subsequent formation and growth of digit bones. *Bone* **2016**, *83*, 256–266. [CrossRef]
30. Amano, K.; Densmore, M.J.; Lanske, B. Conditional Deletion of Indian Hedgehog in Limb Mesenchyme Results in Complete Loss of Growth Plate Formation but Allows Mature Osteoblast Differentiation. *J. Bone Miner. Res.* **2015**, *30*, 2262–2272. [CrossRef]
31. Maeda, Y.; Nakamura, E.; Nguyen, M.T.; Suva, L.J.; Swain, F.L.; Razzaque, M.S.; Mackem, S.; Lanske, B. Indian Hedgehog produced by postnatal chondrocytes is essential for maintaining a growth plate and trabecular bone. *Proc. Natl. Acad. Sci. USA* **2007**, *104*, 6382–6387. [CrossRef] [PubMed]
32. Razzaque, M.S.; Soegiarto, D.W.; Chang, D.; Long, F.; Lanske, B. Conditional deletion of Indian hedgehog from collagen type 2alpha1-expressing cells results in abnormal endochondral bone formation. *J. Pathol.* **2005**, *207*, 453–461. [CrossRef] [PubMed]
33. Long, F.; Zhang, X.M.; Karp, S.; Yang, Y.; McMahon, A.P. Genetic manipulation of hedgehog signaling in the endochondral skeleton reveals a direct role in the regulation of chondrocyte proliferation. *Development* **2001**, *128*, 5099–5108. [PubMed]
34. Vortkamp, A.; Lee, K.; Lanske, B.; Segre, G.V.; Kronenberg, H.M.; Tabin, C.J. Regulation of rate of cartilage differentiation by Indian hedgehog and PTH-related protein. *Science* **1996**, *273*, 613–622. [CrossRef]
35. Karaplis, A.C.; Luz, A.; Glowacki, J.; Bronson, R.T.; Tybulewicz, V.L.; Kronenberg, H.M.; Mulligan, R.C. Lethal skeletal dysplasia from targeted disruption of the parathyroid hormone-related peptide gene. *Genes Dev.* **1994**, *8*, 277–289. [CrossRef]
36. Lee, K.; Lanske, B.; Karaplis, A.C.; Deeds, J.D.; Kohno, H.; Nissenson, R.A.; Kronenberg, H.M.; Segre, G.V. Parathyroid hormone-related peptide delays terminal differentiation of chondrocytes during endochondral bone development. *Endocrinology* **1996**, *137*, 5109–5118. [CrossRef]
37. Lanske, B.; Karaplis, A.C.; Lee, K.; Luz, A.; Vortkamp, A.; Pirro, A.; Karperien, M.; Defize, L.H.; Ho, C.; Mulligan, R.C.; et al. PTH/PTHrP receptor in early development and Indian hedgehog-regulated bone growth. *Science* **1996**, *273*, 663–666. [CrossRef]
38. Weir, E.C.; Philbrick, W.M.; Amling, M.; Neff, L.A.; Baron, R.; Broadus, A.E. Targeted overexpression of parathyroid hormone-related peptide in chondrocytes causes chondrodysplasia and delayed endochondral bone formation. *Proc. Natl. Acad. Sci. USA* **1996**, *93*, 10240–10245. [CrossRef]

39. Hilton, M.J.; Tu, X.; Long, F. Tamoxifen-inducible gene deletion reveals a distinct cell type associated with trabecular bone, and direct regulation of PTHrP expression and chondrocyte morphology by Ihh in growth region cartilage. *Dev. Biol.* **2007**, *308*, 93–105. [CrossRef]

40. Kronenberg, H.M. PTHrP and skeletal development. *Ann. N. Y. Acad. Sci.* **2006**, *1068*, 1–13. [CrossRef]

41. Karp, S.J.; Schipani, E.; St-Jacques, B.; Hunzelman, J.; Kronenberg, H.; McMahon, A.P. Indian hedgehog coordinates endochondral bone growth and morphogenesis via parathyroid hormone related-protein-dependent and -independent pathways. *Development* **2000**, *127*, 543–548. [PubMed]

42. Maeda, Y.; Schipani, E.; Densmore, M.J.; Lanske, B. Partial rescue of postnatal growth plate abnormalities in Ihh mutants by expression of a constitutively active PTH/PTHrP receptor. *Bone* **2010**, *46*, 472–478. [CrossRef] [PubMed]

43. Lai, L.P.; Mitchell, J. Indian hedgehog: Its roles and regulation in endochondral bone development. *J. Cell. Biochem.* **2005**, *96*, 1163–1173. [CrossRef] [PubMed]

44. Marino, R. Growth plate biology: New insights. *Curr. Opin. Endocrinol. Diabetes Obes.* **2011**, *18*, 9–13. [CrossRef]

45. Mak, K.K.; Chen, M.H.; Day, T.F.; Chuang, P.T.; Yang, Y. Wnt/beta-catenin signaling interacts differentially with Ihh signaling in controlling endochondral bone and synovial joint formation. *Development* **2006**, *133*, 3695–3707. [CrossRef]

46. Xie, Y.; Zhou, S.; Chen, H.; Du, X.; Chen, L. Recent research on the growth plate: Advances in fibroblast growth factor signaling in growth plate development and disorders. *J. Mol. Endocrinol.* **2014**, *53*, T11–T34. [CrossRef]

47. Foldynova-Trantirkova, S.; Wilcox, W.R.; Krejci, P. Sixteen years and counting: The current understanding of fibroblast growth factor receptor 3 (FGFR3) signaling in skeletal dysplasias. *Hum. Mutat.* **2012**, *33*, 29–41. [CrossRef]

48. Peters, K.; Ornitz, D.; Werner, S.; Williams, L. Unique expression pattern of the FGF receptor 3 gene during mouse organogenesis. *Dev. Biol.* **1993**, *155*, 423–430. [CrossRef]

49. Colvin, J.S.; Bohne, B.A.; Harding, G.W.; McEwen, D.G.; Ornitz, D.M. Skeletal overgrowth and deafness in mice lacking fibroblast growth factor receptor 3. *Nat. Genet.* **1996**, *12*, 390–397. [CrossRef]

50. Deng, C.; Wynshaw-Boris, A.; Zhou, F.; Kuo, A.; Leder, P. Fibroblast growth factor receptor 3 is a negative regulator of bone growth. *Cell* **1996**, *84*, 911–921. [CrossRef]

51. Chen, L.; Adar, R.; Yang, X.; Monsonego, E.O.; Li, C.; Hauschka, P.V.; Yayon, A.; Deng, C.X. Gly369Cys mutation in mouse FGFR3 causes achondroplasia by affecting both chondrogenesis and osteogenesis. *J. Clin. Investig.* **1999**, *104*, 1517–1525. [CrossRef] [PubMed]

52. Iwata, T.; Chen, L.; Li, C.; Ovchinnikov, D.A.; Behringer, R.R.; Francomano, C.A.; Deng, C.X. A neonatal lethal mutation in FGFR3 uncouples proliferation and differentiation of growth plate chondrocytes in embryos. *Hum. Mol. Genet.* **2000**, *9*, 1603–1613. [CrossRef] [PubMed]

53. Naski, M.C.; Colvin, J.S.; Coffin, J.D.; Ornitz, D.M. Repression of hedgehog signaling and BMP4 expression in growth plate cartilage by fibroblast growth factor receptor 3. *Development* **1998**, *125*, 4977–4988. [PubMed]

54. Segev, O.; Chumakov, I.; Nevo, Z.; Givol, D.; Madar-Shapiro, L.; Sheinin, Y.; Weinreb, M.; Yayon, A. Restrained chondrocyte proliferation and maturation with abnormal growth plate vascularization and ossification in human FGFR-3(G380R) transgenic mice. *Hum. Mol. Genet.* **2000**, *9*, 249–258. [CrossRef] [PubMed]

55. Wang, Y.; Spatz, M.K.; Kannan, K.; Hayk, H.; Avivi, A.; Gorivodsky, M.; Pines, M.; Yayon, A.; Lonai, P.; Givol, D. A mouse model for achondroplasia produced by targeting fibroblast growth factor receptor 3. *Proc. Natl. Acad. Sci. USA* **1999**, *96*, 4455–4460. [CrossRef]

56. Minina, E.; Kreschel, C.; Naski, M.C.; Ornitz, D.M.; Vortkamp, A. Interaction of FGF, Ihh/Pthlh, and BMP signaling integrates chondrocyte proliferation and hypertrophic differentiation. *Dev. Cell* **2002**, *3*, 439–449. [CrossRef]

57. Zhou, S.; Xie, Y.; Tang, J.; Huang, J.; Huang, Q.; Xu, W.; Wang, Z.; Luo, F.; Wang, Q.; Chen, H.; et al. FGFR3 Deficiency Causes Multiple Chondroma-like Lesions by Upregulating Hedgehog Signaling. *PLoS Genet.* **2015**, *11*, e1005214. [CrossRef]

58. Garrison, P.; Yue, S.; Hanson, J.; Baron, J.; Lui, J.C. Spatial regulation of bone morphogenetic proteins (BMPs) in postnatal articular and growth plate cartilage. *PLoS ONE* **2017**, *12*, e0176752. [CrossRef]

59. Jing, Y.; Jing, J.; Ye, L.; Liu, X.; Harris, S.E.; Hinton, R.J.; Feng, J.Q. Chondrogenesis and osteogenesis are one continuous developmental and lineage defined biological process. *Sci. Rep.* **2017**, *7*, 10020. [CrossRef]

60. Salazar, V.S.; Gamer, L.W.; Rosen, V. BMP signalling in skeletal development, disease and repair. *Nat. Rev. Endocrinol.* **2016**, *12*, 203–221. [CrossRef]

61. Minina, E.; Wenzel, H.M.; Kreschel, C.; Karp, S.; Gaffield, W.; McMahon, A.P.; Vortkamp, A. BMP and Ihh/PTHrP signaling interact to coordinate chondrocyte proliferation and differentiation. *Development* **2001**, *128*, 4523–4534. [PubMed]

62. Shu, B.; Zhang, M.; Xie, R.; Wang, M.; Jin, H.; Hou, W.; Tang, D.; Harris, S.E.; Mishina, Y.; O'Keefe, R.J.; et al. BMP2, but not BMP4, is crucial for chondrocyte proliferation and maturation during endochondral bone development. *J. Cell. Sci.* **2011**, *124 Pt 20*, 3428–3440. [CrossRef]

63. Jing, J.; Ren, Y.; Zong, Z.; Liu, C.; Kamiya, N.; Mishina, Y.; Liu, Y.; Zhou, X.; Feng, J.Q. BMP receptor 1A determines the cell fate of the postnatal growth plate. *Int. J. Biol. Sci.* **2013**, *9*, 895–906. [CrossRef] [PubMed]

64. Retting, K.N.; Song, B.; Yoon, B.S.; Lyons, K.M. BMP canonical Smad signaling through Smad1 and Smad5 is required for endochondral bone formation. *Development* **2009**, *136*, 1093–1104. [CrossRef] [PubMed]

65. Yoon, B.S.; Pogue, R.; Ovchinnikov, D.A.; Yoshii, I.; Mishina, Y.; Behringer, R.R.; Lyons, K.M. BMPs regulate multiple aspects of growth-plate chondrogenesis through opposing actions on FGF pathways. *Development* **2006**, *133*, 4667–4678. [CrossRef] [PubMed]

66. Ortega, N.; Wang, K.; Ferrara, N.; Werb, Z.; Vu, T.H. Complementary interplay between matrix metalloproteinase-9, vascular endothelial growth factor and osteoclast function drives endochondral bone formation. *Dis. Model. Mech.* **2010**, *3*, 224–235. [CrossRef]

67. Haimov, H.; Shimoni, E.; Brumfeld, V.; Shemesh, M.; Varsano, N.; Addadi, L.; Weiner, S. Mineralization pathways in the active murine epiphyseal growth plate. *Bone* **2019**, *130*, 115086. [CrossRef]

68. Stickens, D.; Behonick, D.J.; Ortega, N.; Heyer, B.; Hartenstein, B.; Yu, Y.; Fosang, A.J.; Schorpp-Kistner, M.; Angel, P.; Werb, Z. Altered endochondral bone development in matrix metalloproteinase 13-deficient mice. *Development* **2004**, *131*, 5883–5895. [CrossRef]

69. Colnot, C.; de la Fuente, L.; Huang, S.; Hu, D.; Lu, C.; St-Jacques, B.; Helms, J.A. Indian hedgehog synchronizes skeletal angiogenesis and perichondrial maturation with cartilage development. *Development* **2005**, *132*, 1057–1067. [CrossRef]

70. Yoshida, C.A.; Yamamoto, H.; Fujita, T.; Furuichi, T.; Ito, K.; Inoue, K.; Yamana, K.; Zanma, A.; Takada, K.; Ito, Y.; et al. Runx2 and Runx3 are essential for chondrocyte maturation, and Runx2 regulates limb growth through induction of Indian hedgehog. *Genes Dev.* **2004**, *18*, 952–963. [CrossRef]

71. Zelzer, E.; Glotzer, D.J.; Hartmann, C.; Thomas, D.; Fukai, N.; Soker, S.; Olsen, B.R. Tissue specific regulation of VEGF expression during bone development requires Cbfa1/Runx2. *Mech. Dev.* **2001**, *106*, 97–106. [CrossRef]

72. Rossi, M.; Hall, C.M.; Bouvier, R.; Collardeau-Frachon, S.; Le Breton, F.; Bucourt, M.; Cordier, M.P.; Vianey-Saban, C.; Parenti, G.; Andria, G.; et al. Radiographic features of the skeleton in disorders of post-squalene cholesterol biosynthesis. *Pediatric Radiol.* **2015**, *45*, 965–976. [CrossRef] [PubMed]

73. Gofflot, F.; Hars, C.; Illien, F.; Chevy, F.; Wolf, C.; Picard, J.J.; Roux, C. Molecular mechanisms underlying limb anomalies associated with cholesterol deficiency during gestation: Implications of Hedgehog signaling. *Hum. Mol. Genet.* **2003**, *12*, 1187–1198. [CrossRef] [PubMed]

74. Wu, S.; De Luca, F. Role of cholesterol in the regulation of growth plate chondrogenesis and longitudinal bone growth. *J. Biol. Chem.* **2004**, *279*, 4642–4647. [CrossRef] [PubMed]

75. Brown, M.S.; Goldstein, J.L. A proteolytic pathway that controls the cholesterol content of membranes, cells, and blood. *Proc. Natl. Acad. Sci. USA* **1999**, *96*, 11041–11048. [CrossRef]

76. Tsushima, H.; Tang, Y.J.; Puviindran, V.; Hsu, S.C.; Nadesan, P.; Yu, C.; Zhang, H.; Mirando, A.J.; Hilton, M.J.; Alman, B.A. Intracellular biosynthesis of lipids and cholesterol by Scap and Insig in mesenchymal cells regulates long bone growth and chondrocyte homeostasis. *Development* **2018**, *145*, 162396. [CrossRef]

77. Ali, S.A.; Al-Jazrawe, M.; Ma, H.; Whetstone, H.; Poon, R.; Farr, S.; Naples, M.; Adeli, K.; Alman, B.A. Regulation of Cholesterol Homeostasis by Hedgehog Signaling in Osteoarthritic Cartilage. *Arthritis Rheumatol.* **2016**, *68*, 127–137. [CrossRef]

78. Buglino, J.A.; Resh, M.D. Palmitoylation of Hedgehog proteins. *Vitam. Horm.* **2012**, *88*, 229–252.

79. Huang, P.; Nedelcu, D.; Watanabe, M.; Jao, C.; Kim, Y.; Liu, J.; Salic, A. Cellular Cholesterol Directly Activates Smoothened in Hedgehog Signaling. *Cell* **2016**, *166*, 1176–1187. [CrossRef]

80. Ono, N.; Ono, W.; Nagasawa, T.; Kronenberg, H.M. A subset of chondrogenic cells provides early mesenchymal progenitors in growing bones. *Nat. Cell Biol.* **2014**, *16*, 1157–1167. [CrossRef]

81. Yang, G.; Zhu, L.; Hou, N.; Lan, Y.; Wu, X.M.; Zhou, B.; Teng, Y.; Yang, X. Osteogenic fate of hypertrophic chondrocytes. *Cell Res.* **2014**, *24*, 1266–1269. [CrossRef] [PubMed]

82. Yang, L.; Tsang, K.Y.; Tang, H.C.; Chan, D.; Cheah, K.S. Hypertrophic chondrocytes can become osteoblasts and osteocytes in endochondral bone formation. *Proc. Natl. Acad. Sci. USA* **2014**, *111*, 12097–12102. [CrossRef] [PubMed]

83. Zhou, X.; von der Mark, K.; Henry, S.; Norton, W.; Adams, H.; de Crombrugghe, B. Chondrocytes transdifferentiate into osteoblasts in endochondral bone during development, postnatal growth and fracture healing in mice. *PLoS Genet.* **2014**, *10*, e1004820. [CrossRef] [PubMed]

84. Haraguchi, R.; Kitazawa, R.; Imai, Y.; Kitazawa, S. Growth plate-derived hedgehog-signal-responsive cells provide skeletal tissue components in growing bone. *Histochem. Cell Biol.* **2018**, *149*, 365–373. [CrossRef] [PubMed]

85. Shi, Y.; He, G.; Lee, W.C.; McKenzie, J.A.; Silva, M.J.; Long, F. Gli1 identifies osteogenic progenitors for bone formation and fracture repair. *Nat. Commun.* **2017**, *8*, 2043. [CrossRef]

86. Karimian, E.; Chagin, A.S.; Savendahl, L. Genetic regulation of the growth plate. *Front. Endocrinol.* **2011**, *2*, 113. [CrossRef]

87. Makela, E.A.; Vainionpaa, S.; Vihtonen, K.; Mero, M.; Rokkanen, P. The effect of trauma to the lower femoral epiphyseal plate. An experimental study in rabbits. *J. Bone Jt. Surg. Br.* **1988**, *70*, 187–191. [CrossRef]

88. Gao, B.; Guo, J.; She, C.; Shu, A.; Yang, M.; Tan, Z.; Yang, X.; Guo, S.; Feng, G.; He, L. Mutations in IHH, encoding Indian hedgehog, cause brachydactyly type A-1. *Nat. Genet.* **2001**, *28*, 386–388. [CrossRef]

89. Hellemans, J.; Coucke, P.J.; Giedion, A.; De Paepe, A.; Kramer, P.; Beemer, F.; Mortier, G.R. Homozygous mutations in IHH cause acrocapitofemoral dysplasia, an autosomal recessive disorder with cone-shaped epiphyses in hands and hips. *Am. J. Hum. Genet.* **2003**, *72*, 1040–1046. [CrossRef]

90. Mortier, G.R.; Kramer, P.P.; Giedion, A.; Beemer, F.A. Acrocapitofemoral dysplasia: An autosomal recessive skeletal dysplasia with cone shaped epiphyses in the hands and hips. *J. Med. Genet.* **2003**, *40*, 201–207. [CrossRef]

91. Gao, B.; Hu, J.; Stricker, S.; Cheung, M.; Ma, G.; Law, K.F.; Witte, F.; Briscoe, J.; Mundlos, S.; He, L.; et al. A mutation in Ihh that causes digit abnormalities alters its signalling capacity and range. *Nature* **2009**, *458*, 1196–1200. [CrossRef] [PubMed]

92. Ho, R.; McIntyre, A.D.; Kennedy, B.A.; Hegele, R.A. Whole-exome sequencing identifies a novel IHH insertion in an Ontario family with brachydactyly type A1. *SAGE Open Med. Case Rep.* **2018**, *6*, 2050313X18818711. [CrossRef] [PubMed]

93. Ma, G.; Yu, J.; Xiao, Y.; Chan, D.; Gao, B.; Hu, J.; He, Y.; Guo, S.; Zhou, J.; Zhang, L.; et al. Indian hedgehog mutations causing brachydactyly type A1 impair Hedgehog signal transduction at multiple levels. *Cell Res.* **2011**, *21*, 1343–1357. [CrossRef] [PubMed]

94. Shen, L.; Ma, G.; Shi, Y.; Ruan, Y.; Yang, X.; Wu, X.; Xiong, Y.; Wan, C.; Yang, C.; Cai, L.; et al. p. E95K mutation in Indian hedgehog causing brachydactyly type A1 impairs IHH/Gli1 downstream transcriptional regulation. *BMC Genet.* **2019**, *20*, 10. [CrossRef] [PubMed]

95. Bovee, J.V.; Hogendoorn, P.C.; Wunder, J.S.; Alman, B.A. Cartilage tumours and bone development: Molecular pathology and possible therapeutic targets. *Nat. Rev. Cancer* **2010**, *10*, 481–488. [CrossRef] [PubMed]

96. Garcia, R.A.; Inwards, C.Y.; Unni, K.K. Benign bone tumors—Recent developments. *Semin. Diagn. Pathol.* **2011**, *28*, 73–85. [CrossRef]

97. Romeo, S.; Hogendoorn, P.C.; Dei Tos, A.P. Benign cartilaginous tumors of bone: From morphology to somatic and germ-line genetics. *Adv. Anat. Pathol.* **2009**, *16*, 307–315. [CrossRef]

98. Bowen, M.E.; Boyden, E.D.; Holm, I.A.; Campos-Xavier, B.; Bonafe, L.; Superti-Furga, A.; Ikegawa, S.; Cormier-Daire, V.; Bovee, J.V.; Pansuriya, T.C.; et al. Loss-of-function mutations in PTPN11 cause metachondromatosis, but not Ollier disease or Maffucci syndrome. *PLoS Genet.* **2011**, *7*, e1002050. [CrossRef]

99. Deng, Q.; Li, P.; Che, M.; Liu, J.; Biswas, S.; Ma, G.; He, L.; Wei, Z.; Zhang, Z.; Yang, Y.; et al. Activation of hedgehog signaling in mesenchymal stem cells induces cartilage and bone tumor formation via Wnt/beta-Catenin. *Elife* **2019**, *8*, e50208. [CrossRef]

100. Hopyan, S.; Gokgoz, N.; Poon, R.; Gensure, R.C.; Yu, C.; Cole, W.G.; Bell, R.S.; Juppner, H.; Andrulis, I.L.; Wunder, J.S.; et al. A mutant PTH/PTHrP type I receptor in enchondromatosis. *Nat. Genet.* **2002**, *30*, 306–310. [CrossRef]

101. Huegel, J.; Sgariglia, F.; Enomoto-Iwamoto, M.; Koyama, E.; Dormans, J.P.; Pacifici, M. Heparan sulfate in skeletal development, growth, and pathology: The case of hereditary multiple exostoses. *Dev. Dyn.* **2013**, *242*, 1021–1032. [CrossRef] [PubMed]

102. Jennes, I.; Pedrini, E.; Zuntini, M.; Mordenti, M.; Balkassmi, S.; Asteggiano, C.G.; Casey, B.; Bakker, B.; Sangiorgi, L.; Wuyts, W. Multiple osteochondromas: Mutation update and description of the multiple osteochondromas mutation database (MOdb). *Hum. Mutat.* **2009**, *30*, 1620–1627. [CrossRef] [PubMed]

103. Sobreira, N.L.; Cirulli, E.T.; Avramopoulos, D.; Wohler, E.; Oswald, G.L.; Stevens, E.L.; Ge, D.; Shianna, K.V.; Smith, J.P.; Maia, J.M.; et al. Whole-genome sequencing of a single proband together with linkage analysis identifies a Mendelian disease gene. *PLoS Genet.* **2010**, *6*, e1000991. [CrossRef] [PubMed]

104. Tiet, T.D.; Hopyan, S.; Nadesan, P.; Gokgoz, N.; Poon, R.; Lin, A.C.; Yan, T.; Andrulis, I.L.; Alman, B.A.; Wunder, J.S. Constitutive hedgehog signaling in chondrosarcoma up-regulates tumor cell proliferation. *Am. J. Pathol.* **2006**, *168*, 321–330. [CrossRef] [PubMed]

105. Koziel, L.; Kunath, M.; Kelly, O.G.; Vortkamp, A. Ext1-dependent heparan sulfate regulates the range of Ihh signaling during endochondral ossification. *Dev. Cell* **2004**, *6*, 801–813. [CrossRef]

106. Stickens, D.; Zak, B.M.; Rougier, N.; Esko, J.D.; Werb, Z. Mice deficient in Ext2 lack heparan sulfate and develop exostoses. *Development* **2005**, *132*, 5055–5068. [CrossRef]

107. Zak, B.M.; Schuksz, M.; Koyama, E.; Mundy, C.; Wells, D.E.; Yamaguchi, Y.; Pacifici, M.; Esko, J.D. Compound heterozygous loss of Ext1 and Ext2 is sufficient for formation of multiple exostoses in mouse ribs and long bones. *Bone* **2011**, *48*, 979–987. [CrossRef]

108. Yang, W.; Wang, J.; Moore, D.C.; Liang, H.; Dooner, M.; Wu, Q.; Terek, R.; Chen, Q.; Ehrlich, M.G.; Quesenberry, P.J.; et al. Ptpn11 deletion in a novel progenitor causes metachondromatosis by inducing hedgehog signalling. *Nature* **2013**, *499*, 491–495. [CrossRef]

109. Kaplan, F.S.; Hahn, G.V.; Zasloff, M.A. Heterotopic Ossification: Two Rare Forms and What They Can Teach Us. *J. Am. Acad. Orthop. Surg.* **1994**, *2*, 288–296. [CrossRef]

110. Kaplan, F.S.; Shore, E.M. Progressive osseous heteroplasia. *J. Bone Miner. Res.* **2000**, *15*, 2084–2094. [CrossRef]

111. Regard, J.B.; Cherman, N.; Palmer, D.; Kuznetsov, S.A.; Celi, F.S.; Guettier, J.M.; Chen, M.; Bhattacharyya, N.; Wess, J.; Coughlin, S.R.; et al. Wnt/beta-catenin signaling is differentially regulated by Galpha proteins and contributes to fibrous dysplasia. *Proc. Natl. Acad. Sci. USA* **2011**, *108*, 20101–20106. [CrossRef] [PubMed]

112. Regard, J.B.; Malhotra, D.; Gvozdenovic-Jeremic, J.; Josey, M.; Chen, M.; Weinstein, L.S.; Lu, J.; Shore, E.M.; Kaplan, F.S.; Yang, Y. Activation of Hedgehog signaling by loss of GNAS causes heterotopic ossification. *Nat. Med.* **2013**, *19*, 1505–1512. [CrossRef] [PubMed]

International Journal of
Molecular Sciences

Review

Sonic Hedgehog Signaling and Tooth Development

Akihiro Hosoya [1,*]**, Nazmus Shalehin** [1]**, Hiroaki Takebe** [1]**, Tsuyoshi Shimo** [2] **and Kazuharu Irie** [1]

[1]   Division of Histology, Department of Oral Growth and Development, School of Dentistry, Health Sciences
University of Hokkaido, Ishikari-Tobetsu, Hokkaido 061-0293, Japan; shalehin@hoku-iryo-u.ac.jp (N.S.);
takebeh@hoku-iryo-u.ac.jp (H.T.); irie@hoku-iryo-u.ac.jp (K.I.)

[2]   Division of Reconstructive Surgery for Oral and Maxillofacial Region, Department of Human Biology and
Pathophysiology, School of Dentistry, Health Sciences University of Hokkaido, Ishikari-Tobetsu,
Hokkaido 061-0293, Japan; shimotsu@hoku-iryo-u.ac.jp

*   Correspondence: hosoya@hoku-iryo-u.ac.jp; Tel.: +81-133-23-1938; Fax: +81-133-23-1236

Received: 14 January 2020; Accepted: 19 February 2020; Published: 26 February 2020

Abstract: Sonic hedgehog (Shh) is a secreted protein with important roles in mammalian embryogenesis. During tooth development, Shh is primarily expressed in the dental epithelium, from initiation to the root formation stages. A number of studies have analyzed the function of Shh signaling at different stages of tooth development and have revealed that Shh signaling regulates the formation of various tooth components, including enamel, dentin, cementum, and other soft tissues. In addition, dental mesenchymal cells positive for Gli1, a downstream transcription factor of Shh signaling, have been found to have stem cell properties, including multipotency and the ability to self-renew. Indeed, Gli1-positive cells in mature teeth appear to contribute to the regeneration of dental pulp and periodontal tissues. In this review, we provide an overview of recent advances related to the role of Shh signaling in tooth development, as well as the contribution of this pathway to tooth homeostasis and regeneration.

Keywords: sonic hedgehog; tooth development; epithelial and mesenchymal interaction; Gli1; mesenchymal stem cell; lineage tracing analysis; stem cell marker

1. Introduction

Hedgehog (Hh) signaling has been reported to have important roles in the development of many organs including craniofacial tissues such as palate, lip, salivary gland [1–5], as well as tooth [6]. This signaling requires primary cilia that function in intraflagellar transport (IFT) [7]. Disruption of IFT trafficking from the base to the tip of the cilium in *kif3a*-deficient mice results in phenotypes similar to the loss of Hh signaling, such as tooth dysplasia [8]. Under quiescent conditions, when Hh signaling is not activated, Patched (Ptch), a receptor of three hedgehog orthologs, including Sonic hedgehog (Shh), Indian hedgehog, and Desert hedgehog, represses Smoothened (Smo). Canonical Hh signaling is mediated via Smo activation. When the hedgehog ligand binds Ptch, it relieves this suppression and Smo accumulates in the tip of the primary cilium. Accordingly, Gli becomes dissociated from Suppressor of Fused (Sufu), a negative regulator of the Shh signaling. It then leads to the activation of Gli transcription factors and the downstream hedgehog signaling pathway [9]. Gli transcription factors have DNA-binding zinc finger domains that bind to sequences on their target genes to initiate or inhibit their transcription [10]. In contrast, non-canonical Hh signaling occurs through Patched1, independently of Smo and Gli [11].

Tooth germ is composed of both epithelial and mesenchymal tissues, with dental epithelial tissue originating from the oral epithelium. However, unlike the nearby oral epithelium, the dental epithelium expresses *Shh* [12–16]. During the period of tooth crown formation, Shh-expressing cells are strictly localized in the dental epithelium, including the enamel knot that corresponds to future cusps, as well as ameloblast-lineage cells [17–19]. On the other hand, Ptch-positive cells and its downstream proteins are located in the dental mesenchyme in the absence of *Shh* expression [20]. Therefore, it is believed that an epithelial-mesenchymal interaction exists in which *Shh* expressed in the epithelium acts on Ptch-positive mesenchymal cells during tooth development. Conversely, several reports have demonstrated that cells in the dental mesenchyme regulate Shh expression in the dental epithelium [21–30]. It has been shown that expression levels of *Shh* in the dental epithelium are decreased in *runt-related transcription factor 2* (*Runx2*) mutant mice [31]. Runx2 is an essential transcription factor for osteoblast differentiation and is expressed in both osteogenic- and odontogenic-lineage cells, indicating that dental mesenchymal cells may regulate Shh expression in the epithelium. Furthermore, it has been reported that Shh signaling is strictly regulated in certain types of cells and is required for cellular proliferation and differentiation during different stages of tooth development (Table 1, Figure 1).

Table 1. Roles of Shh signaling during tooth development.

Stage	Expressing Cells		Function	References
	Shh	*Ptch, Smo, Gli*		
Initiation	Epithelium	Dental mesenchyme	Epithelial invagination	[32–34]
Crown formation	Enamel organ	Enamel organ	Ameloblast differentiation	[35–38]
Calcification	Enamel organ	Dental papillae	Tooth morphogenesis	[37–41]
Root formation	HERS	Dental mesenchyme	Root elongation	[35,42–44]

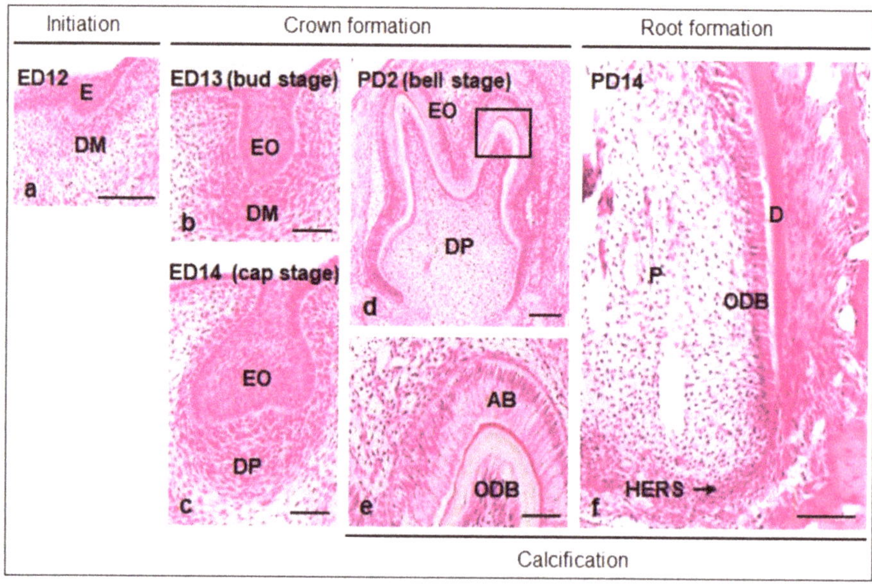

Figure 1. Process of molar tooth development in mouse. (**a–e**) Tooth development begins with thickening of the oral epithelium (E) and progresses to crown (bud, cap, and bell stages) and root formation stages. Calcification of enamel, dentin (D), and cementum occurs after the bell stage. The formation stages "initiation," "crown formation," and "root formation" correspond to the terms in Table 1. Higher magnification of the boxed region in "d" is shown in "e." AB, ameloblast; DM, dental mesenchyme; DP, dental papillae; ED, embryonic day; EO, enamel organ; HERS, Hertwig's epithelial root sheath; ODB, odontoblast; P, pulp; PD, postnatal day. Scale bars = 100 µm (**a**), 50 µm (**b–d,f**), 25 µm (**e**).

In this review, we focus on the functions of Shh signaling related to tooth development. In addition, we introduce recent findings concerning the relationship between Shh signaling and stem cell maintenance, with an emphasis on the potential of Shh signaling for the regeneration of dental tissues.

2. Shh Is Important for Epithelial Invagination at the Initiation of Tooth Development

Tooth development is regulated by reciprocal interactions occurring between epithelial and mesenchymal tissues. These interactions are controlled by several conserved signaling molecules, including bone morphogenetic proteins (BMPs), fibroblast growth factor, Wnts, and Shh [24,25]. At the beginning of tooth development, the oral epithelium actively grows and invaginates toward the mesenchyme. The expression of *Shh* has been demonstrated in the thickening epithelium at the site where tooth formation will occur [31,45–47]. Shh signaling related molecules such as *Ptch*, *Smo*, *Gli1*, *Gli2*, and *Gli3* are also expressed in the dental mesenchyme around the thickening epithelium [32,48]. The implantation of Shh-soaked beads into the dental mesenchyme has been shown to enhance the expression of *Ptch* and *Gli1* at this site, subsequently resulting in an irregular shape of the thickening epithelium. On the other hand, the implantation of Shh-soaked beads into the oral epithelium, but not around the thickening dental epithelium, induces an ectopic epithelial invagination [33]. Enhancement of *Shh* expression in the dental epithelium using the Keratin 14 promoter inhibits cellular proliferation and arrests tooth development during the early stage [34]. Conversely, the inhibition of Shh signaling by cyclopamine, an antagonist of Smo, inhibits the invagination and extension of the oral epithelium into the dental mesenchyme [33]. In *Gli2* and *Gli3* double-mutant mice, although epithelial thickening is observed in the oral epithelium, the epithelium does not proceed to form the enamel organ [32]. Therefore, Shh signaling appears to have an important role in dental epithelial cellular proliferation and invagination.

3. Shh Regulates Enamel Formation

During tooth development, the invaginated dental epithelium extends and forms the enamel organ. Epithelial cells in this tissue can be divided into three types of tissues, namely, the inner and outer epithelia and the stellate reticulum. During tooth crown formation, the cells in the inner enamel epithelium differentiate into ameloblasts that form the enamel. The inner enamel epithelium at this stage expresses both *Shh* and *Ptch* [24,49–54], and suppression of these expressions results in the inhibition of the proliferative activity of the epithelial cells [35]. In addition, it has been demonstrated that inhibition of Shh signaling in tooth germ using a neutralizing antibody suppresses ameloblast differentiation [36].

Shh is also expressed in enamel-secreting ameloblasts [55–57]. As such, the loss of Shh signaling in ameloblast-lineage cells using genetic modification techniques has been shown to cause unpolarized ameloblast differentiation and enamel hypoplasia, resulting in the disruption of normal tooth morphology [37]. Therefore, Shh signaling appears to have multiple roles, which include the proliferation and differentiation of cells in the inner enamel epithelium and in differentiated ameloblasts.

4. Shh Signaling Functions in the Dental Mesenchyme and Is Involved in Tooth Morphogenesis

It has been reported that the inactivation of Shh signaling in the dental epithelium results in the formation of small teeth with the disappearance of Ptch1- and Gli1-positive cells in the dental mesenchyme [39]. Suppression of Sufu in dental mesenchymal cells results in deletion of primary enamel knot in the enamel organ as well as retardation of transition from bud to cap stage of tooth development [40]. It has also been demonstrated that crown size depends on the contact area between the *Shh*-expressing inner enamel epithelium and the dental mesenchyme [54]. These findings indicate that Shh signaling may regulate cellular proliferation in the dental mesenchyme, thereby controlling tooth morphogenesis [38,41,58].

5. Deletion of Shh Signaling in Hertwig's Epithelial Root Sheath (HERS) Suppresses Tooth Root Elongation

After crown formation, the inner and outer enamel epithelium fuse at the lower edge of the enamel organ, forming a bilayered tissue referred to as HERS. Morphologically, the HERS bends inward during the early stages of root formation and grows between the dental papilla and dental follicle. In general, the HERS has been accepted as the principal structure controlling root formation, as this tissue disappears upon completion of root formation. Recent studies have demonstrated that growth factors, including BMPs and transforming growth factor-beta, mediate reciprocal epithelial-mesenchymal interactions during tooth root development [21,45,46,59]. It has also been shown that the epithelial cells of the HERS secrete Shh [42,43,60]. In this process, via Shh signaling, dental mesenchymal cells expressing Ptch are stimulated to form the root dentin [35,36].

Nuclear factor Ic (*Nfic*) knockout mice have normal tooth crowns, but a defect of tooth root formation can be observed in the molars [61]. This suggests that *Nfic* has an essential role in tooth root formation. The loss of Shh in the HERS has been shown to inhibit the expression of *Nfic* in the dental mesenchyme around the HERS [43]. Therefore, it is considered that Shh is an important signaling molecule of the epithelial-mesenchymal interaction and regulates tooth root formation.

6. Signaling Pathways of BMP-SHH and SHH-BMP Regulate Tooth Root Formation

While evidence suggests that Shh signaling has an important role for tooth root development [42,44], the mechanisms of this process remain controversial. In the process of tooth root development, as mentioned above, BMPs are important signaling molecules that regulate epithelial-mesenchymal tissue interactions [45,46,62]. In particular, BMPs principally function via receptor complexes consisting of BMP receptor types I (BMPR-I) and II (BMPR-II) [63]. BMPs activate these receptors upon binding, which then leads to the phosphorylation of R-Smads. Phosphorylated R-Smads subsequently interacts with Smad4 to form a complex, which is translocated to the nucleus [64,65]. This complex then induces the expression of downstream proteins, including Runx2, which are essential transcription factors for hard tissue-forming cell differentiation [66].

The inactivation of Smad4 in the dental epithelium using Keratin 14-Cre; Smad4fl/fl mice have been shown to cause the absence of *Shh* expression in the HERS, resulting in the formation of short tooth roots [43]. In addition, a similar phenotype is observed in mice with mutated BMPR-I in the dental epithelium [67]. In the dental mesenchyme, some positive cells for downstream proteins of Shh signaling are known to be present, including Gli1. The inhibition of BMP signaling in these Gli1-positive cells results in a failure of root dentin formation [43,67]. Therefore, it can be speculated that certain key molecules regulated by Shh signaling may be closely associated with tooth root development, suggesting that BMP and Shh signaling pathways may be regulators of tooth root formation.

7. Gli1-Expressing Cells Possess Stem Cell Properties in Mature Tooth

Multipotent mesenchymal stem cells have been described in a variety of tissues with varying developmental origins and physiological functions [68,69]. Although human permanent and deciduous teeth are known to contain mesenchymal stem cells in the periodontal ligament and dental pulp [70–72], visualization of these cells has yet to be achieved. Recently, iGli1/Tomato mice, which are transgenic for the *Gli1CreERT2; R26RtDTomato* gene [73,74], have been used for lineage tracing analysis of Gli1-positive cells in various organs [75–81]. In this mouse model, Gli1-positive cells were shown to express the Cre recombinase-mutated estrogen receptor (CreERT2). Since CreERT2 is only active in the presence of tamoxifen, Gli1-positive cells start to express Tomato red fluorescence after tamoxifen administration. Tomato red fluorescence is also observed in the daughter cells of Gli1-positive cells after cell division. Therefore, this system can be used to continuously trace Gli1-positive cells and their daughter cells (Figure 2a).

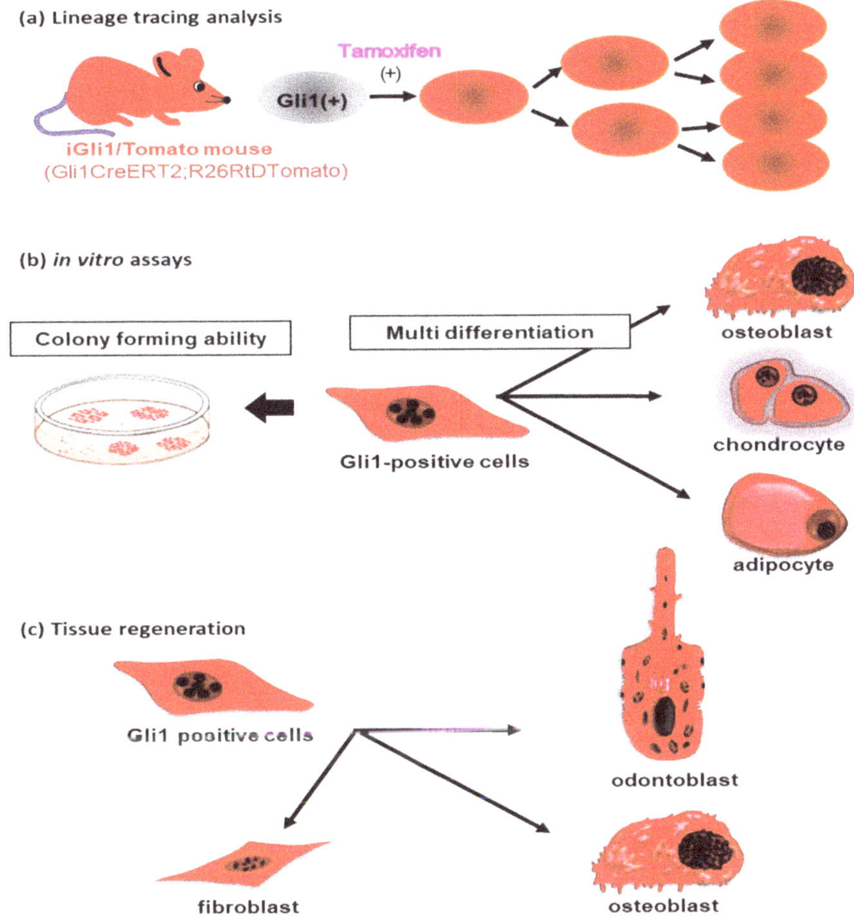

Figure 2. Schematic illustrations of characteristics of Gli1-positive cells in mature teeth. (**a**) After tamoxifen administration in iGli1/Tomato mice, Gli1-positive cells are shown to express Tomato red fluorescence. Cells that once expressed Tomato red fluorescence continuously emit this fluorescence even after cell division. Using this system, it is possible to trace the differentiation process of Gli1-positive cells and their progeny cells. (**b**) Gli1-positive cells exhibit high colony-forming unit fibroblast (CFU-F) activity. These cells also have trilineage potential to form osteoblasts, chondrocytes, and adipocytes in vitro. (**c**) After tooth transplantation into subcutaneous tissue, Gli1-positive cells differentiate into odontoblasts, osteoblasts, and fibroblasts during tissue regeneration.

In a previous study, we revealed that Gli1-positive cells are present in the dental pulp and the periodontal ligament in mature teeth [82]. These cells are barely detected around the blood vessels in mature tooth (Figure 3a–e). In addition, Gli1-positive cells have been identified as mesenchymal stem cells with the ability to self-renew and with trilineage differentiation potential (Figure 2b). Although Gli1-positive cells are quiescent under normal conditions after the completion of tooth formation, they can proliferate after tissue injury, contributing to tissue repair (Figure 2c). In the following chapters, recent studies demonstrating the stem cell abilities of Gli1-positive cells during tooth development will be discussed.

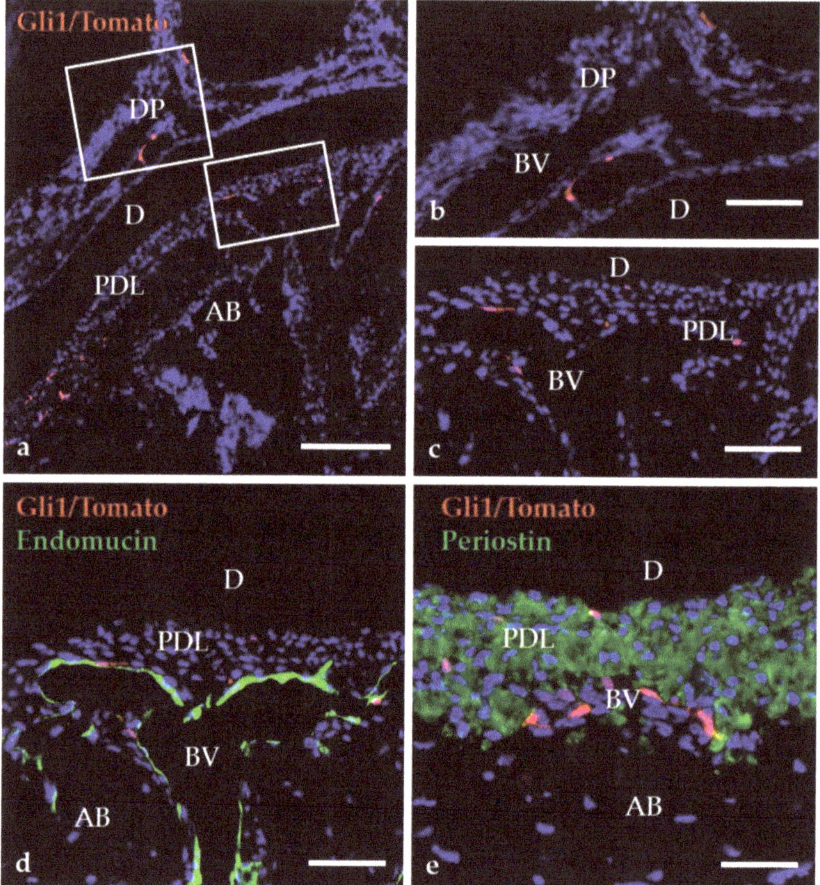

Figure 3. Distribution of Gli1-positive cells in mature teeth. Higher magnification of the boxed region in "a" are shown in "b"–"e." (**a–c**) Gli1-positive cells are present in the dental pulp (DP) and the periodontal ligament (PDL). (**d–e**) The merged image of Endomucin and Periostin with Gli1/Tomato fluorescence demonstrate that most Gli1/Tomato-positive cells are distributed near blood vessels (BV). AB, alveolar bone; D, dentin. Scale bars = 100 μm (**a**), 25 μm (**b–e**).

8. Gli1-Positive Cells Supply Ameloblast-Lineage Cells in the Rodent Incisor

Since rodent incisors erupt continuously throughout the life of the animal, epithelial stem cells that differentiate into enamel-forming ameloblasts are present in the dental epithelium at the posterior apex of the incisor [83]. In addition, Gli1-positive cells are distributed in proximity to *Shh*-expressing cells in the cervical loop of the incisor. These cells have been shown to be co-localized with bromodeoxyuridine label-retaining cells, suggesting the presence of both stem cells and transit-amplifying cells [84]. Using lineage tracing analysis, Gli1-positive cells in the dental epithelium of the mouse incisor have been shown to proliferate and differentiate into ameloblasts [85]. Furthermore, since the formation of enamel can be blocked in the mouse incisor by the administration of hedgehog pathway inhibitors [36,37,85], Shh signaling may contribute to both the maintenance of epithelial stem cells and ameloblast differentiation.

9. Gli1-Positive Cells Are Mesenchymal Stem Cells in Developing Tooth

Mesenchymal cells in tooth germ have been considered to originate from the cranial neural crest [86]. In the mouse incisor, it has been reported that most cells originating from the cranial neural crest express *Gli1* and are localized at the posterior apex of the dental mesenchyme without a high proliferation ability. These cells expand and populate the entire dental pulp, as well as the periodontal ligament [87,88]. Zhao et al. [87] suggested that Shh secreted by sensory nerves, not the dental epithelium, is important for the maintenance of these Gli1-positive cells in the dental mesenchyme of the mouse incisor. Similarly, it has been reported that the nerve-derived Shh is involved in supporting the stem cell niche in hair follicle for its development and regeneration [89].

Just after the beginning of root formation stage of the mouse molar, the HERS secrets Shh [42,43]. Gli1-positive cells are then distributed in the dental mesenchyme around the HERS [44,67,90] and proliferate as the tooth root elongates, differentiating into root-forming cells such as odontoblasts, cementoblasts, and fibroblasts in the dental pulp and the periodontal ligament [44,91]. These Gli1-positive cells have also been shown to have multilineage potential and high colony-forming unit fibroblast (CFU-F) activity *in vitro* [90]. Furthermore, root elongation is not observed in tooth germ lacking Gli1-positive cells during the root formation stage [44]. Therefore, Gli1-positive cells are believed to supply the cells involved in tooth root formation. These results also indicate that Gli1 may be a useful marker of mesenchymal stem cells in the developing tooth (Table 2).

Table 2. Differentiatial ability of Gli1-positive cells in mouse developing tooth.

Tooth	Localization of Gli1-Positive Cells	Differentiating Cells	References
Incisor	Epithelium in cervical loop	Ameloblasts	[85]
	Mesenchyme around cervical loop	Crown forming cells without ameloblasts	[87,88]
Molar	Mesenchyme around HERS	Root forming cells	[44,67,90]

10. Can Shh Signaling Be a Target for Tooth Regeneration Therapy?

Cell replacement therapies using undifferentiated cells are considered to be one of the most effective methods for cellular and tissue regeneration. As such, regenerative therapies using stem cells have been widely studied in a variety of organs [92,93]. This approach is considered to be constructive as it promotes healing in the original cells. In vivo studies have shown that, after tooth transplantation into the subcutaneous tissue, stem and undifferentiated cells can differentiate into odontoblasts [94,95], cementoblasts, and osteoblasts [96,97]. Interestingly, the majority of cells with this regenerative ability have been shown to express *Gli1* (Figure 2c).The collection of dental pulp and periodontal ligament cells containing Gli1-positive cells from teeth extracted for orthodontic reasons or from nonfunctional third molars is possible. In addition, the elucidation of mechanisms concerning stem and undifferentiated cell maintenance by Shh signaling may lead to the application of Gli1-positive cells for tooth regeneration. However, in practical terms, a large number of replacement cells would be required because stem cells in tooth are present only in a limited number. Furthermore, in vitro culture systems to expand these stem cells, while maintaining their unique characteristics, have not been established. Therefore, a better understanding of the mechanisms underlying the maintenance of stemness, as well as tooth cell differentiation in Gli1-positive cells, may lead to more effective biologically activating therapies than are currently offered by traditional dental treatments.

11. Conclusions

Shh signaling is deeply involved in tooth formation and has different functions at each stage of tooth development. Therefore, a greater understanding of tooth formation may accelerate the development of novel regenerative and restorative therapies. Indeed, recent studies have shown that cells expressing *Gli1*, a downstream factor of Shh signaling, are mesenchymal cells in both developing and mature teeth. Thus, it is expected that additional functions of Shh signaling in tooth formation, as well as the regulatory mechanism of stem cell properties in the dental mesenchyme, will be elucidated and lead to the development of new dental therapies.

Author Contributions: Conceptualization, A.H. and K.I.; writing—original draft preparation, A.H.; writing—review and editing, A.H., N.S., H.T., T.S. and K.I.; visualization, A.H. and N.S.; funding acquisition, A.H., H.T., T.S. and K.I. All authors have read and agreed to the published version of the manuscript.

Funding: This research was funded by Grant-in-Aid from Japan Society for the Promotion of Science (JSPS).

Acknowledgments: We thank Toshihide Mizoguchi, Hiroaki Nakamura, Kunihiko Yoshiba, and Nagako Yoshiba for their valuable advice, comments, and discussion.

Conflicts of Interest: The authors declare no potential conflicts of interest with respect to the authorship and/or publication of this article.

Abbreviations

BMP	Bone Morphogenetic Protein
BMPR	Bone Morphogenetic Protein Receptor
CreERT2	Cre Recombinase-mutated Estrogen Receptor
HERS	Hertwig's Epithelial Root Sheath
Hh	Hedgehog
IFT	Intraflagellar Transport
Ptch	Patched
Runx2	Runt-related Transcription Factor 2
Shh	Sonic Hedgehog
Smo	Smoothend

References

1. Kurosaka, H. The roles of hedgehog signaling in upper lip formation. *Biomed. Res. Int.* **2015**, 901041. [CrossRef] [PubMed]
2. Xavier, G.M.; Seppala, M.; Barrell, W.; Birjandi, A.A.; Geoghegan, F.; Cobourne, M.T. Hedgehog receptor function during craniofacial development. *Dev. Biol.* **2016**, *415*, 198–215. [CrossRef] [PubMed]
3. Dworkin, S.; Boglev, Y.; Owens, H.; Goldie, S.J. The role of Sonic hedgehog in craniofacial patterning, morphogenesis and cranial neural crest survival. *J. Dev. Biol.* **2016**, *4*, E3. [CrossRef] [PubMed]
4. Elliott, K.H.; Millington, G.; Brugmann, S.A. A novel role for cilia-dependent Sonic hedgehog signaling during submandibular gland development. *Dev. Dyn.* **2018**, *247*, 818–831. [CrossRef]
5. Abramyan, J. Hedgehog signaling and embryonic craniofacial disorders. *J. Dev. Biol.* **2019**, *7*, E2. [CrossRef]
6. Seppala, M.; Fraser, G.J.; Birjandi, A.A.; Xavier, G.M.; Cobourne, M.T. Sonic hedgehog signaling and development of the dentition. *J. Dev. Biol.* **2017**, *5*, E6. [CrossRef]
7. Goetz, S.C.; Anderson, K.V. The Primary Cilium: A signaling center during vertebrate development. *Nat. Rev. Genet.* **2010**, *11*, 331–344. [CrossRef]
8. Liu, B.; Chen, S.; Cheng, D.; Jing, W.; Helms, J.A. Primary cilia integrate hedgehog and Wnt signaling during tooth development. *J. Dent. Res.* **2014**, *93*, 475–482. [CrossRef]
9. Briscoe, J.; Therond, P.P. The mechanisms of hedgehog signaling and its roles in development and disease. *Nat. Rev. Mol. Cell Biol.* **2013**, *14*, 418–431. [CrossRef]
10. Sasaki, H.; Hui, C.; Nakafuku, M.; Kondoh, H. A binding site for Gli proteins is essential for HNF-3beta floor plate enhancer activity in transgenics and can respond to Shh in vitro. *Development* **1997**, *124*, 1313–1322.
11. Yang, J.; Andre, P.; Ye, L.; Yang, Y.Z. The hedgehog signaling pathway in bone formation. *Int. J. Oral Sci.* **2015**, *7*, 73. [CrossRef] [PubMed]
12. Jernvall, J.; Keränen, S.V.; Thesleff, I. Evolutionary modification of development in mammalian teeth: Quantifying gene expression patterns and topography. *Proc. Natl. Acad. Sci. USA* **2000**, *97*, 14444–14448. [CrossRef] [PubMed]
13. Koyama, E.; Wu, C.; Shimo, T.; Iwamoto, M.; Ohmori, T.; Kurisu, K.; Ookura, T.; Bashir, M.M.; Abrams, W.R.; Tucker, T.; et al. Development of stratum intermedium and its role as a Sonic hedgehog-signaling structure during odontogenesis. *Dev. Dyn.* **2001**, *222*, 178–191. [CrossRef] [PubMed]

14. Miyado, M.; Ogi, H.; Yamada, G.; Kitoh, J.; Jogahara, T.; Oda, S.; Sato, I.; Miyado, K.; Sunohara, M. Sonic hedgehog expression during early tooth development in suncus murinus. *Biochem. Biophys. Res. Commun.* **2007**, *363*, 269–275. [CrossRef]

15. Hovorakova, M.; Smrckova, L.; Lesot, H.; Lochovska, K.; Peterka, M.; Peterkova, R. Sequential Shh expression in the development of the mouse upper functional incisor. *J. Exp. Zool. B Mol. Dev. Evol.* **2013**, *320*, 455–464.

16. Hu, X.; Zhang, S.; Chen, G.; Lin, C.; Huang, Z.; Chen, Y.; Zhang, Y. Expression of SHH signaling molecules in the developing human primary dentition. *BMC Dev. Biol.* **2013**, *13*, 11. [CrossRef]

17. Yamanaka, A.; Yasui, K.; Sonomura, T.; Iwai, H.; Uemura, M. Development of deciduous and permanent dentitions in the upper jaw of the house shrew (Suncus murinus). *Arch. Oral Biol.* **2010**, *55*, 279–287. [CrossRef]

18. Hovorakova, M.; Prochazka, J.; Lesot, H.; Smrckova, L.; Churava, S.; Boran, T.; Kozmik, Z.; Klein, O.; Peterkova, R.; Peterka, M. Shh expression in a rudimentary tooth offers new insights into development of the mouse incisor. *J. Exp. Zool. B Mol. Dev. Evol.* **2011**, *316*, 347–358. [CrossRef]

19. Nakatomi, C.; Nakatomi, M.; Saito, K.; Harada, H.; Ohshima, H. The enamel knot-like structure is eternally maintained in the apical bud of postnatal mouse incisors. *Arch. Oral Biol.* **2015**, *60*, 1122–1130. [CrossRef]

20. Khan, M.; Seppala, M.; Zoupa, M.; Cobourne, M.T. Hedgehog pathway gene expression during early development of the molar tooth root in the mouse. *Gene Expr. Patterns* **2007**, *7*, 239–243. [CrossRef]

21. Zhang, Y.; Zhang, Z.; Zhao, X.; Yu, X.; Hu, Y.; Geronimo, B.; Fromm, S.H.; Chen, Y.P. A new function of BMP4: Dual role for BMP4 in regulation of Sonic hedgehog expression in the mouse tooth germ. *Development* **2000**, *127*, 1431–1443. [PubMed]

22. Sarkar, L.; Cobourne, M.; Naylor, S.; Smalley, M.; Dale, T.; Sharpe, P.T. Wnt/Shh interactions regulate ectodermal boundary formation during mammalian tooth development. *Proc. Natl. Acad. Sci. USA* **2000**, *97*, 4520–4524. [CrossRef] [PubMed]

23. Ten Berge, D.; Brouwer, A.; Korving, J.; Reijnen, M.J.; van Raaij, E.J.; Verbeek, F.; Gaffield, W.; Meijlink, F. Prx1 and Prx2 are upstream regulators of sonic hedgehog and control cell proliferation during mandibular arch morphogenesis. *Development* **2001**, *128*, 2929–2938. [PubMed]

24. Handrigan, G.R.; Richman, J.M. A network of Wnt, hedgehog and BMP signaling pathways regulates tooth replacement in snakes. *Dev. Biol.* **2010**, *318*, 130–141. [CrossRef] [PubMed]

25. Ahn, Y.; Sanderson, B.W.; Klein, O.D.; Krumlauf, R. Inhibition of Wnt signaling by Wise (Sostdc1) and negative feedback from Shh controls tooth number and patterning. *Development* **2010**, *137*, 3221–3231. [CrossRef] [PubMed]

26. Cho, S.W.; Kwak, S.; Woolley, T.E.; Lee, M.J.; Kim, E.J.; Baker, R.E.; Kim, H.J.; Shin, J.S.; Tickle, C.; Maini, P.K.; et al. Interactions between Shh, Sostdc1 and Wnt signaling and a new feedback loop for spatial patterning of the teeth. *Development* **2011**, *138*, 1807–1816. [CrossRef] [PubMed]

27. Tokita, M.; Chaeychomsri, W.; Siruntawineti, J. Developmental basis of toothlessness in turtles: Insight into convergent evolution of vertebrate morphology. *Evolution* **2013**, *67*, 260–273. [CrossRef]

28. Aurrekoetxea, M.; Irastorza, I.; García-Gallastegui, P.; Jiménez-Rojo, L.; Nakamura, T.; Yamada, Y.; Ibarretxe, G.; Unda, F.J. Wnt/β-catenin regulates the activity of epiprofin/Sp6, SHH, FGF, and BMP to coordinate the stages of odontogenesis. *Front. Cell Dev. Biol.* **2016**, *4*, 25. [CrossRef]

29. Sagai, T.; Amano, T.; Maeno, A.; Kiyonari, H.; Seo, H.; Cho, S.W.; Shiroishi, T. SHH signaling directed by two oral epithelium-specific enhancers controls tooth and oral development. *Sci. Rep.* **2017**, *7*, 13004. [CrossRef]

30. Seo, H.; Amano, T.; Seki, R.; Sagai, T.; Kim, J.; Cho, S.W.; Shiroishi, T. Upstream enhancer elements of Shh regulate oral and dental patterning. *J. Dent. Res.* **2018**, *97*, 1055–1063. [CrossRef]

31. Wang, X.P.; Aberg, T.; James, M.J.; Levanon, D.; Groner, Y.; Thesleff, I. Runx2 (Cbfa1) inhibits Shh signaling in the lower but not upper molars of mouse embryos and prevents the budding of putative successional teeth. *J. Dent. Res.* **2005**, *84*, 138–143. [CrossRef] [PubMed]

32. Hardcastle, Z.; Mo, R.; Hui, C.C.; Sharpe, P.T. The Shh signalling pathway in tooth development: Defects in Gli2 and Gli3 mutants. *Development* **1998**, *125*, 2803–2811. [PubMed]

33. Li, J.; Chatzeli, L.; Panousopoulou, E.; Tucker, A.S.; Green, J.B. Epithelial stratification and placode invagination are separable functions in early morphogenesis of the molar tooth. *Development* **2016**, *143*, 670–681. [CrossRef] [PubMed]

34. Cobourne, M.T.; Xavier, G.M.; Depew, M.; Hagan, L.; Sealby, J.; Webster, Z.; Sharpe, P.T. Sonic hedgehog signalling inhibits palatogenesis and arrests tooth development in a mouse model of the nevoid basal cell carcinoma syndrome. *Dev. Biol.* **2009**, *331*, 38–49. [CrossRef] [PubMed]

35. Wu, C.; Shimo, T.; Liu, M.; Pacifici, M.; Koyama, E. Sonic hedgehog functions as a mitogen during bell stage of odontogenesis. *Connect. Tissue Res.* **2003**, *44*, 92–96. [CrossRef]

36. Koyama, E.; Wu, C.; Shimo, T.; Pacifici, M. Chick Limbs With Mouse Teeth: An effective in vivo culture system for tooth germ development and analysis. *Dev. Dyn.* **2003**, *226*, 149–154. [CrossRef]

37. Gritli-Linde, A.; Bei, M.; Maas, R.; Zhang, X.M.; Linde, A.; McMahon, A.P. Shh signaling within the dental epithelium is necessary for cell proliferation, growth and polarization. *Development* **2002**, *129*, 5323–5337. [CrossRef]

38. Yu, J.C.; Fox, Z.D.; Crimp, J.L.; Littleford, H.E.; Jowdry, A.L.; Jackman, W.R. Hedgehog signaling regulates dental papilla formation and tooth size during zebrafish odontogenesis. *Dev. Dyn.* **2015**, *244*, 577–590. [CrossRef]

39. Dassule, H.R.; Lewis, P.; Bei, M.; Maas, R.; McMahon, A.P. Sonic hedgehog regulates growth and morphogenesis of the tooth. *Development* **2000**, *127*, 4775–4785.

40. Li, J.; Xu, J.; Cui, Y.; Wang, L.; Wang, B.; Wang, Q.; Zhang, X.; Qiu, M.; Zhang, Z. Mesenchymal Sufu regulates development of mandibular molars via Shh signaling. *J. Dent. Res.* **2019**, *98*, 1348–1356. [CrossRef]

41. Ishida, K.; Murofushi, M.; Nakao, K.; Morita, R.; Ogawa, M.; Tsuji, T. The regulation of tooth morphogenesis is associated with epithelial cell proliferation and the expression of Sonic hedgehog through epithelial-mesenchymal interactions. *Biochem. Biophys. Res. Commun.* **2011**, *405*, 455–461. [CrossRef] [PubMed]

42. Nakatomi, M.; Morita, I.; Eto, K.; Ota, M.S. Sonic hedgehog signaling is important in tooth root development. *J. Dent. Res.* **2006**, *85*, 427–431. [CrossRef] [PubMed]

43. Huang, X.; Xu, X.; Bringas, P., Jr.; Hung, Y.P.; Chai, Y. Smad4-Shh-Nfic signaling cascade-mediated epithelial-mesenchymal interaction is crucial in regulating tooth root development. *J. Bone Miner. Res.* **2010**, *25*, 1167–1178. [CrossRef] [PubMed]

44. Liu, Y.; Feng, J.; Li, J.; Zhao, H.; Ho, T.V.; Chai, Y. An Nfic-hedgehog signaling cascade regulates tooth root development. *Development* **2015**, *142*, 3374–3382. [CrossRef]

45. Thesleff, I. Epithelial-mesenchymal signalling regulating tooth morphogenesis. *J. Cell Sci.* **2003**, *116*, 1647–1648. [CrossRef]

46. Tucker, A.; Sharpe, P. The cutting-edge of mammalian development; how the embryo makes teeth. *Nat. Rev. Genet.* **2004**, *5*, 499–508. [CrossRef]

47. Buchtová, M.; Handrigan, G.R.; Tucker, A.S.; Lozanoff, S.; Town, L.; Fu, K.; Diewert, V.M.; Wicking, C.; Richman, J.M. Initiation and patterning of the snake dentition are dependent on Sonic hedgehog signaling. *Dev. Biol.* **2008**, *319*, 132–145. [CrossRef]

48. Cobourne, M.T.; Miletich, I.; Sharpe, P.T. Restriction of sonic hedgehog signalling during early tooth development. *Development* **2004**, *131*, 2875–2885. [CrossRef]

49. Koyama, E.; Yamaai, T.; Iseki, S.; Ohuchi, H.; Nohno, T.; Yoshioka, H.; Hayashi, Y.; Leatherman, J.L.; Golden, E.B.; Noji, S.; et al. Polarizing activity, Sonic hedgehog, and tooth development in embryonic and postnatal mouse. *Dev. Dyn.* **1996**, *206*, 59–72. [CrossRef]

50. Peterková, R.; Peterka, M.; Viriot, L.; Lesot, H. Dentition development and budding morphogenesis. *J. Craniofac. Genet. Dev. Biol.* **2000**, *20*, 158–172.

51. Kriangkrai, R.; Iseki, S.; Eto, K.; Chareonvit, S. Dual odontogenic origins develop at the early stage of rat maxillary incisor development. *Anat. Embryol. Berl.* **2006**, *211*, 101–108. [CrossRef] [PubMed]

52. Nunes, F.D.; Valenzuela Mda, G.; Rodini, C.O.; Massironi, S.M.; Ko, G.M. Localization of Bmp-4, Shh and Wnt-5a transcripts during early mice tooth development by in situ hybridization. *Braz. Oral Res.* **2007**, *21*, 127–133. [CrossRef] [PubMed]

53. Zhang, L.; Hua, F.; Yuan, G.H.; Zhang, Y.D.; Chen, Z. Sonic hedgehog signaling is critical for cytodifferentiation and cusp formation in developing mouse molars. *J. Mol. Histol.* **2008**, *39*, 87–94. [CrossRef] [PubMed]

54. Handrigan, G.R.; Richman, J.M. Autocrine and paracrine Shh signaling are necessary for tooth morphogenesis, but not tooth replacement in snakes and lizards (Squamata). *Dev. Biol.* **2010**, *337*, 171–186. [CrossRef]

55. Iseki, S.; Araga, A.; Ohuchi, H.; Nohno, T.; Yoshioka, H.; Hayashi, F.; Noji, S. Sonic hedgehog is expressed in epithelial cells during development of whisker, hair, and tooth. *Biochem. Biophys. Res. Commun.* **1996**, *218*, 688–693. [CrossRef]

56. Kumamoto, H.; Ohki, K.; Ooya, K. Expression of Sonic hedgehog (SHH) signaling molecules in ameloblastomas. *J. Oral Pathol. Med.* **2004**, *33*, 185–190. [CrossRef]

57. Takahashi, S.; Kawashima, N.; Sakamoto, K.; Nakata, A.; Kameda, T.; Sugiyama, T.; Katsube, K.; Suda, H. Differentiation of an ameloblast-lineage cell line (ALC) is induced by Sonic hedgehog signaling. *Biochem. Biophys. Res. Commun.* **2007**, *353*, 405–411. [CrossRef]

58. Jackman, W.R.; Yoo, J.J.; Stock, D.W. Hedgehog signaling is required at multiple stages of zebrafish tooth development. *BMC Dev. Biol.* **2010**, *10*, 119. [CrossRef]

59. Thesleff, I.; Sharpe, P. Signalling networks regulating dental development. *Mech Dev* **1997**, *67*, 111–123. [CrossRef]

60. Bae, W.J.; Auh, Q.S.; Lim, H.C.; Kim, G.T.; Kim, H.S.; Kim, E.C. Sonic hedgehog promotes cementoblastic differentiation via activating the BMP pathways. *Calcif. Tissue Int.* **2016**, *99*, 396–407. [CrossRef]

61. Steele-Perkins, G.; Butz, K.G.; Lyons, G.E.; Zeichner-David, M.; Kim, H.J.; Cho, M.I.; Gronostajski, R.M. Essential role for NFI-C/CTF transcription-replication factor in tooth root development. *Mol. Cell Biol.* **2003**, *23*, 1075–1084. [CrossRef] [PubMed]

62. Nie, X.; Luukko, K.; Kettunen, P. BMP signalling in craniofacial development. *Int. J. Dev. Biol.* **2006**, *50*, 511–521. [CrossRef] [PubMed]

63. Derynck, R.; Zhang, Y.E. Smad-dependent and Smad-independent pathways in TGF-beta family signalling. *Nature* **2003**, *425*, 577–584. [CrossRef] [PubMed]

64. Derynck, R.; Zhang, Y.; Feng, X.H. Smads: Transcriptional activators of TGF-beta responses. *Cell* **1998**, *95*, 737–740. [CrossRef]

65. Shi, Y.; Massague, J. Mechanisms of TGF-beta signaling from cell membrane to the nucleus. *Cell* **2003**, *113*, 685–700. [CrossRef]

66. Komori, T.; Yagi, H.; Nomura, S.; Yamaguchi, A.; Sasaki, K.; Deguchi, K.; Shimizu, Y.; Bronson, R.T.; Gao, Y.H.; Inada, M.; et al. Targeted disruption of Cbfa1 results in a complete lack of bone formation owing to maturational arrest of osteoblasts. *Cell* **1997**, *89*, 755–764. [CrossRef]

67. Li, J.; Feng, J.; Liu, Y.; Ho, T.V.; Grimes, W.; Ho, H.A.; Park, S.; Wang, S.; Chai, Y. BMP-SHH signaling network controls epithelial stem cell fate via regulation of its niche in the developing tooth. *Dev. Cell* **2015**, *33*, 125–135. [CrossRef]

68. Friedenstein, A.J.; Petrakova, K.V.; Kurolesova, A.I.; Frolova, G.P. Heterotopic of bone marrow. Analysis of precursor cells for osteogenic and hematopoietic tissues. *Transplantation* **1968**, *6*, 230–247. [CrossRef]

69. Bianco, P.; Robey, P.G.; Simmons, P.J. Mesenchymal stem cells: Revisiting history, concepts, and assays. *Cell Stem Cell* **2008**, *2*, 313–319. [CrossRef]

70. Gronthos, S.; Mankani, M.; Brahim, J.; Robey, P.G.; Shi, S. Postnatal human dental pulp stem cells (DPSCs) in vitro and in vivo. *Proc. Natl. Acad. Sci. USA* **2000**, *97*, 13625–13630. [CrossRef]

71. Miura, M.; Gronthos, S.; Zhao, M.; Lu, B.; Fisher, L.W.; Robey, P.G.; Shi, S. SHED: Stem cells from human exfoliated deciduous teeth. *Proc. Natl. Acad. Sci. USA* **2003**, *100*, 5807–5812. [CrossRef] [PubMed]

72. Seo, B.M.; Miura, M.; Gronthos, S.; Bartold, P.M.; Batouli, S.; Brahim, J.; Young, M.; Robey, P.G.; Wang, C.Y.; Shi, S. Investigation of multipotent postnatal stem cells from human periodontal ligament. *Lancet* **2004**, *364*, 149–155. [CrossRef]

73. Ahn, S.; Joyner, A.L. Dynamic changes in the response of cells to positive hedgehog signaling during mouse limb patterning. *Cell* **2004**, *118*, 505–516. [CrossRef] [PubMed]

74. Madisen, L.; Zwingman, T.A.; Sunkin, S.M.; Oh, S.W.; Zariwala, H.A.; Gu, H.; Ng, L.L.; Palmiter, R.D.; Hawrylycz, M.J.; Jones, A.R.; et al. A robust and high-throughput Cre reporting and characterization system for the whole mouse brain. *Nat. Neurosci.* **2010**, *13*, 133–140. [CrossRef] [PubMed]

75. Kramann, R.; Schneider, R.K.; DiRocco, D.P.; Machado, F.; Fleig, S.; Bondzie, P.A.; Henderson, J.M.; Ebert, B.L.; Humphreys, B.D. Perivascular Gli1+ progenitors are key contributors to injury-induced organ fibrosis. *Cell Stem Cell* **2015**, *16*, 51–66. [CrossRef] [PubMed]

76. Zhao, H.; Feng, J.; Ho, T.V.; Grimes, W.; Urata, M.; Chai, Y. The suture provides a niche for mesenchymal stem cells of craniofacial bones. *Nat. Cell. Biol.* **2015**, *17*, 386–396. [CrossRef] [PubMed]

77. Liu, C.; Rodriguez, K.; Yao, H.H. Mapping lineage progression of somatic progenitor cells in the mouse fetal testis. *Development* **2016**, *143*, 3700–3710. [CrossRef]

78. Kramann, R.; Goettsch, C.; Wongboonsin, J.; Iwata, H.; Schneider, R.K.; Kuppe, C.; Kaesler, N.; Chang-Panesso, M.; Machado, F.G.; Gratwohl, S.; et al. Adventitial MSC-like cells are progenitors of vascular smooth muscle cells and drive vascular calcification in chronic kidney disease. *Cell Stem Cell* **2016**, *19*, 628–642. [CrossRef]

79. Shi, Y.; He, G.; Lee, W.C.; McKenzie, J.A.; Silva, M.J.; Long, F. Gli1 identifies osteogenic progenitors for bone formation and fracture repair. *Nat. Commun.* **2017**, *8*, 2043. [CrossRef]

80. Degirmenci, B.; Valenta, T.; Dimitrieva, S.; Hausmann, G.; Basler, K. GLI1-expressing mesenchymal cells form the essential Wnt-secreting niche for colon stem cells. *Nature* **2018**, *558*, 449–453. [CrossRef]

81. Ó hAinmhire, E.; Wu, H.; Muto, Y.; Donnelly, E.L.; Machado, F.G.; Fan, L.X.; Chang-Panesso, M.; Humphreys, B.D. A conditionally immortalized Gli1-positive kidney mesenchymal cell line models myofibroblast transition. *Am. J. Physiol. Renal. Physiol.* **2019**, *316*, F63–F75. [CrossRef] [PubMed]

82. Pang, P.; Shimo, T.; Takada, H.; Matsumoto, K.; Yoshioka, N.; Ibaragi, S.; Sasaki, A. Expression pattern of sonic hedgehog signaling and calcitonin gene-related peptide in the socket healing process after tooth extraction. *Biochem. Biophys. Res. Commun.* **2015**, *467*, 21–26. [CrossRef] [PubMed]

83. Harada, H.; Kettunen, P.; Jung, H.S.; Mustonen, T.; Wang, Y.A.; Thesleff, I. Localization of putative stem cells in dental epithelium and their association with Notch and FGF signaling. *J. Cell Biol.* **1999**, *147*, 105–120. [CrossRef] [PubMed]

84. Ishikawa, Y.; Nakatomi, M.; Ida-Yonemochi, H.; Ohshima, H. Quiescent adult stem cells in murine teeth are regulated by Shh signaling. *Cell Tissue Res.* **2017**, *369*, 497–512. [CrossRef]

85. Seidel, K.; Ahn, C.P.; Lyons, D.; Nee, A.; Ting, K.; Brownell, I.; Cao, T.; Carano, R.A.D.; Curran, T.; Schober, M.; et al. Hedgehog signaling regulates the generation of ameloblast progenitors in the continuously growing mouse incisor. *Development* **2010**, *137*, 3753–3761. [CrossRef]

86. Yamazaki, H.; Tsuneto, M.; Yoshino, M.; Yamamura, K.; Hayashi, S. Potential of dental mesenchymal Cells in developing teeth. *Stem Cells* **2007**, *25*, 78–87. [CrossRef]

87. Zhao, H.; Feng, J.; Seidel, K.; Shi, S.; Chai, Y. Secretion of Shh by a neurovascular bundle niche supports mesenchymal stem cell homeostasis in the adult mouse incisor. *Cell Stem Cell* **2014**, *14*, 160–173. [CrossRef]

88. Shi, C.; Yuan, Y.; Guo, Y.; Jing, J.; Ho, T.V.; Han, X.; Li, J.; Feng, J.; Chai, Y. BMP signaling in regulating mesenchymal stem cells in incisor homeostasis. *J. Dent. Res.* **2019**, *98*, 904–911. [CrossRef]

89. Brownell, I.; Guevara, E.; Bai, C.B.; Loomis, C.A.; Joyner, A.L. Nerve-derived Sonic hedgehog defines a niche for hair follicle stem cells capable of becoming epidermal stem cells. *Cell Stem Cell* **2011**, *8*, 552–565. [CrossRef]

90. Feng, J.; Jing, J.; Li, J.; Zhao, H.; Punj, V.; Zhang, T.; Xu, J.; Chai, Y. BMP signaling orchestrates a transcriptional network to control the fate of mesenchymal stem cells in mice. *Development* **2017**, *144*, 2560–2569. [CrossRef]

91. Li, C.; Jing, Y.; Wang, K.; Ren, Y.; Liu, X.; Wang, X.; Wang, Z.; Zhao, H.; Feng, J.Q. Dentinal mineralization is not limited in the mineralization front but occurs along with the entire odontoblast process. *Int. J. Biol. Sci.* **2018**, *14*, 693–704. [CrossRef] [PubMed]

92. Mimeault, M.; Hauke, R.; Batra, S.K. Stem cells: A revolution in therapeutics-recent advances in stem cell biology and their therapeutic applications in regenerative medicine and cancer therapies. *Clin. Pharmacol. Ther.* **2007**, *82*, 252–264. [CrossRef] [PubMed]

93. Picinich, S.C.; Mishra, P.J.; Mishra, P.J.; Glod, J.; Banerjee, D. The therapeutic potential of mesenchymal stem cells. Cell- & tissue-based therapy. *Expert. Opin. Biol. Ther.* **2007**, *7*, 965–973.

94. Hosoya, A.; Yoshiba, K.; Yoshiba, N.; Hoshi, K.; Iwaku, M.; Ozawa, H. An immunohistochemical study on hard tissue formation in a subcutaneously transplanted rat molar. *Histochem. Cell Biol.* **2003**, *119*, 27–35. [CrossRef] [PubMed]

95. Hosoya, A.; Yukita, A.; Yoshiba, K.; Yoshiba, N.; Takahashi, M.; Nakamura, H. Two distinct processes of bone-like tissue formation by dental pulp cells after tooth transplantation. *J. Histochem. Cytochem.* **2012**, *60*, 861–873. [CrossRef]

96. Hosoya, A.; Ninomiya, T.; Hiraga, T.; Zhao, C.; Yoshiba, K.; Nakamura, H. Alveolar bone regeneration of subcutaneously transplanted rat molar. *Bone* **2008**, *42*, 350–357. [CrossRef]

97. Hasan, M.R.; Takebe, H.; Shalehin, N.; Obara, N.; Saito, T.; Irie, K. Effects of tooth storage media on periodontal ligament preservation. *Dent. Traumatol.* **2017**, *33*, 383–392. [CrossRef]

Communication

Sonic Hedgehog Regulates Bone Fracture Healing

Hiroaki Takebe [1],*, Nazmus Shalehin [1], Akihiro Hosoya [1], Tsuyoshi Shimo [2] and Kazuharu Irie [1]

[1]   Division of Histology, Department of Oral Growth and Development, School of Dentistry, Health Sciences University of Hokkaido, 1757 Kanazawa, Ishikari-Tobetsu Hokkaido 061-0293, Japan; shalehin@hoku-iryo-u.ac.jp (N.S.); hosoya@hoku-iryo-u.ac.jp (A.H.); irie@hoku-iryo-u.ac.jp (K.I.)

[2]   Division of Reconstructive Surgery for Oral and Maxillofacial Region, Department of Human Biology and Pathophysiology, School of Dentistry, Health Sciences University of Hokkaido, 1757 Kanazawa, Ishikari-Tobetsu Hokkaido 061-0293, Japan; shimotsu@hoku-iryo-u.ac.jp

*   Correspondence: takebeh@hoku-iryo-u.ac.jp; Tel.: +81-0133-23-1211

Received: 25 December 2019; Accepted: 16 January 2020; Published: 20 January 2020

Abstract: Bone fracture healing involves the combination of intramembranous and endochondral ossification. It is known that Indian hedgehog (Ihh) promotes chondrogenesis during fracture healing. Meanwhile, Sonic hedgehog (Shh), which is involved in ontogeny, has been reported to be involved in fracture healing, but the details had not been clarified. In this study, we demonstrated that Shh participated in fracture healing. Six-week-old Sprague–Dawley rats and Gli-CreER^{T2}; tdTomato mice were used in this study. The right rib bones of experimental animals were fractured. The localization of Shh and Gli1 during fracture healing was examined. The localization of Gli1 progeny cells and osterix (Osx)-positive cells was similar during fracture healing. Runt-related transcription factor 2 (Runx2) and Osx, both of which are osteoblast markers, were observed on the surface of the new bone matrix and chondrocytes on day seven after fracture. Shh and Gli1 were co-localized with Runx2 and Osx. These findings suggest that Shh is involved in intramembranous and endochondral ossification during fracture healing.

Keywords: sonic hedgehog; stem cell; animal experiment; fracture healing

---

1. Introduction

The fracture healing process consists of four overlapping phases, namely, inflammation, proliferation, callus formation, and bone remodeling. Immediately following fracture, the injury initiates an inflammatory response that is necessary to promote healing. The response induces the development of a hematoma, which consists of cells from both peripheral blood vessels and bone marrow. The hematoma coagulates between and around the fracture site and within the bone marrow, providing a template for callus formation [1]. Vascularization supplies mesenchymal stem cells (MSCs), which differentiate into chondrocytes or osteoblasts simultaneously with cartilage tissue development (proliferation phase) [2,3]. The cartilage matrix begins to form at the fractured bone gap during the callus formation phase. Meanwhile, intramembranous ossification occurs internal to the periosteum adjacent to the fracture line and forms the bone matrix [4]. MSCs directly differentiate into osteoblasts at the fracture site along the proximal and distal edges of fractured bone during intramembranous ossification. After cartilage tissue maturation, new bone formation is initiated as the cartilage tissue is resorbed and vascularization is induced to replace the cartilage tissue with bone. It has also been reported that primary bone formation is initiated peripheral to the newly formed cartilage region at the fractured bone site [5]. The bone remodeling phase recapitulates embryonic bone development with a combination of cellular proliferation and differentiation, increasing the cellular volume and matrix deposition [1]. Finally, remodeling of the hard callus into a lamellar bone structure occurs (bone remodeling phase).

The biological process occurring during bone fracture healing is regulated by several signaling molecules. Hedgehog (HH) proteins are among the signaling molecules required for endochondral bone formation during embryonic development, and they regulate bone homeostasis by controlling MSC proliferation [6,7]. HH signaling is also involved in the regulation of MSC proliferation in adult tissues. Aberrant activation of HH pathways has been linked to multiple types of human cancer [7]. These pathways are also activated during intramembranous and endochondral ossification in the fracture healing process, but it is not clear if they are involved in the healing process [5]. HH signaling pathways play critical roles in developmental processes and in the postnatal homeostasis of many tissues, including bone and cartilage. The HH family of intercellular signaling proteins plays important roles in regulating the development of many tissues and organs. Their name is derived from the observation of a hedgehog-like appearance in *Drosophila* embryos with genetic mutations that block their action. Three types of HH proteins have been reported in mammals, namely Sonic HH (Shh), Indian HH (Ihh), and Desert HH (Dhh). Ihh is up-regulated during the initial stage of fracture repair, and it regulates differentiation indirectly by controlling cartilage development at the fracture site. Ihh regulates osteoblast differentiation indirectly by controlling cartilage development [8]. In general, Shh acts in the early stages of development to regulate patterning and growth [9]. Recently, several studies reported that Shh might be related to fracture healing [10,11]. Following the inactivation of HH signaling, the activity of Smo is inhibited by a receptor known as Patched (Ptch). Binding of the HH ligand Ptch relieves the inhibition of Smo, and activated Smo blocks the proteolysis of Gli proteins in the cytoplasm and promotes their dissociation from suppressor of fused (SuFu). Following dissociation from SuFu, activated Gli proteins translocate into the nucleus and promote the expression of Hh target genes, including *Gli1* [9,12]. Gli1 positivity has been identified as a marker for MSCs [13]. Another study uncovered that Gli1 is involved in osteoblast differentiation [14]. However, it is unclear that whether Shh proteins are involved in fracture healing. In this study, we demonstrated that Shh protein and the related proteins Smo and Gli1 were involved in osteoblast differentiation at the fracture healing site via immunohistochemical analysis.

2. Results and Discussion

In this study, we hypothesized that Shh is related to the healing process of fractures and investigated and compared the positive localization of Runx2 and Osx, which appear during the fracture repair process, with that of Shh and its downstream factor Gli1. Runt-related transcription factor 2 (Runx2), which is an essential factor for bone formation, is expressed very early in skeletal development. Osterix (Osx) is activated downstream of Runx2 during osteoblastic lineage differentiation [15,16]. On the day of fracture (day 0), a few Runx2-positive and Osx-positive cells were observed on the bone surface in the periosteum (Figure 1a,c). Shh-positive cells were rarely observed in the periosteum on day 0 (Figure 1b). Furthermore, Gli1-positive cells were also rarely observed (Figure 1d). However, Shh and Gli1 positivity were localized to osteocytes in the bone matrix. These results indicated that Shh signaling occurred in osteocytes but not in undifferentiated cells in the periosteum. Moreover, in this study, we traced the fate of Gli1-positive cells in Gli1-Cre recombinase-mutated estrogen receptors (CreER^{T2}); tdTomato mice on day seven after fracture by administering tamoxifen. Previous reports demonstrated that 3 days are required for Cre activation after tamoxifen administration [8]. In our genetically modified mouse system, both Gli1-positive cells and their progeny were permanently marked by red fluorescent protein expression. Gli1-CreER^{T2}; tdTomato mice, which are transgenic for the *Gli1-CreER^{T2}/Rosa26-loxP-stop-loxP-tdTomato* gene, were used to generate Gli1-positive and progeny cells through lineage-tracing analysis. Gli1-positive cells expressed the CreER^{T2}. CreER^{T2} has the function of specifically recognizing and removing the LoxP site. Moreover, CreER^{T2} binds to tamoxifen but not to natural estrogens. Gli1-positive cells were found to express tomato red fluorescence after tamoxifen administration. Since tomato fluorescence is expressed permanently, not only Gli1-positive cells but also progeny cells were found to express tomato red fluorescence [17].

New cartilage matrix formed around the fracture site (Figure 1e). The localization of Gli1 progeny cells and Osx-positive cells was examined at the fracture healing site (Figure 1e*). Gli1 positivity was observed in chondrocytes and the perichondrium around the cartilage matrix at the fracture site (Figure 1e*). Osx positivity was localized in cells surrounding new cartilage matrix. The merged image of Gli1-positive and Osx-positive areas indicated that most Osx-positive cells were co-localized with Gli1-positive cells (Figure 1e*). These results indicate that Gli-positive cells and their progeny cells might differentiate into osteoblasts after bone fracture. This result was consistent with another report that Gli1 marked a major skeletal progenitor pool contributing to both bone and cartilage formation during bone fracture healing in postnatal mice [8].

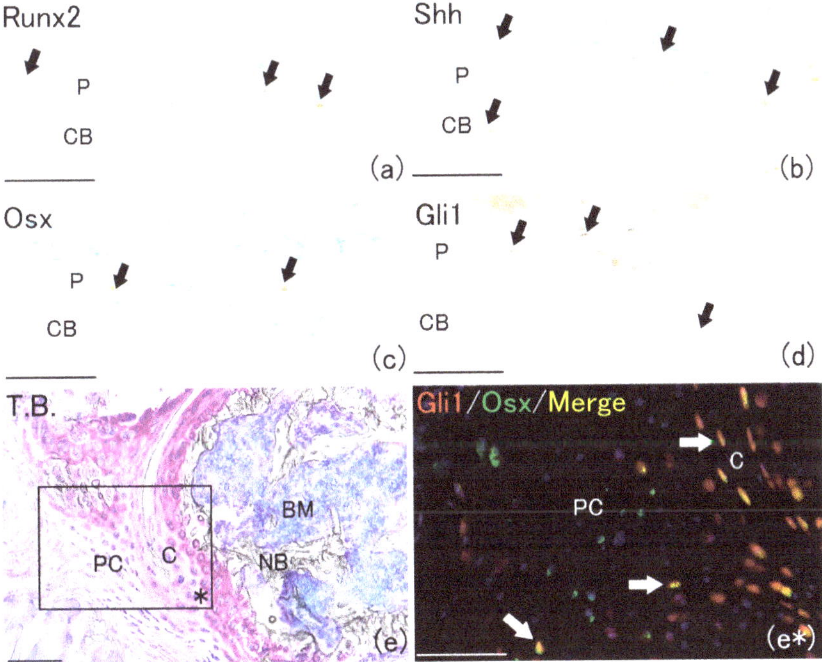

Figure 1. Histological analysis at day 0 on rat lib bone and at day seven on mouse lib bone fracture. (**a**) Runx2-positive cells were rarely observed at the surface of bone matrix in the periosteum (arrows). Scale bar: 50 μm. (**b**) Shh-positive expression was observed at the surface of bone matrix and osteocyte (arrows). Scale bar: 50 μm. (**c**) Osx-positive cells were rarely observed at the surface of bone matrix in the periosteum and same localization as Runx2. Scale bar: 50 μm. (**d**) Gli1-positive expression was observed at the surface of bone matrix and osteocyte and same as Shh localization (arrows). Scale bar: 50 μm. (**e**) Cross section pictures of fracture site. Newly formed cartilage matrix was observed around new bone and bone marrow. Scale bar: 100 μm. (**e***) Gli1-positive cells were observed in perichondrium and formed new cartilage at the fracture site (red fluorescent cell). Osx-positive cells were localized around the new cartilage matrix (green fluorescent cell). The merged image of Gli1- and Osx-positive areas demonstrated that the most Osx-positive cells merged on Gli1-positive cells (arrows). Scale bar: 50 μm. (P: periosteum; CB: cortical bone; PC: perichondrium; C: cartilage; NB: new bone; BM: bone marrow).

On day 1, hematoma and granulation tissue were observed at the bone fracture gap. Runx2-positive and Osx-positive cells were observed extensively in the remaining periosteum near the fracture site (Figure 2a*,b*). However, few Runx2-positive and Osx-positive cells were noted in the intact periosteum far from the fracture site (Figure 2a**,b**). Runx2-positive and Osx-positive cell numbers in the periosteum near the fracture site were remarkably higher than those far from the fracture site (Figure 3A,B). These results indicated that MSCs committed to osteoblast or chondroblast differentiation participated in intramembranous and endochondral ossification only near the fracture site.

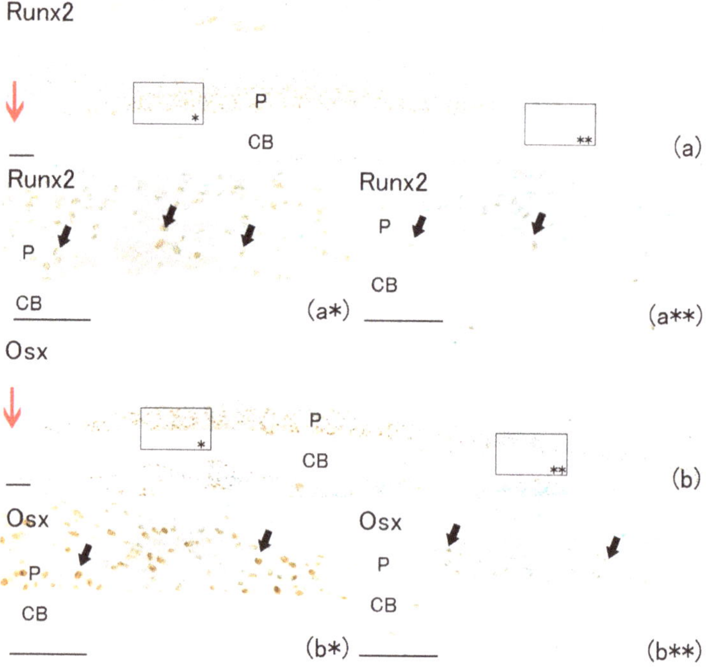

Figure 2. Histological analysis at day 1 on rat lib bone fracture (red arrow). (**a**) Runx2-positive cells were observed in remained periosteum. Scale bar: 100 μm. (**a***) Runx2-positive cells near the fracture site (arrows). Scale bar: 50 μm. (**a****) Runx2-positive cells were rarely observed far from bone fracture site. Scale bar: 50 μm. (**b**) Osx-positive cells were observed in remaining periosteum. Scale bar: 100 μm. (**b***) Osx-positive cells near the fracture site (arrows). Scale bar: 50 μm. (**b****) Osx-positive cells were rarely observed far from the bone fracture site. Scale bar: 50 μm. (P: periosteum; CB: cortical bone).

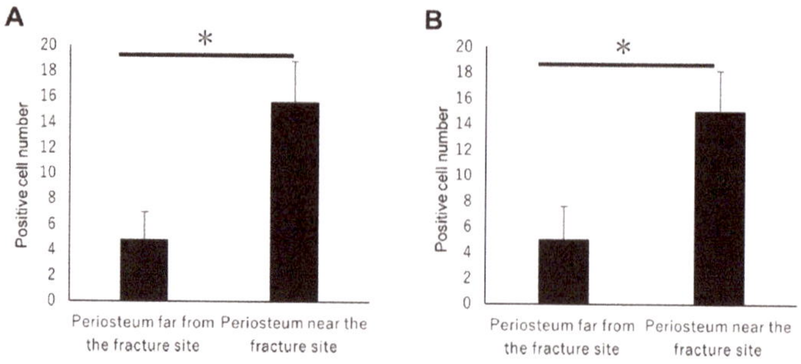

Figure 3. The positive cell number count on day 1. (**A**) Runx2-positive and (**B**) Osx-positive cell numbers in the periosteum near the fracture site were remarkably higher than those far from the fracture site. Data represent the mean ± SE (n = 3/group). Asterisks indicate the statistical significance of the differences (* $p < 0.05$, *t*-test).

Large numbers of Shh-positive and Gli1-positive cells were also found in the periosteum only near the fracture site (Figure 4a*,b*). Their numbers increased on day one in the periosteum adjacent to the fracture site compared with the number of Shh-positive and Gli1-positive cells on day 0 (Figure 5A,B). Shh-positive and Gli1-positive cell numbers in the periosteum near the fracture site were remarkably higher than those far from the fracture site (Figure 5A,B). These results indicate that Shh and Gli1, which emerge after bone fracture in the periosteum, might be associated with osteoblast differentiation.

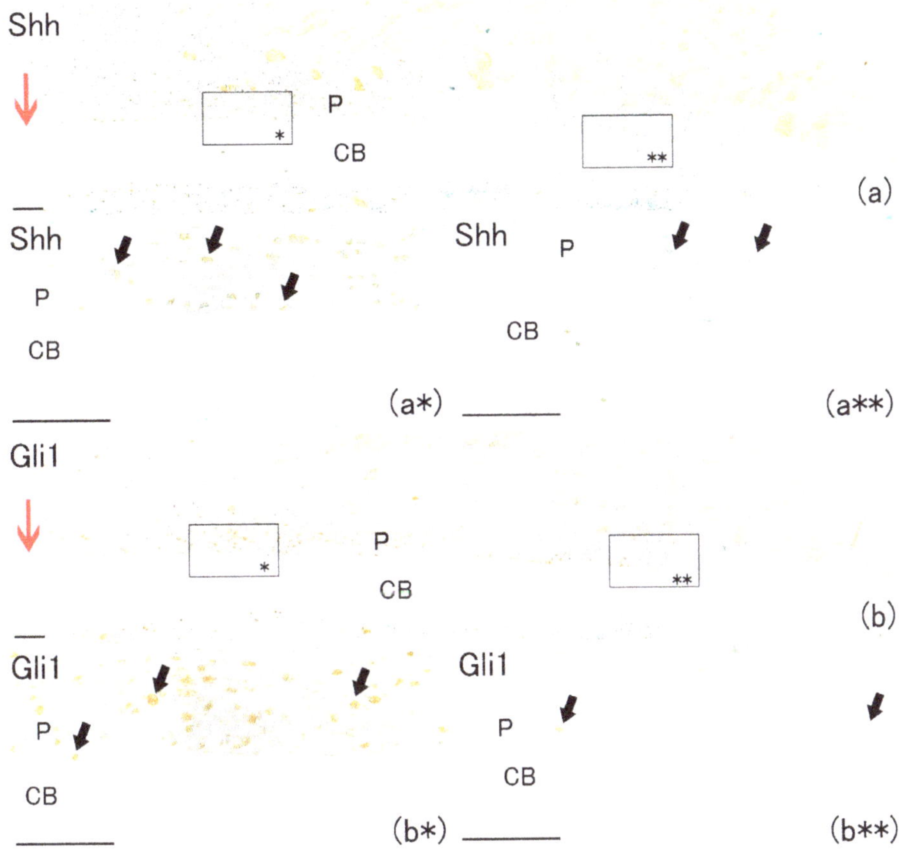

Figure 4. Histological analysis at day 1 on rat lib bone fracture (red arrow). (**a**) Shh-positive cells were observed in remaining periosteum. Scale bar: 100 μm. (**a***) Shh-positive expression near the fracture site (arrows). Scale bar: 50 μm. (**a****) Shh-positive cells were rarely observed far from the bone fracture site. Scale bar: 50 μm. (**b**) Gli1-positive cells were observed in remaining periosteum. Scale bar: 100 μm. (**b***) Gli1-positive cells near the fracture site (arrows). Scale bar: 50 μm. (**b****) Gli1-positive cells were rarely observed far from the bone fracture site. Scale bar: 50 μm.

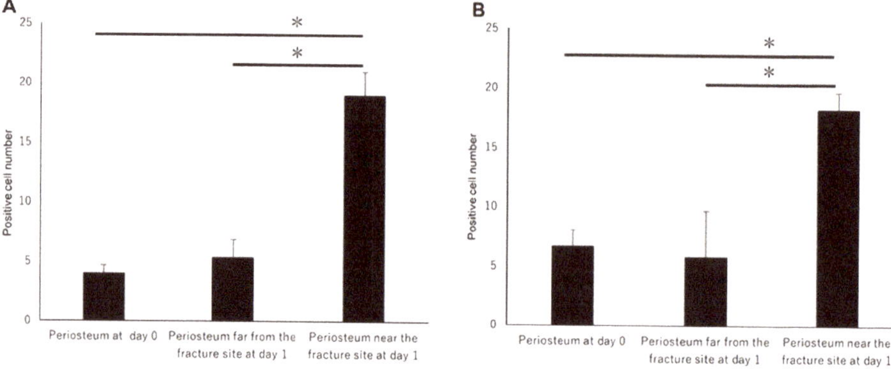

Figure 5. The positive cell number count on day 1. (**A**) Shh-positive and (**B**) Gli1-positive cell numbers in the periosteum near the fracture site were remarkably higher than those far from the fracture site. (A) Shh-positive and (B) Gli1-positive cell numbers in the periosteum near the fracture site at day 1 were higher than those at day 0. Data represent the mean ± SE (n = 3/group). Asterisks indicate the statistical significance of the differences (*p < 0.01, Tukey's test).

On day 7, a newly formed cartilage matrix was observed in the fracture gap (Figure 6a). It has been reported that endochondral ossification is observed in the fracture gap at that time [18,19]. Sox9 positivity was observed in chondrocytes as well as on the bone matrix surface (Figure 6b*). Runx2-positive and Osx-positive cells were observed on the surfaces of newly formed bone matrix and chondrocytes (Figure 6b**,c*,c**). Shh-positive and Gli1-positive cells were also observed on the surfaces of newly formed bone matrix and chondrocytes (Figure 6d*,d**,e*,e**). In addition, new bone matrix extending from the proximal and distal edges of the fractured bone surface, which is termed intramembranous ossification, was observed (Figure 6a). A large number of Osx-positive cells were also observed on the surface of newly formed bone extending from the proximal and distal edges. Osx-positive cells were localized on the new bone surface around the newly formed cartilage matrix (Figure 6c*). This result is consistent with another report in which MSCs directly differentiated into osteoblasts in the perichondrium around the cartilage matrix after bone fracture, resulting in bone formation [20]. Shh-positive and Gli1-positive cells localized along the surface of newly formed bone (Figure 6d*,d**,e*,e**). These results indicate that the Shh–Gli1 signaling pathway might regulate intramembranous and endochondral ossification at the fracture site.

On day 14, newly formed cartilage matrix at the fracture site began to resorb, and it was replaced by newly formed bone known as primary bone (Figure 7a) [21]. In the resorbed cartilage area, many Osx-positive and cathepsin K (CathK)-positive cells were observed (Figure 7b,c). Positivity for osteopontin (OPN), which is a bone matrix component, was observed around the resorbed cartilage matrix (Figure 7d) [22]. Shh and Gli1 positivity localized around the resorbed cartilage matrix (Figure 7e,f). These results indicate that the Shh–Gli1 signaling pathway participates in new bone formation by osteoblasts.

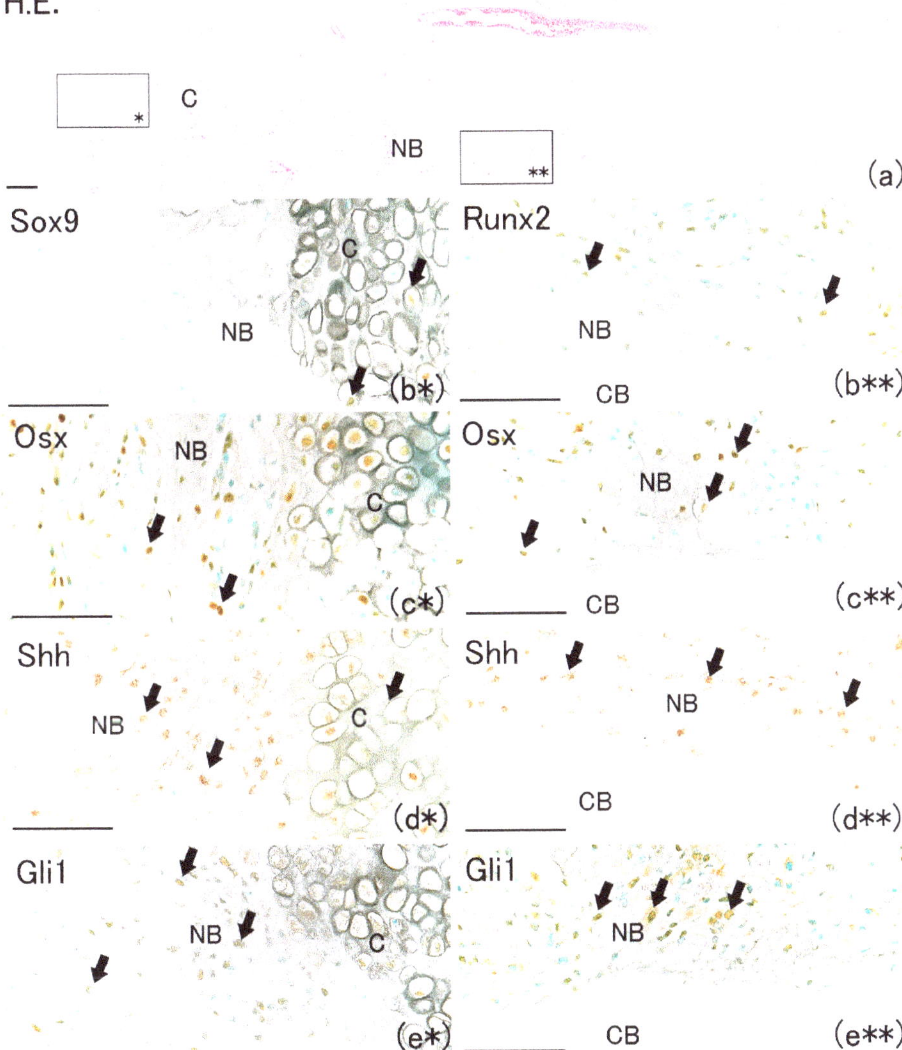

Figure 6. Newly formed cartilage and bone matrix were observed in the fracture site. (**a**) Newly formed cartilage matrix was observed in fracture site at day 7 on rat lib bone fracture. Scale bar: 100 µm. (**b***) Sox9-positive cells were observed in chondrocyte but not on the cell surface of bone matrix. Scale bar: 50 µm. (**c***) Osx-positive cells were observed at the surface of newly formed bone matrix and in chondrocyte (arrows). Scale bar: 50 µm. (**d***) Shh-positive expression was also around the new cartilage matrix (arrows). Scale bar: 50 µm. (**e***) Gli1-positive cells were also around the new cartilage matrix (arrows). Scale bar: 50 µm. (**b****) Runx2 positive cells at the surface of new bone but in the existing bone (arrows). Scale bar: 50 µm. (**c****) Osx-positive cells at the surface of new bone but in the existing bone (arrows). Scale bar: 50 µm; (**d****) Shh-positive expression at the surface of new bone but in the existing bone (arrows). Scale bar: 50 µm. (**e****) Gli1-positive cells at the surface of new bone but in the existing bone (arrows). Scale bar: 50 µm. (CB: cortical bone; NB: newly formed bone; C: cartilage).

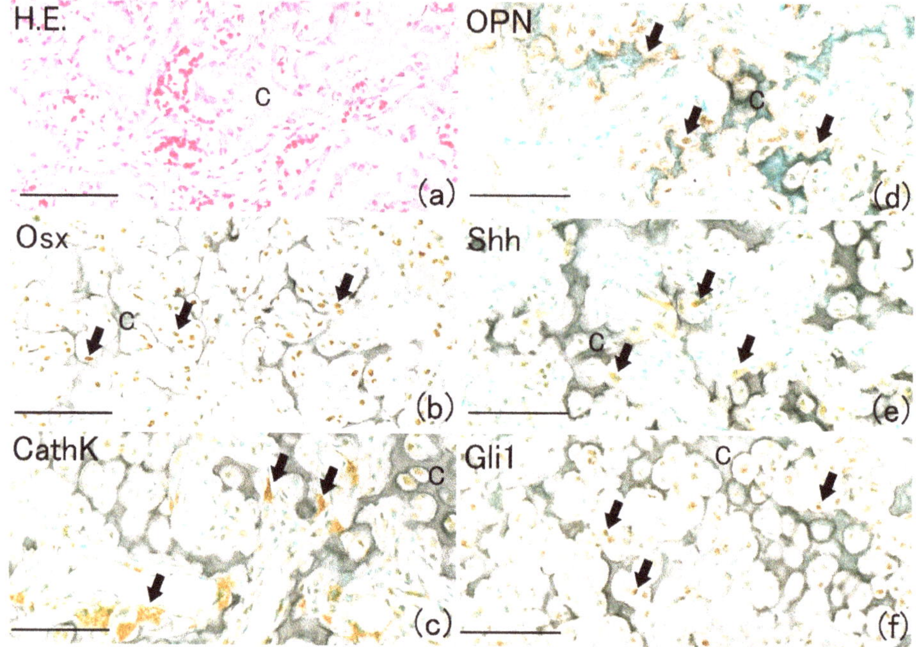

Figure 7. Histological analysis at day 14 on rat lib bone fracture. (**a**) Reduction of formed cartilage matrix began to resorb and be replaced by new bone formation on rat lib bone fracture. (**b**) Osx-positive cells were observed around reduced cartilage matrix (arrows). (**c**) Localization of CathK-positive cells was around remaining cartilage matrix. (**d**) OPN-positive area was observed around the cartilage matrix. (**e**) Shh-positive areas localized around cartilage matrix (arrows). (**f**) Gli1-positive areas found around cartilage matrix (arrows). (NB: newly formed bone; C: cartilage; Scale bar: 50 μm).

3. Materials and Methods

3.1. Experimental Animals

Twenty-six-week-old male Sprague–Dawley rats (Hokudo, Sapporo, Japan) and three Gli-CreER^{T2}; tdTomato male mice (Jackson Laboratory, Bar Harbor, ME, USA) were used. All experimental animals were maintained in a specific pathogen-free facility. All experiments were approved and performed according to guidelines set forth by the Animal Ethics Committee of the Health Sciences University of Hokkaido (The ethical permission code and permission date: 19-028, 8 March, 2019 and 19-045, 29 March 2019).

3.2. Tamoxifen Administration

Gli1-CreER^{T2}; tdTomato mice were injected intraperitoneally with tamoxifen (Sigma-Aldrich, St. Louis, MO, USA) once daily for 3 consecutive days (40 mg/mL, dissolved in corn oil).

3.3. Fracture Experiment

The right eighth rib of each experimental animal was fractured as previously described [10]. Briefly, each experimental animal was anesthetized, and the eighth rib on the right side was exposed and cut vertical to the axis with scissors. As a control, the right eighth rib of select animals was similarly exposed but not fractured.

3.4. Tissue Preparation

The animals were anesthetized subcutaneously with pentobarbital sodium (40 mg/kg) and killed via cervical dislocation. The ribs of Gli1-CreER^{T2}; tdTomato mice were collected 7 days after fracture and immediately frozen at $-80\,^{\circ}$C. Each sample was embedded in 5 % carboxymethyl cellulose (CMC) gel (Section-Lab Co. Ltd., Tokyo, Japan). Each frozen CMC sample was covered with polyvinylidene chloride film (Section-Lab Co. Ltd.) and sagittally sectioned at a thickness of 5 µm. The ribs of rats were collected 0, 1, 7, and 14 days after fracture and fixed in 4.0% paraformaldehyde in 0.1 M phosphate buffer (pH 7.4) overnight at 4 $^{\circ}$C. Specimens were demineralized via immersion in 10% ethylenediaminetetraacetic acid (pH 7.4) for 4 weeks at 4 $^{\circ}$C. Following demineralization, the specimens were embedded in paraffin and sectioned at a thickness of 5 µm.

3.5. Immunohistochemistry

For immunohistochemistry, the dehydrated sections were treated with 0.3% H_2O_2 in phosphate-buffered saline (PBS; pH 7.4) for 30 min at room temperature to inactivate endogenous peroxidase. Sections were pretreated with 3% bovine serum albumin in PBS for 30 min at room temperature, followed by incubation with primary antibodies against Shh (1:100, Bioss, Woburn, MA, USA), Gli1 (1:100, Novus Biologicals, Centennial, CO, USA), Osx (1:1000, Abcam, Cambridge, MA, USA), Runx2 (1:1000, MBL, Nagoya, Japan), and OPN (1:100, antibody was kindly provided by Dr. Hiroaki Nakamura, Matsumoto Dental University, Japan) overnight at 4 $^{\circ}$C. Sections were reacted with Histofine Simple Stain rat MAX-PO (MULTI; Nichirei, Tokyo, Japan) for 1 h at room temperature. Color was developed using liquid diaminobenzidine substrate-chromogen system (Dako, Carpinteria, CA, USA). Immunostained sections were then counterstained with methylene green.

3.6. Image Analysis

The 500 µm portion from the fracture line of the cortical bone to the midshaft was defined as the periosteum near the fracture site, and the 2000 µm portion from the fracture line of the cortical bone to the midshaft was defined as the periosteum far from the fracture site. The number of positive cells of Runx2, Osx, Shh, and Gli1 present in the periosteum near and far from the fracture were counted by defining a square (100 × 100 µm^2).

3.7. Statistical Analysis

The statistical analysis of the data gathered from Runx2-, Osx-, Shh-, and Gli1-positive cell counting was performed using SPSS version 23 (SPSS Inc., Chicago, IL, USA). Analyses of variance were followed by the *t*-test and Tukey's test to determine significance.

4. Conclusions

To date, it has been demonstrated that Ihh participates in fracture healing by promoting chondrocyte differentiation. Ihh signaling critically regulates osteoblast differentiation during endochondral bone development after bone fracture [19,23]. Our results demonstrate that the Shh–Gli1 signaling pathway is involved in intramembranous and endochondral ossification during the fracture healing process.

Author Contributions: Conceptualization, K.I. and T.S.; methodology, H.T. and N.S.; validation, A.H., N.S. and T.S.; formal analysis, H.T.; investigation, H.T.; resources, T.S.; writing—original draft preparation, H.T. and A.H.; writing—review and editing, H.T. All authors have read and agreed to the published version of the manuscript.

Funding: This work was supported by JSPS KAKENHI Grant Number JP18K17090.

Acknowledgments: The authors would like to thank Hiroaki Nakamura, Professor of Dentistry, Matsumoto Dental University.

Abbreviations

Ihh	Indian hedgehog
Shh	Sonic hedgehog
Osx	Osterix
Runx2	Runt-related transcription factor 2
MSCs	Mesenchymal stem cells
HH	Hedgehog
Ptch	Patched
SuFu	Suppressor of fused
P	Periosteum
CB	Cortical bone
C	Cartilage
T.B.	Toluidine blue
PC	Perichondrium
NB	Newly formed bone
BM	Bone marrow
CreER^{T2}	Cre recombinase-mutated estrogen receptor
CathK	Cathepsin K
OPN	Osteopontin

References

1. Marsell, R.; Einhorn, T.A. The biology of fracture healing. *Injury* **2011**, *42*, 551–555. [CrossRef]
2. Williams, J.N.; Kambrath, A.V.; Patel, R.B.; Kang, K.S.; Mével, E.; Li, Y.; Cheng, Y.H.; Pucylowski, A.J.; Hassert, M.A.; Voor, M.J. Inhibition of CaMKK2 enhances fracture healing by stimulating Indian hedgehog signaling and accelerating endochondral ossification. *J. Bone Miner. Res.* **2018**, *33*, 930–944. [CrossRef] [PubMed]
3. Ai-Aql, Z.; Alagl, A.S.; Graves, D.T.; Gerstenfeld, L.C.; Einhorn, T.A. Molecular mechanisms controlling bone formation during fracture healing and distraction osteogenesis. *J. Dent. Res.* **2008**, *87*, 107–118. [CrossRef] [PubMed]
4. Dimitriou, R.; Tsiridis, E.; Giannoudis, P.V. Current concepts of molecular aspects of bone healing. *Injury* **2005**, *36*, 1392–1404. [CrossRef] [PubMed]
5. Einhorn, T.A.; Gerstenfeld, L.C. Fracture healing: Mechanisms and interventions. *Nat. Rev. Rheumatol.* **2015**, *11*, 45–54. [CrossRef]
6. Mak, K.K.; Bi, Y.; Wan, C.; Chuang, P.-T.; Clemens, T.; Young, M.; Yang, Y. Hedgehog signaling in mature osteoblasts regulates bone formation and resorption by controlling PTHrP and RANKL expression. *Dev. Cell* **2008**, *14*, 674–688. [CrossRef]
7. Varjosalo, M.; Taipale, J. Hedgehog signaling. *J. Cell Sci.* **2007**, *120*, 3–6. [CrossRef]
8. Shi, Y.; He, G.; Lee, W.-C.; McKenzie, J.A.; Silva, M.J.; Long, F. Gli1 identifies osteogenic progenitors for bone formation and fracture repair. *Nat. Commun.* **2017**, *8*, 2043. [CrossRef]
9. Yang, J.; Andre, P.; Ye, L.; Yang, Y.-Z. The Hedgehog signalling pathway in bone formation. *Int. J. Oral Sci.* **2015**, *7*, 73–79. [CrossRef]
10. Horikiri, Y.; Shimo, T.; Kurio, N.; Okui, T.; Matsumoto, K.; Iwamoto, M.; Sasaki, A. Sonic hedgehog regulates osteoblast function by focal adhesion kinase signaling in the process of fracture healing. *PLoS ONE* **2013**, *8*, e76785. [CrossRef]
11. Edwards, P.C.; Ruggiero, S.; Fantasia, J.; Burakoff, R.; Moorji, S.; Paric, E.; Razzano, P.; Grande, D.; Mason, J. Sonic hedgehog gene-enhanced tissue engineering for bone regeneration. *Gene Ther.* **2005**, *12*, 75–86. [CrossRef]
12. Niewiadomski, P.; Niedziolka, S.M.; Markiewicz, L.; Uspienski, T.; Baran, B.; Chojnowska, K. Gli Proteins: Regulation in Development and Cancer. *Cells* **2019**, *8*, 147. [CrossRef]
13. Zhao, H.; Feng, J.; Ho, T.-V.; Grimes, W.; Urata, M.; Chai, Y. The suture provides a niche for mesenchymal stem cells of craniofacial bones. *Nat. Cell Biol.* **2015**, *17*, 386–396. [CrossRef]

14. Kitaura, Y.; Hojo, H.; Komiyama, Y.; Takato, T.; Chung, U.-I.; Ohba, S. Gli1 haploinsufficiency leads to decreased bone mass with an uncoupling of bone metabolism in adult mice. *PLoS ONE* **2014**, *9*, e109597. [CrossRef]

15. Franceschi, R.; Ge, C.; Xiao, G.; Roca, H.; Jiang, D. Transcriptional regulation of osteoblasts. *Ann. N. Y. Acad. Sci.* **2007**, *1116*, 196–207. [CrossRef]

16. Hirata, A.; Sugahara, T.; Nakamura, H. Localization of runx2, osterix, and osteopontin in tooth root formation in rat molars. *J. Histochem. Cytochem.* **2009**, *57*, 397–403. [CrossRef]

17. Kos, C.H. Methods in nutrition science: Cre/loxP system for generating tissue-specific knockout mouse models. *Nutr. Rev.* **2004**, *62*, 243–246. [CrossRef]

18. Hu, D.P.; Ferro, F.; Yang, F.; Taylor, A.J.; Chang, W.; Miclau, T.; Marcucio, R.S.; Bahney, C.S. Cartilage to bone transformation during fracture healing is coordinated by the invading vasculature and induction of the core pluripotency genes. *Development* **2017**, *144*, 221–234. [CrossRef]

19. Murakami, S.; Noda, M. Expression of Indian hedgehog during fracture healing in adult rat femora. *Calcif. Tissue Int.* **2000**, *66*, 272–276. [CrossRef]

20. Asaumi, K.; Nakanishi, T.; Asahara, H.; Inoue, H.; Takigawa, M. Expression of neurotrophins and their receptors (TRK) during fracture healing. *Bone* **2000**, *26*, 625–633. [CrossRef]

21. Fazzalari, N. Bone fracture and bone fracture repair. *Osteoporos. Int.* **2011**, *22*, 2003–2006. [CrossRef]

22. Sodek, J.; Ganss, B.; McKee, M. Osteopontin. *Crit. Rev. Oral Biol. Med.* **2000**, *11*, 279–303. [CrossRef]

23. Li, S.; Xiang, C.; Wei, X.; Li, H.; Li, K.; Sun, X.; Wang, S.; Zhang, M.; Deng, J.; Wang, X. Knockdown Indian Hedgehog (Ihh) does not delay Fibular Fracture Healing in genetic deleted Ihh mice and pharmaceutical inhibited Ihh Mice. *Sci. Rep.* **2018**, *8*, 10351. [CrossRef]

 International Journal of
Molecular Sciences

Review

Sonic Hedgehog Signaling in Organogenesis, Tumors, and Tumor Microenvironments

Kuo-Shyang Jeng *, Chiung-Fang Chang and Shu-Sheng Lin

Department of Surgery and Medical Research, Far Eastern Memorial Hospital, New Taipei 220, Taiwan;
cfchang.gina@gmail.com (C.-F.C.); ffcc840922@gmail.com (S.-S.L.)
* Correspondence: kevin.ksjeng@gmail.com; Tel.: +886-2-8966-7000 (ext. 1611)

Received: 27 December 2019; Accepted: 20 January 2020; Published: 23 January 2020

Abstract: During mammalian embryonic development, primary cilia transduce and regulate several signaling pathways. Among the various pathways, Sonic hedgehog (SHH) is one of the most significant. SHH signaling remains quiescent in adult mammalian tissues. However, in multiple adult tissues, it becomes active during differentiation, proliferation, and maintenance. Moreover, aberrant activation of SHH signaling occurs in cancers of the skin, brain, liver, gallbladder, pancreas, stomach, colon, breast, lung, prostate, and hematological malignancies. Recent studies have shown that the tumor microenvironment or stroma could affect tumor development and metastasis. One hypothesis has been proposed, claiming that the pancreatic epithelia secretes SHH that is essential in establishing and regulating the pancreatic tumor microenvironment in promoting cancer progression. The SHH signaling pathway is also activated in the cancer stem cells (CSC) of several neoplasms. The self-renewal of CSC is regulated by the SHH/Smoothened receptor (SMO)/Glioma-associated oncogene homolog I (GLI) signaling pathway. Combined use of SHH signaling inhibitors and chemotherapy/radiation therapy/immunotherapy is therefore key in targeting CSCs.

Keywords: hedgehog; smoothened; cancer stem cells

---

1. Introduction

During mammalian embryonic development, primary cilia with microtubule-based cellular organelles protrude from the surface of the cell [1]. Primary cilia defects cause "ciliopathies", adversely affecting the development of brain, kidneys, eyes, liver, and other organs. Acting as cellular antenna, primary cilia transduce and regulate several signaling pathways such as Sonic hedgehog (SHH) and Wingless-activated (WNT) [1]. Among the various pathways, SHH is one of the most significant.

SHH signaling remains quiescent in adult mammalian tissues. However, in multiple adult tissues, it becomes active during differentiation, proliferation, and maintenance [2]. Moreover, aberrant activation of SHH signaling occurs in cancers of skin, brain, liver, gallbladder, pancreas, stomach, colon, breast, lung, prostate, and hematological malignancies [3].

Tumor microenvironment/stroma can affect tumor development and metastasis [4,5]. The tumor microenvironment/stroma includes endothelial cells, immune cells, adipocytes, and activated fibroblasts (the so-called "cancer-associated fibroblasts" (CAFs)) [6]. CAFs fuel cancer cells via secreting soluble factors to trigger metastasis and chemoresistance. These triggers include extracellular acidification, inflammation, activation of matrix metalloproteases, and decreased efficacy of chemotherapeutic drugs [7–12]. In addition, SHH produced by CAFs could regulate the microenvironment for cancer progression [13]. The details of tumor microenvironment in each organ are discussed in the following sections.

This article reviews the recent studies of the role of SHH in organogenesis, tumors, and tumor microenvironments.

2. Sonic Hedgehog Signaling Pathway Includes Canonical and Non-Canonical Pathways

The SHH signaling pathway could be categorized into canonical and non-canonical pathways. The canonical SHH signal transduction pathway consists of main components such as the Patched receptor (PTCH1, PTCH2), a 12-domain transmembrane receptor, the Smoothened receptor (SMO), a 7-domain transmembrane receptor coupled to G protein-coupled receptor (GPCR), and negative regulatory protein suppressor of fused homolog (SUFU) and the Glioma-associated oncogene homolog (GLI) family of transcription factors (GLI1, GLI2 and GLI3) [14]. Tumors produce ligands to activate the SHH pathway in an autocrine/juxtacrine manner. SHH ligands remove the inhibition of SMO by PTCH. SMO can activate GLI to regulate target gene expression and affect migration/invasion, cell cycle, tumor growth, and cancer stem cells. Moreover, paracrine Hedgehog (HH) signaling is important for epithelial cancers [15]. HH ligands secreted by tumor cells activate the signaling in the surrounding stroma, which provides a favorable microenvironment for tumor growth.

The non-canonical SHH signaling could be classified into three types, including (1) PTCH-mediated, (2) SMO-dependent/GLI-independent, and (3) SMO-independent GLI activation [16,17]. A simplified, broad, non-canonical SHH signal is defined as any related SHH signaling pathway component that differs from the usual canonical SHH signaling pattern [18]. For example, activation of SMO or GLI may occur via other signaling pathways such as Protein kinase A (PKA), Guanosine triphosphatase (GTPase), Phosphoinositide 3-kinase (P13K)/mammalian target of rapamycin (mTOR), or Rho, to drive target gene expression. The detailed mechanism of non-canonical SHH signal transduction still remains elusive, but probably acts as an alternative activation pathway when the canonical SHH signal transduction does not work functionally, or when transduction needs to "escape" the canonical SHH signaling pathway during cytotoxic or inflammatory stress [16].

3. SHH in Organogenesis, Vasculogenesis, and Angiogenesis

SHH signaling is essential for cell growth and tissue patterning. The pathway involves the development of neural tube, lung, skin, axial skeleton, gastrointestinal tract, pancreas, and other organs, as well as the regulation of tissue homeostasis and stem cell behavior [19]. During brain development, enhancer Shh brain enhancer 7 (SBE7) could not only initiate Shh expression but also induce Shh, which controls craniofacial morphogenesis and the etiology of holoprosencephaly [20].

SHH signaling induces endothelial cells and connective tissue support cells to release proangiogenic factors (Ang1, Ang2, and VEGF) [21,22]. However, some studies suggest that the response of SHH is limited to mesothelial and smooth muscle cells but not to endothelial cells [23]. Geng et al. [24] mentioned that SHH signaling affects vasculogenesis and angiogenesis. Others have found that the lung displays decreased vascularization in SHH-deficient mice [25]. Overexpression of SHH induces hypervascularization of the neuroectoderm during the development of mouse embryos [26]. Mutant zebrafish with deficient SHH signaling develop abnormal circulation and vascularization such as a single axial vessel with no arterial markers [27,28]. Though SHH is important in embryonic vessel formation, its role in tumor vasculature remains unclear.

4. SHH in Organogenesis of Forebrain and Cerebellum and in Medulloblastoma

4.1. SHH in Organogenesis of Forebrain and Cerebellum

During brain development, SHH signaling plays an essential role in two phases. The prechordal plate elicits early ShHH signaling to overlay the prechordal plate. Then, the triggered SHH signaling affects the late neuronal differentiation of the forebrain. The prechordal enhancer *Shh brain enhancer 7* (*SEB7*) regulates SHH signaling to affect the development and growth of the forebrain [20].

Primary cilia are essential for cerebellar differentiation, and the proliferation of neuronal granule precursors is SHH-dependent [29]. Additionally, SHH orchestrates the development and maturation of the cerebellum [30]. During embryogenesis, activation of SHH signaling occurs in the ventricular germinal zone (VZ) and regulates the proliferation of VZ-derived progenitors. Purkinje cells also secret

SHH, and SHH can sustain the amplifications of the postnatal neurogenic niches (consisting of the external granular layer, the white matter, the excitatory granule cells, and inhibitory interneurons). During development, SHH signaling plays a role in Bergmann glial differentiation and facilitates the foliation of the cerebellum [30]. The mammalian GPCR 37-like 1 exists specifically in cerebellar Bergmann glia astrocytes and affects the proliferation and differentiation of postnatal cerebellar granules as well as Bergmann glia and Purkinje neuron maturation [29].

SHH signaling affects fissure formation. GLI2 activates SHH-induced granule cell progenitor proliferation and drives initial fissuring [29]. In the late prenatal or postnatal stages, SHH sustains the expansion of the external granular layer [30,31].

4.2. SHH in Medulloblastoma

Grausam et al. reported that medulloblastoma (MB) is the most common brain malignancy in pediatric patients, carrying a high mortality of up to 30% and high heterogeneity [32]. This disease arises from the cerebellum and is associated with early leptomeningeal metastasis, recurrence, and poor prognosis. According to their transcriptional profiles, MB can be classified into four subtypes: (1) WNT-MBs, (2) SHH-MBs, (3) Group C with abnormal transforming growth factor 1 beta (TGF1β) pathway, and (4) Group D with tandem duplication of a-synuclein-interacting protein [33–38]. SHH-MBs occur in both children and adults [39,40]. The transcriptional and genetic profiles are differently expressed in infant MBs and adult MBs. SHH-MBs are correlated with aberrations of the components of the SHH pathway (PTCH1, SUFU, GLI transcription factors and SMO) [30,41–46]. In particular, several SMO mutations involved in MB tumorigenesis have been found (L225R, N223D, S391N, D338N, D477G, D473H, G457S) [47]. Grausam et al. found that SHH pathway inhibitors may decrease both the proliferation and metastasis of tumors in a mouse MB model [32]. Inhibition HH signaling by sonidegib (LED225) and vismodegib (GDC-0449) have anti-tumor activity in SHH-driven MB [48]. Sonidegib led to better objective response rates than vismodegib against SHH-driven MB among five different clinical trials.

4.3. SHH in the Microenvironment of Medulloblastoma (MB)

SHH signaling activity alone is not sufficient for advanced development of medulloblastoma. Brain tumors consist of tumors, stem-like cells, and tumor-associated components (stroma) including vascular cells, immune cells, astrocytes, microglia, and extracellular martrix. Some epithelial cancer tumors can trigger SHH signaling to the stroma which enhances tumor growth [49]. The SHH subgroup of MB could significantly increase the gene expression of tumor-associated macrophages [50]. Tumor-associated macrophages are abundantly present in SHH MBs. Patients with decreased macrophage count usually have significantly worse prognosis [51].

5. SHH in Organogenesis, Tumors, and Tumor Microenvironments of the Liver

5.1. SHH in Organogenesis of Liver

During embryogenesis, the SHH pathway plays an essential role in hepatic specification of endodermic progenitors [52]. Conversely, there is no activation of SHH pathway in the adult mature hepatocyte [53], and the presence of SHH, PTCH, and GLI1 in the normal adult liver is minimal. Moreover, during liver repair, both myofibroblasts and progenitors can produce and respond to SHH ligands. In addition, Hippo/Yes-associated protein (YAP1) is a downstream effector of HH signaling pathway for liver regeneration [54].

5.2. SHH in Liver Injury and Hepatocarcinogenesis

SMO regulates adult liver repair by enhancing epithelial–mesenchymal transition [55]. From animal studies, chronic liver injury has been found to activate the SHH pathway. After Fas-induced liver injury, SMO is upregulated in hepatocytes. HBx (HBV gene product HBx protein) transformation induces the activation of the SHH signaling pathway. Cai et al. found that the SHH signaling pathway is activated during hepatocarcinogenesis [52]. Activated SHH signaling facilitates the cell proliferation after enhancing the G2/M transition via an increase in cyclin B1 and cyclin-dependent kinase 1 (CDK1) [52]. SMO plays an essential role during early hepatocarcinogenesis [55]. Overexpression of SMO-mediated c-Myc affects hepatocarcinogenesis significantly [56]. In hepatocellular carcinoma (HCC) patients, SMO mutation at the C-terminal lysine (K575M) involves the binding between PTCH and SMO to alleviate SMO from PTCH suppression to activate the downstream signaling [57]. Liu et al. found that hypoxia-inducing oxidative stress, epithelial–mesenchymal transition (EMT), and activation of the non-canonical SHH signaling pathway aggravates the invasiveness of HCC cells [58]. Chen et al. found that the SHH pathway induces the migration and the invasion of HCC cells via activation of focal adhesion kinase (FAK)/P13K/AKT signaling-mediated matrix metalloproteinase-2, as well as matrix metalloproteinase-9 production [59]. Other authors found that the SMO inhibitor GDC-0499 could inhibit hepatocarcinogenesis in HBx transgenic mice [60].

The expression of SMO affects the prognosis of HCC patients [61]. The overexpression of SMO and an increased ratio of *SMO* mRNA to *PTCH* mRNA correlates with HCC size in patients [62]. Jeng et al. reported that the high expression of SHH signaling pathway molecules affects the risk of post-resection recurrence with HCC [62]. Wang et al. found that *SMO* polymorphisms in transplant recipients are associated with an increased risk of postoperative HCC recurrence [63].

5.3. SHH in the Microenvironment of Primary Liver Cancer

The most common primary liver cancers are hepatocellular carcinoma (HCC) and cholangiocarcinoma. HCC originates from hepatocytes and cholangiocarcinoma originates from bile duct cells.

The disruption or change of the liver microenvironment and immune cell composition mainly promotes the malignant transformation and progression of HCC [64]. When the liver regeneration microenvironment deteriorates, inflammation and vascular changes can occur to enhance hepatocarcinogenesis [65]. To improve the microenvironment in liver regeneration through a regulation of multi-component, multi-target, multi-level, multi-channel, and multi-timed factors, an updated strategy for liver cancer prevention or inhibition is required. Hedgehog signaling could promote tumor-associated macrophages (TAMs) and thereby lead to immunosuppression [66]. SMO expression in myeloid is required not only for HCC growth but also for M2 polarization of TAMs.

Cholangiocarcinoma remains the second most common primary malignancy of the liver. Razumilava et al. found that cholangiocarcinoma cells could express non-canonical SHH signaling with chemotaxis even when cilia function is impaired. The non-canonical SHH signaling pathway contributes to the progression of cholangiocarcinoma [67]. Fingas et al. described the use of cyclopamine (SMO inhibitor) as able to increase the apoptosis of cholangiocarcinoma cells. Cyclopamine also inhibited tumor growth and metastasis in a rodent model study [68]. El et al. noted that SHH signaling pathway inhibitors can enhance the necrosis of cholangiocarcinoma cell [69]. Moreover, myofibroblast-derived platelet-derived growth factor (PDGF)-BB-mediated cyto-protection in cholangiocarcinoma is dependent on the HH signaling pathway [68]. PDGF-BB could induce translocations of SMO to the plasma membrane. Therefore, SMO inhibitor could promote the apoptosis of cholangiocarcinoma cells as well as their metastasis.

6. SHH in Gallbladder Organogenesis, Gallbladder Cancer, and Tumor Microenvironment

6.1. SHH in Gallbladder Organogenesis

The genetic and in vitro studies found that the SHH signaling pathway is essential for the proper formation of smooth muscles downstream of *Sox17* in the development of the gallbladder during the late organogenesis periods [70].

6.2. SHH in Gallbladder Cancer

Matsushita et al. found a higher expression of SHH in human gallbladder cancer specimens compared to normal gallbladder tissue [71]. SMO inhibitors could inhibit the proliferation of cancer cells. Inhibition of gallbladder cancer cell invasiveness functions via the suppression of matrix metalloproteinase-2 (MMP-2) and MMP-9 and epithelial-mesenchymal transition [71]. SMO si-RNA-transfected gallbladder cancer cells underwent a decrease in tumor volume according to a xenograft study [71]. The expressions of SHH, PTCH, and GLI1 are upregulated in gallbladder cancer. Aberrant activation of SHH signaling protein could be found in chronic cholecystitis and gallbladder cancer. SHH expression increased in severe chronic cholecystitis but decreased after the progression to gallbladder cancer. High GLI1 expression is correlated with worse prognosis of gallbladder cancer [72]. Moreover, high expression of SHH-signaling molecules SHH, PTCH, and GLI are associated with poor survival in the gallbladder [73]. Several mutations of the *SHH* gene in gallbladder carcinoma could be identified and associated with the carcinogenesis [74].

6.3. SHH in the Microenvironment of Gallbladder Cancer

Patients with high levels of SHH-signaling molecules were found to be associated with unfavorable survival outcomes. It could be associated with inflammation states [75]. Inflammatory responses could drive cancer progression such as EMT, angiogenesis, and metastasis. However, the evidence for SHH in the microenvironment of gallbladder cancer is still required for the investigation.

7. SHH in Organogenesis of the Pancreas, Pancreatic Cancer, and Pancreatic Cancer Microenvironment and Cancer Stem Cells

7.1. SHH in Pancreas Organogenesis

During the development of the human pancreas, SHH signaling remains low in pancreatic progenitor cells. Between embryonic week 14 and 18, both SMO and GLI2 gradually start to accumulate in primary cilia [76,77]. Then, GLI3 becomes gradually lower in the nucleus and cytoplasm of ductal epithelial cells during pancreas development. SHH signaling is necessary for both proliferation and maturation of the pancreas [76,77].

7.2. SHH in Pancreatic Cancer

The primary cilium and receptor SMO have been demonstrated in the vessels and stromal fibroblasts of tumors, providing evidence of SHH signal pathway activation. One hypothesis has been proposed that states that the behavior of mesenchymal and endothelial cells is affected by SHH, which is secreted from pancreatic cancer in a paracrine manner [78,79]. Aberrant SHH expression occurs in the early stages and during the progression of pancreatic cancer. Expression then increases from pre-malignant to malignant lesions of the pancreas [78]. Kumar et al. also emphasized the essential role of SHH signaling in the development and metastasis of pancreatic cancer cells [80]. Niyaz et al. mentioned that dysregulated SMO could be a therapy target in pancreatic cancers [81].

Conversely, Tian et al. found that SMO expression in epithelial cells has no role in affecting the development of pancreatic cancer [82].

7.3. SHH in the Microenvironment of Pancreatic Cancer

Li et al. emphasized that the activity of the SHH pathway is low in normal pancreatic tissue, whereas in pancreatic adenocarcinoma, the activity of the SHH pathway signaling in tumor epithelia and surrounding stromal tissue becomes higher [79]. Saini et al. found that SHH is upregulated in both the stroma and epithelial compartments in poorly differentiated pancreatic ductal carcinoma [83]. Wang et al. suggested that tumor necrosis factor alpha and interleukin-1 beta in stromal hyperplasia could promote the growth of pancreatic ductal adenocarcinoma after the SHH pathway activation for both canonical and non-canonical models [84]. Bailey et al. supported the hypothesis that the pancreatic epithelia secretes SHH, which is essential in establishing and regulating the pancreatic tumor microenvironment to affect cancer progression [78]. A high expression of SHH enhances the size and metastasis of primary pancreatic tumors [78]. SHH ligands exhibited by pancreatic cancers promote tumor growth indirectly via SHH signaling activation in the surrounding stroma. This paracrine activation of SHH signaling in the tumor microenvironment affords an environment favorable for the proliferation, metastasis, and drug resistance of cancer cells [79]. Mouse models (subcutaneous and orthotopic implantation) using a pancreatic tumor cell line showed that cells can secrete SHH, and that expression of SHH was high in a transformed primary cell line. SHH significantly affects both microenvironment and tumor progression, and is a potential target to suppress the desmoplastic and metastatic processes involved in pancreatic cancer [78]. Rucki et al. suggested that dual inhibition of the SHH and hepatocyte growth factor (HGF) pathways in the stroma could significantly suppress cancer cell growth and metastasis [85].

The activated SHH in pancreatic tumors enhances angiogenesis, lymphangiogenesis, and metastasis to produce a pro-angiogenic effect and to promote metastasis in the stroma [78]. The effects on lymphangiogenesis on SHH are important and could affect the metastasis of pancreatic tumor cells to the lymph nodes [86]. Targeting the stroma of pancreatic cancer could improve drug delivery and inhibit both angiogenesis and lymphangiogenesis. Pitarresi et al. [87] proposed a mechanism of stromal fibroblasts enhancing pancreatic tumor cell growth. Tumors secrete SHH, which activates the SHH pathway in pancreatic fibroblasts [88]. The fibroblasts in the tumor microenvironment then promote tumor growth via the disruption of paracrine SHH signaling. Some investigators found that SHH antagonists may successfully suppress tumor growth in xenograft tumors [88], whereas Pitarresi et al. knocked out SMO in fibroblasts to enhance tumor growth. Pitarres et al. found that the SMO gene in stromal fibroblasts affected the proliferation of pancreatic cancer cells [87]. In turn, deletion of SMO could activate oncogenic protein kinase B in the fibroblasts [87].

However, Tian et al. found that only the tumor stroma is competent in transducing the SHH signal, given that SMO is activated in the mesenchyme [82]. These researchers used a mouse model of pancreatic cancer and found that SHH signaling activation is present in the SHH-expressing tumor epithelium surrounding the stroma. Using quantitative RT-PCR to examine tissue samples of both primary or metastatic human pancreatic cancer, activation of the SHH pathway in the tumor stroma was found. Researchers have suggested that SHH-mediated tumorigenesis is a paracrine model. Pancreatic tumor cells could secrete SHH ligand to induce the SHH target genes in the adjacent stroma, thus promoting tumor growth [82].

Rhim et al. found that SHH-deficient tumors with reduced stromal content became more aggressive. These tumors presented undifferentiated histology, increased vascularity, and heightened proliferation. The administration of vascular endothelial growth factor receptor (VEGFR) blocking antibody could improve the survival of SHH-deficient tumors, meaning that SHH-driven stroma inhibits tumor growth partly via restraining tumor angiogenesis [89]. Pitarresi et al. analyzed fibroblasts in a sample of patients with pancreatic cancer, demonstrating heterogeneous patterns of expression in the components of the stromal fibroblast [87]. They also found that patients with decreased stromal phosphatase and tensin homologs usually led to a worse prognosis. These data established the potential to modulate pancreatic cancer stroma for targeted therapy [87].

After SMO deletion, fibroblasts overexpressed *transforming growth factor-alpha* (*TGF-α*) mRNA, and TGF-α protein, resulting in activation of epidermal growth factor receptor signaling in acinar cells and in acinar-ductal metaplasia [90]. This means that a non-cell-autonomous mechanism could modulate Kras G12D-driven acinar-ductal metaplasia. Such a phenomena could be balanced through cross-talk between the SHH/SMO pathway and alpha serine/threonine-protein kinase/GLI2 pathways in the stromal fibroblasts [90]. Zhou et al. demonstrated that using SMO-positive pancreatic cancer cells, GDC-0449 could downregulate SHH signaling genes and reverse fibroblast-induced resistance to doxorubicin [91]. Liu et al. found that genetic ablation of SMO in stromal fibrosis could disrupt the paracrine SHH signaling with acinar-ductal metaplasia under a Kras G12D mouse model [90].

Kumar et al. designed a novel GDC-0449 analog 2-chloro-N ^{1}-[4-chloro-3-(2-pyridinyl)phenyl]-N ^{4},N ^{4}-bis(2-pyridinylmethyl)-1,4-benzenedicarboxamide (MDB5) to inhibit pancreatic cancer [80]. Using a mouse model, Olive et al. reported that SMO inhibitor enhances the vasculature within the tumor and facilitates the delivery of chemotherapy agents to pancreatic cancer [92]. Von Ahrens et al. emphasized that targeting SHH to act on the stroma and exploit the secretory capability of CAFs could enhance drug delivery and prevent chemoresistance in cancer cells [93].

7.4. SHH in Cancer Stem Cells of Pancreatic Cancer

SMO could affect epithelial–mesenchymal transition, invasion, and migration of cancer stem cells in the pancreas [94]. Wang et al. knocked down SMO to inhibit pancreas cancer stem cells that possessed characteristics of self-renewal, epithelial-mesenchymal transition, invasion, migration, lung metastasis, chemoresistance to gemcitabine, and tumorigenesis [94]. The inhibition of SHH signaling pathway by sulforaphane could alter the expression of stem cell-related genes such as *Nanog* and *Oct-4* [95]. Therefore, targeting cancer stem cells by SHH pathway could improve the outcomes of pancreatic cancer patients.

8. SHH in Organogenesis of the Gastrointestinal (GI) System, GI Cancer, the Microenvironment, and Stem Cells of GI Cancer

8.1. SHH in Organogenesis of the Stomach

SHH signaling affects foregut development [96]. Among the three SHH ligands in the mammalian genome, SHH levels are highest in the mucosa of the embryonic foregut [96]. Ranakho–Santos et al. emphasized that the SHH signal plays an important role in organogenesis of the mammalian gastrointestinal tract [97]. SHH plays a significant role during epithelial development and differentiation, homeostasis, and neoplastic transformation of the stomach [19,96,98]. Van den Brink et al. reported that in humans, there are abundant *SHH* mRNA and SHH proteins in the gastric fundus, but no SHH protein is present in the esophagus or intestines [97,98]. SHH is needed during the growth and differentiation of the esophagus [99]. High expression of SHH in parietal cells contributes to gastric acid production [96]. Myofibroblasts are the predominant cell to respond to SHH ligand in normal stomach tissue. SHH induces the epithelial phenotype in gastric organogenesis. SHH null mice show an overgrowth of gastric epithelium as patterned into glandular and nonglandular regions [100].

8.2. SHH in GI Cancer and its Microenvironment

8.2.1. Stomach: Gastritis, Gastric Ulcer, Gastric Carcinogenesis, and Gastric Cancer Stem Cells

Ranakho–Santos et al. suggested that mutations to the SHH signaling pathway affect human gastrointestinal function [100]. Chronic inflammation caused by *Helicobacter pylori* (HP) infection causes parietal cell atrophy and metaplastic cell proliferation (a precursor to human gastric cancer) [96]. In a mouse study following HP infection, canonical SHH signaling-induced inflammatory cells were recruited from the bone marrow to the stomach along with metaplasia [96]. Gastric parietal cells secreting SHH affect the regeneration of the epithelium after gastritis following HP infection [101]. Dysregulation of the SHH signaling pathway causes the disruption of gastric differentiation, loss of gastric acid secretion, and the development of cancer [101]. Merchant et al. showed that overexpression of SHH in parietal cells induces gastric acid production. In an uninfected stomach, myofibroblasts are the predominant cells that respond to SHH ligand. Xiao et al. used a mouse model to find that ulcer healing occurs with decreased ulcer size, angiogenesis, macrophage infiltration, and granulation tissue formation upon re-expression of SHH within ulcerated tissue [102]. Re-expression of SHH affects gastric regeneration as well. In a mouse model following HP infection, canonical SHH signaling induces bone marrow to recruit inflammatory cells to the stomach, leading to metaplastic development. The transcription factor GLI1 regulates the polarization of invading myeloid cells and myeloid-derived suppressor cells to afford a microenvironment that favors wound healing and neoplastic transformation. In mice, GLI1 mediates a shift in phenotype to gastric myeloid-derived suppressor cells via inducing Schlafen 4 (slfn4) directly. These could be taken as biomarkers to predict gastric cancer progression and determine benefit after SHH antagonist treatments [96].

Atrophic change with loss of parietal cells also causes loss of SHH expression, indicating an early sign in the mucosa before cellular transformation [103]. The change of SHH expression induces gastric cancer development. SHH is primarily located in parietal cells of the gastric body. However, the intestinal type of gastric cancer mainly develops in the antrum. The method by which fundic SHH regulates proliferation in the antrum remains elusive. SHH regulates downstream targets, including PTCH and the TGF-beta family members (bone morphogenic proteins (BMPs)) [97]. The latter targets of the SHH pathway are present in the mesenchyme rather than the epithelium, suggesting that SHH regulates epithelial–mesenchymal crosstalk. It is likely that these mesenchymal factors are present preferentially in the antrum, whereas gastric atrophy and subsequent loss of SHH could remove the inhibitory signal that suppresses antral proliferation. Loss of SHH in the mucosa during HP-associated atrophic gastritis becomes an early change prior to cellular transformation. SHH plays an important role in sustaining gastric epithelial differentiation, and the loss of SHH favors early carcinogenesis. Yang et al. reported that a high expression of SMO and GLI1 is correlated with gastric carcinogenesis [104]. Other researchers found that SMO or GLI1 inhibitors impair the migration and invasion of gastric cancer cells [104]. Chong et al. mentioned that in gastric cancer cells, galectin-1 promotes cancer invasion and epithelial–mesenchymal transition via activation of the non-canonical SHH pathway [105]. Wu et al. found that GDC-0499 inhibits the proliferation of gastric cancer cell line SGC-7901 and accelerates apoptosis [106].

Surface markers of gastric cancer stem cells CD133 and CD44 were found to be significantly decreased in the SGC-7901 gastric cancer cell line following GDC-0499 treatment [106]. This SMO antagonist could affect the maintenance and other properties of gastric cancer stem cells [106]. In paclitaxel-treated gastric cancer cells, overexpression of SMO could reduce activated caspase 3, thus decreasing cancer cell death [107]. Ma et al. found that in paclitaxel-resistant gastric cancer cell lines, there was an overexpression of SMO. SMO overexpression upregulates 5-Bromo-2′-Deoxyuridine (BrdU) in gastric cancer cells [107].

8.2.2. Colon Cancer

Zhang et al. found that when compared with normal colon tissue, overexpression of SMO and GLI protein is noted in colon cancer tissue and colonic adenoma tissue [108]. Li et al. reported that in colorectal cancer, SMO expression corresponds with tumor status and patient prognosis [109]. Ding et al. found that SMO expression is an independent biomarker for postoperative liver metastasis. Similarly, SMO plays an important role in colon cancer progression [110].

Colon cancer driven by cancer stem cells forms a heterogeneous tumor, and whole-transcriptome analysis has revealed enhancement of WNT and Hedgehog signaling in cancer stem cells. Canonical GLI-dependent SHH signaling negatively affects WNT signaling in intestinal tumors. Regan et al. found that the SHH signaling in colon cancer stem cells includes SHH-dependent, non-canonical PTCH1-dependent, and GLI-independent pathways, suggesting that non-canonical SHH signaling positively affects WNT signaling and is essential for the survival of colon cancer stem cells [111]. Niyaz et al. suggested that dysregulated SMO could be as a treatment target of colon cancer [81].

Wu et al. found that GDC-0449 inhibits the replication of colon cancer cells and triggers apoptosis via downregulating B-cell lymphoma 2 (Bcl-2) [112]. Magistri et al. found that GDC-0499 could suppress and modulate cellular plasticity and invasiveness of colorectal cancer [113].

9. The Role of the SHH/SMO Pathway in Breast Organogenesis, Breast Cancer, and Its Microenvironment

9.1. SHH in Organogenesis of the Breast

SHH Signaling During Normal Mammary Gland Development

Riobo-Del Galdo et al. emphasized that SHH signaling is essential in breast development and homeostasis. The expression of SHH component pathways in mammary tissue differ at different stages of development [114–119]. During embryonic development, the canonical SHH signaling pathway is inhibited in breast tissues [4,111], and SHH gene expression is affected temporally and spatially via genetic and epigenetic mechanisms [4,111]. In a mouse study, early mammary bud formation was found to require active repression of GLI1 by GLI3R [4]. During puberty, ductal morphogenesis is affected by canonical and non-canonical SHH signaling, for which type I non-canonical SHH signaling plays an essential role [4,111]. During puberty, the elongation of the terminal buds is affected via activation of cellular Src kinase (c-Src), estrogen receptor alpha (ERα), and extracellular signal-regulated kinase (ERK) cascades in mammary luminal epithelial cells [120–122]. At this stage, a decrease of the expression of SHH ligands GLI1, GLI2, GLI3, and PTCH1 in the mature mammary gland is also found [114,118,123]. In normal adult mammary tissue, this pathway becomes downregulated.

9.2. SHH in Breast Cancer

In transgenic mice, active SMO with high canonical signaling activity may be involved in the development to mammary ductal dysplasia [4,124,125].

SHH contributes to tumorigenesis and progression with some types of breast cancer [111]. SMO expression is present in ductal carcinoma in situ (DCIS) and invasive breast cancer (IBC), but is absent in normal breast tissue. SMO expression affects tumor size, lymph node involvement, and tumor recurrence. However, it does not affect histological grade or other oncology markers [111]. SMO expression does not correlate with PTCH1 expression in either DCIS or IBC. This means that the activation of the SHH pathway cannot regulate SMO. Such evidence suggests that targeting downstream molecules of SMO when treating breast cancer may not be effective [111].

Guerrini et al. found that the SHH signaling pathway regulates breast cancer cell migration and invasion through carbonic anhydrase (CA) II [126]. SHH pathway activation affects breast cancer metastasis [127]. Many studies support the claim that target genes *GLI1* and *GLI2* are involved in breast cancer cell proliferation, survival, migration, invasion, EMT, angiogenesis, and osteolytic metastasis [4,128–135]. Benvenuto et al. used an SMO inhibitor (GDC-0449) and GLI inhibitor (GANT-61) to target the SHH/GLI pathway to inhibit breast cancer cell growth in both in vitro and in vivo studies [136]. The researchers found that in breast cancer, downstream SMO targeting is better than upstream SMO when attempting to interrupt SHH signaling [136]. The development of highly vascularized tumors is regulated by overexpression of SHH, which affects the pro-angiogenic transcription factor cysteine-rich angiogenic induced 61 (CYR61) in a GLI-dependent manner [4,131]. Han et al. found cancer stem cells to be heavily present in breast cancer via non-canonical SMO-independent SHH signaling activation [137]. SHH inhibitors are therefore another therapeutic option [138].

9.3. SHH in Estrogen Receptor-Positive Breast Cancer

Recently, studies of estrogen receptor (ER)-positive breast cancer (BC) cell lines have revealed that estrogen can increase GLI1 and GLI2 [4,139]. However, GANT61, which inhibits GLI1 and GLI2 activity, could reduce the proliferation of cancer stem cells in culture. GLI transcription factors as mediators were also able to affect estrogen in BC [139]. Some authors found that estrogen affects overexpression of SHH and GLI1, activating SHH signaling and enhancing invasiveness of the ER-positive T47D (HER2-) and BT-474 (HER2+) cells [140]. These results suggest that cross-talk between ER- and SHH-signaling pathways facilitate the invasiveness of ER-positive BC cells [4]. The association between GLI1 and ER (luminal subtype marker) remains elusive [4]. In ER-positive breast cancer, overexpression of GLI1 affects early disease onset, higher SHH expression, higher Ki-67 index, higher histological grade, advanced stage, lymph node metastasis, and both shorter disease-free survival and overall survival. Overexpression of GLI1 acts as a predictor in age, ER-positive expression, distant metastasis, short disease-free survival, and short overall survival. However, it does not correlate with the tumor size [4,139,140].

9.4. SHH in Triple-Negative Breast Cancer (TNBC)

Canonical SHH signaling plays a role in triple-negative breast cancer (TNBC) [4]. The SHH signaling pathway is a regulator of angiogenesis in TNBC [138]. Mauro et al. identified that angiogenesis of TNBC is regulated by the SHH pathway [138]. In TNBC, some researchers found a correlation between SMO expression and histological grade or tumor stage. Riaz et al. found that expression of SMO corresponds with early onset and subtype of TNBC. Canonical SHH signaling enhances tumor angiogenesis via mechanisms including metalloproteases, CYR61, and VEGF receptor 2 (VEGFR2), resulting in TNBC growth and metastasis [4,131,138,141]. The osteolytic bone metastasis of TNBC is also affected by the SHH pathway [4]. TNBC has a high proportion of basal-like progenitors, which retain primary cilia and GLI1 expression. The ligand-dependent stimulation of canonical SHH pathways affect TNBC [138,142]. In vitro studies reveal that overexpression of SHH enhances cell proliferation, colony formation, migration, and invasion of TNBC [131,143]. Likewise, an in vivo study revealed that such overexpression enhances the growth of orthotopic xenograft and promotes lung metastasis [131].

Some investigators emphasize that GLI1 upregulation mainly affects the maintenance and proliferation of breast CSC. GLI1 activation could upregulate the multidrug-resistant protein-1 (MDR-1), resulting in resistance to doxorubicin, paclitaxel, and cisplatin [4,144]. Recently, Ruiz-Borrego et al. conducted a phase Ib clinical trial study using combined sonidegib (LDE225) (a small molecular oral inhibitor of the SMO/SHH pathway) and docetaxel to treat advanced TNBC patients [145].

9.5. SHH in the Microenvironment of Breast Cancer

Aberrant upregulation of SHH affects changes in the tumor microenvironment of breast cancer [4,130], whereas type II non-canonical SHH signaling plays a role in the tumor stroma of breast cancer [4]. The tumor microenvironment/stroma affects tumor development and metastasis [4,5], and the tumor microenvironment/stroma of breast cancer includes endothelial cells, immune cells, adipocytes, and activated fibroblasts (the so-called "cancer-associated fibroblasts" (CAFs)) [6]. CAFs fuel tumor cells via secreting soluble factors [7–10] to induce metastasis and chemoresistance. This process involves extracellular acidification, inflammation, activation of matrix metalloproteases, and decreased effects of chemotherapeutic drugs [7,11,12]. Tumor microenvironment cells also include tumor-associated macrophages with aberrant genetic and epigenetic changes that may induce a high expression of signaling molecules to enhance the survival of tumor cells [146]. Inhibitors targeting SHH, Notch, CDKs, mTOR, and WNT are promising and are involved in ongoing clinical trials, either in single use or combined use in therapy [146]. Such microenvironment remodeling also activates an antioxidant response in SHH signaling to enhance the CSC in ER-positive BC [147]. A hypoxic microenvironment affects the upregulation of GLI1. In hypoxia, hypoxia-inducible factor 1-alpha (HIF-1α) induces SHH expression in fibroblasts to affect GLI1 induction in a paracrine manner [4,143–148].

10. SHH in Organogenesis of the Lung and Lung Cancer

10.1. Organogenesis of the Lung

SHH is necessary for the growth and differentiation of the trachea and lungs. During lung development, SHH plays an essential role in lung development, specifically for lung specification, primary bud formation, and branching morphogenesis. Mutations in SHH or associated signaling components result in foregut defects in humans [99]. Abnormal secretion of SHH causes severe foregut defects and lung hypoplasia. Pulmonary morphogenesis depends deeply on SHH activation and molecular interactions with other signaling pathways [149]. SHH signaling pathway molecules are required for embryonic lung development [150]. Hypoplastic lungs were found in *SHH, GLI1, GLI2,* or *GLI3* knockout. PTCH knockout is lethal before lung development begins.

10.2. Suppressing the SHH/SMO Pathway to Inhibit Lung Cancer

SHH expression is negatively correlated with tumor differentiation in lung cancer [151]. Patients with higher SHH expression could have a poorer prognosis and worse overall survival [152]. Therefore, SHH could be a prognostic marker. Szczepny et al. reported that an autocrine, ligand-dependent model of the SHH signaling pathway contributes to the pathogenesis of small cell lung cancer [153]. They also found a novel role of non-canonical SHH signaling in producing chromosomal instability [153]. Sun et al. reported that hyperactivated SMO could facilitate the proliferation of non-small cell lung cancer cells [154], and found that HECT and RLD domain containing E3 ubiquitin ligase 4 (HERC4) is inhibited after destabilizing oncoprotein SMO [154].

11. Targeting SHH/SMO/GLI Signaling Pathway for Cancer Stem Cells

Cancer stem cells (CSC), a subpopulation of cancer cells with self-sustaining characteristics, play an essential role in tumorigenesis, cancer progression, metastasis, recurrence, and drug resistance [155,156]. The SHH signaling pathway activates in cancer stem cells of several neoplasms such as glioblastoma, as well as cancers of the colon, liver, breast, pancreas, and blood neoplasms (chronic myeloid leukemia and multiple myeloma) [14]. The pathway not only triggers tumorigenesis with uncontrolled cell growth, but also promotes cell migration, mitosis, and can sustain cancer cell survival [14]. Moreover, self-renewal of CSC regulated by the SHH/SMO/GLI signaling pathway has been observed [156].

PTCH1-dependent and SMO-independent (type I non-canonical Hedgehog signaling) paths are both necessary for the survival of CSC [111].

Regan et al. proposed that PTCH1-dependent (non-canonical SHH signaling) positively affects WNT to maintain CSCs with an undifferentiated state [111]. PTCH1 is a dependence receptor that can induce apoptosis even when SHH ligand is absent [157]. However, canonical SMO-dependent SHH signaling, as mediated by GLI1 nuclear localization, downregulates WNT signaling and tumor cell differentiation. Targeting non-canonical SHH signaling to induce CSC differentiation may provide a strategy to eliminate the therapy-resistant CSCs. SHH is therefore proposed as a target in the treatment of SHH-dependent pancreatic cancer and breast cancers [158,159].

SHH/SMO/GLI affecting epithelial–mesenchymal transition allows the transformation of polarized epithelial cells into motile mesenchymal cells, enhancing invasive growth and metastasis [111,155]. Some investigators found that drug transport pump expression in cancer stem cells enabling cytotoxic drug resistance were upregulated by SHH signaling [156]. This is important in the combined use of SHH/SMO/GLI signaling inhibitors and chemotherapy, radiation therapy, or immunotherapy to target CSCs. SMO receptor antagonists may also be able to inhibit this process [14]. Using pharmacological inhibitors that target the SHH/SMO/GLI pathway to inhibit CSC is therefore a promising strategy [155].

12. Concluding Remarks

SHH plays an important role in organogenesis, cancer, and the cancer microenvironment of some organs. Combined use of SHH signaling inhibitors and chemotherapy/radiation therapy/immunotherapy could be key in targeting cancer stem cells. Better understanding of these mechanisms could help us better target the SHH pathway against cancer.

Funding: This research was supported by Far Eastern Memorial Hospital (FEMH-2019-C-002, FEMH-2019-C-078) and Ministry of Science and Technology, Taiwan (MOST 108-2314-B-418-001).

Acknowledgments: We appreciate all support from the department of Medical Research and the Core Laboratory of Far Eastern Memorial Hospital, Taiwan.

Conflicts of Interest: The authors declare no conflict of interest.

Abbreviations

SHH	Sonic Hedgehog
CSC	Cancer stem cells
GLI	Glioma-associated oncogene homolog
WNT	Wingless-activated
CAFs	Cancer-associated fibroblasts
SMO	Smoothened
GPCR	G protein-coupled receptor
SUFU	Suppressor of fused homolog
HH	Hedgehog
PKA	Protein kinase A
GTPase	Guanosine triphosphatase
PI3K	Phosphoinositide 3-kinase
mTOR	mammalian target of rapamycin
SBE7	Shh brain enhancer 7
VZ	Ventricular zone
TGF1β	Transforming growth factor 1 beta
YAP1	Hippo/Yes-associated protein
HBx	HBV gene product HBx protein
CDK1	cyclin-dependent kinase 1
HCC	Hepatocellular carcinoma
EMT	epithelial-mesenchymal transition
FAK	focal adhesion kinase
TAMs	Tumor-associated macrophages
PDGF	Platelet-derived growth factor
MMP-2	Matrix metalloproteinase-2
HGF	hepatocyte growth factor
VEGFR	Vascular endothelial growth factor receptor
TGF-α	transforming growth factor-alpha
MDB5	2-chloro-N^1-[4-chloro-3-(2-pyridinyl)phenyl]-N^4,N^4-bis(2-pyridinylmethyl)-1,4-benzenedicarboxamide
HP	Helicobacter pylori
slfn4	Schlafen 4
BMPs	bone morphogenic proteins
BrdU	5-Bromo-2′-Deoxyuridine
Bcl-2	B-cell lymphoma 2
c-Src	cellular Src kinase
ERα	Estrogen receptor alpha
ERK	Extracellular signal-regulated kinase
DICS	ductal carcinoma in situ
IBC	invasive breast cancer
CA	carbonic anhydrase
CYR61	Cysteine-rich angiogenic induced 61
ER	estrogen receptor
BC	breast cancer
TNBC	triple negative breast cancer
MDR-1	multidrug resistant protein-1
HIF-1α	Hypoxia-inducible factor 1-alpha
HERC4	HECT and RLD domain containing E3 ubiquitin ligase 4

References

1. Park, S.M.; Jang, H.J.; Lee, J.H. Roles of primary cilia in the developing brain. *Front. Cell. Neurosci.* **2019**, *13*, 218. [CrossRef]

2. Saqui-Salces, M.; Merchant, J.L. Hedgehog signaling and gastrointestinal cancer. *Biochim. Biophys. Acta* **2010**, *1803*, 786–795. [CrossRef]

3. Montagnani, V.; Stecca, B. Role of protein kinases in hedgehog pathway control and implications for cancer therapy. *Cancers (Basel)* **2019**, 11. [CrossRef]

4. Riobo-Del Galdo, N.A.; Lara Montero, A.; Wertheimer, E.V. Role of hedgehog signaling in breast cancer: pathogenesis and therapeutics. *Cells* **2019**, 8. [CrossRef]

5. Hanahan, D.; Coussens, L.M. Accessories to the crime: Functions of cells recruited to the tumor microenvironment. *Cancer Cell* **2012**, *21*, 309–322. [CrossRef]

6. Petersen, O.W.; Ronnov-Jessen, L.; Howlett, A.R.; Bissell, M.J. Interaction with basement membrane serves to rapidly distinguish growth and differentiation pattern of normal and malignant human breast epithelial cells. *Proc. Natl. Acad. Sci. USA* **1992**, *89*, 9064–9068. [CrossRef]

7. Bartling, B.; Hofmann, H.S.; Silber, R.E.; Simm, A. Differential impact of fibroblasts on the efficient cell death of lung cancer cells induced by paclitaxel and cisplatin. *Cancer Biol.* **2008**, *7*, 1250–1261. [CrossRef]

8. Martinez-Outschoorn, U.; Sotgia, F.; Lisanti, M.P. Tumor microenvironment and metabolic synergy in breast cancers: Critical importance of mitochondrial fuels and function. *Semin. Oncol.* **2014**, *41*, 195–216. [CrossRef]

9. Nieman, K.M.; Kenny, H.A.; Penicka, C.V.; Ladanyi, A.; Buell-Gutbrod, R.; Zillhardt, M.R.; Romero, I.L.; Carey, M.S.; Mills, G.B.; Hotamisligil, G.S.; et al. Adipocytes promote ovarian cancer metastasis and provide energy for rapid tumor growth. *Nat. Med.* **2011**, *17*, 1498–1503. [CrossRef]

10. Zhang, W.; Trachootham, D.; Liu, J.; Chen, G.; Pelicano, H.; Garcia-Prieto, C.; Lu, W.; Burger, J.A.; Croce, C.M.; Plunkett, W.; et al. Stromal control of cystine metabolism promotes cancer cell survival in chronic lymphocytic leukaemia. *Nat. Cell Biol.* **2012**, *14*, 276–286. [CrossRef]

11. Laberge, R.M.; Awad, P.; Campisi, J.; Desprez, P.Y. Epithelial-mesenchymal transition induced by senescent fibroblasts. *Cancer Microenviron.* **2012**, *5*, 39–44. [CrossRef]

12. Martinez-Outschoorn, U.E.; Lin, Z.; Ko, Y.H.; Goldberg, A.F.; Flomenberg, N.; Wang, C.; Pavlides, S.; Pestell, R.G.; Howell, A.; Sotgia, F.; et al. Understanding the metabolic basis of drug resistance: Therapeutic induction of the Warburg effect kills cancer cells. *Cell Cycle* **2011**, *10*, 2521–2528. [CrossRef]

13. Sun, Q.; Zhang, B.; Hu, Q.; Qin, Y.; Xu, W.; Liu, W.; Yu, X.; Xu, J. The impact of cancer-associated fibroblasts on major hallmarks of pancreatic cancer. *Theranostics* **2018**, *8*, 5072–5087. [CrossRef]

14. Espinosa-Bustos, C.; Mella, J.; Soto-Delgado, J.; Salas, C.O. State of the art of Smo antagonists for cancer therapy: Advances in the target receptor and new ligand structures. *Future Med. Chem.* **2019**, *11*, 617–638. [CrossRef]

15. Theunissen, J.W.; de Sauvage, F.J. Paracrine hedgehog signaling in cancer. *Cancer Res.* **2009**, *69*, 6007–6010. [CrossRef]

16. Xin, M.; Ji, X.; De La Cruz, L.K.; Thareja, S.; Wang, B. Strategies to target the hedgehog signaling pathway for cancer therapy. *Med. Res. Rev.* **2018**, *38*, 870–913. [CrossRef]

17. Jenkins, D. Hedgehog signalling: Emerging evidence for non-canonical pathways. *Cell Signal.* **2009**, *21*, 1023–1034.

18. Marini, K.D.; Payne, B.J.; Watkins, D.N.; Martelotto, L.G. Mechanisms of hedgehog signalling in cancer. *Growth Factors* **2011**, *29*, 221–234. [CrossRef]

19. Merchant, J.L. Hedgehog signalling in gut development, physiology and cancer. *J. Physiol.* **2012**, *590*, 421–432. [CrossRef]

20. Sagai, T.; Amano, T.; Maeno, A.; Ajima, R.; Shiroishi, T. SHH signaling mediated by a prechordal and brain enhancer controls forebrain organization. *Proc. Natl. Acad. Sci. USA* **2019**, *116*, 23636–23642. [CrossRef]

21. Byrd, N.; Grabel, L. Hedgehog signaling in murine vasculogenesis and angiogenesis. *Trends Cardiovasc. Med.* **2004**, *14*, 308–313. [CrossRef]

22. D'Amore, P.A.; Ng, Y.S. Won't you be my neighbor? Local induction of arteriogenesis. *Cell* **2002**, *110*, 289–292. [CrossRef]

23. Byrd, N.; Becker, S.; Maye, P.; Narasimhaiah, R.; St-Jacques, B.; Zhang, X.; McMahon, J.; McMahon, A.; Grabel, L. Hedgehog is required for murine yolk sac angiogenesis. *Development* **2002**, *129*, 361–372.

24. Geng, L.; Cuneo, K.C.; Cooper, M.K.; Wang, H.; Sekhar, K.; Fu, A.; Hallahan, D.E. Hedgehog signaling in the murine melanoma microenvironment. *Angiogenesis* **2007**, *10*, 259–267. [CrossRef]

25. Pepicelli, C.V.; Lewis, P.M.; McMahon, A.P. Sonic hedgehog regulates branching morphogenesis in the mammalian lung. *Curr. Biol.* **1998**, *8*, 1083–1086. [CrossRef]

26. Rowitch, D.H.; B, S.J.; Lee, S.M.; Flax, J.D.; Snyder, E.Y.; McMahon, A.P. Sonic hedgehog regulates proliferation and inhibits differentiation of CNS precursor cells. *J. Neurosci.* **1999**, *19*, 8954–8965.

27. Brown, L.A.; Rodaway, A.R.; Schilling, T.F.; Jowett, T.; Ingham, P.W.; Patient, R.K.; Sharrocks, A.D. Insights into early vasculogenesis revealed by expression of the ETS-domain transcription factor Fli-1 in wild-type and mutant zebrafish embryos. *Mech. Dev.* **2000**, *90*, 237–252.

28. Lawson, N.D.; Vogel, A.M.; Weinstein, B.M. Sonic hedgehog and vascular endothelial growth factor act upstream of the Notch pathway during arterial endothelial differentiation. *Dev. Cell* **2002**, *3*, 127–136.

29. Di Pietro, C.; Marazziti, D.; La Sala, G.; Abbaszadeh, Z.; Golini, E.; Matteoni, R.; Tocchini-Valentini, G.P. Primary cilia in the murine cerebellum and in mutant models of medulloblastoma. *Cell. Mol. Neurobiol.* **2017**, *37*, 145–154. [CrossRef]

30. De Luca, A.; Cerrato, V.; Fuca, E.; Parmigiani, E.; Buffo, A.; Leto, K. Sonic hedgehog patterning during cerebellar development. *Cell. Mol. Life Sci.* **2016**, *73*, 291–303. [CrossRef]

31. Corrales, J.D.; Rocco, G.L.; Blaess, S.; Guo, Q.; Joyner, A.L. Spatial pattern of sonic hedgehog signaling through Gli genes during cerebellum development. *Development* **2004**, *131*, 5581–5590. [CrossRef]

32. Grausam, K.B.; Dooyema, S.D.R.; Bihannic, L.; Premathilake, H.; Morrissy, A.S.; Forget, A.; Schaefer, A.M.; Gundelach, J.H.; Macura, S.; Maher, D.M.; et al. ATOH1 promotes leptomeningeal dissemination and metastasis of sonic hedgehog subgroup medulloblastomas. *Cancer Res.* **2017**, *77*, 3766–3777. [CrossRef]

33. Leto, K.; Arancillo, M.; Becker, E.B.; Buffo, A.; Chiang, C.; Ding, B.; Dobyns, W.B.; Dusart, I.; Haldipur, P.; Hatten, M.E.; et al. Consensus paper: Cerebellar development. *Cerebellum* **2016**, *15*, 789–828. [CrossRef]

34. Kool, M.; Koster, J.; Bunt, J.; Hasselt, N.E.; Lakeman, A.; van Sluis, P.; Troost, D.; Meeteren, N.S.; Caron, H.N.; Cloos, J.; et al. Integrated genomics identifies five medulloblastoma subtypes with distinct genetic profiles, pathway signatures and clinicopathological features. *PLoS ONE* **2008**, *3*, e3088. [CrossRef]

35. Northcott, P.A.; Korshunov, A.; Witt, H.; Hielscher, T.; Eberhart, C.G.; Mack, S.; Bouffet, E.; Clifford, S.C.; Hawkins, C.E.; French, P.; et al. Medulloblastoma comprises four distinct molecular variants. *J. Clin. Oncol.* **2011**, *29*, 1408–1414. [CrossRef]

36. Northcott, P.A.; Hielscher, T.; Dubuc, A.; Mack, S.; Shih, D.; Remke, M.; Al-Halabi, H.; Albrecht, S.; Jabado, N.; Eberhart, C.G.; et al. Pediatric and adult sonic hedgehog medulloblastomas are clinically and molecularly distinct. *Acta Neuropathol.* **2011**, *122*, 231–240. [CrossRef]

37. Northcott, P.A.; Shih, D.J.; Peacock, J.; Garzia, L.; Morrissy, A.S.; Zichner, T.; Stutz, A.M.; Korshunov, A.; Reimand, J.; Schumacher, S.E.; et al. Subgroup-specific structural variation across 1000 medulloblastoma genomes. *Nature* **2012**, *488*, 49–56. [CrossRef]

38. Aref, D.; Moffatt, C.J.; Agnihotri, S.; Ramaswamy, V.; Dubuc, A.M.; Northcott, P.A.; Taylor, M.D.; Perry, A.; Olson, J.M.; Eberhart, C.G.; et al. Canonical TGF-beta pathway activity is a predictor of SHH-driven medulloblastoma survival and delineates putative precursors in cerebellar development. *Brain Pathol.* **2013**, *23*, 178–191. [CrossRef]

39. Oliver, T.G.; Read, T.A.; Kessler, J.D.; Mehmeti, A.; Wells, J.F.; Huynh, T.T.; Lin, S.M.; Wechsler-Reya, R.J. Loss of patched and disruption of granule cell development in a pre-neoplastic stage of medulloblastoma. *Development* **2005**, *132*, 2425–2439. [CrossRef]

40. Crawford, J.R.; MacDonald, T.J.; Packer, R.J. Medulloblastoma in childhood: New biological advances. *Lancet Neurol.* **2007**, *6*, 1073–1085. [CrossRef]

41. Wechsler-Reya, R.J.; Scott, M.P. Control of neuronal precursor proliferation in the cerebellum by sonic hedgehog. *Neuron* **1999**, *22*, 103–114. [CrossRef]

42. Raffel, C. Medulloblastoma: Molecular genetics and animal models. *Neoplasia* **2004**, *6*, 310–322. [CrossRef]

43. Taylor, M.D.; Liu, L.; Raffel, C.; Hui, C.C.; Mainprize, T.G.; Zhang, X.; Agatep, R.; Chiappa, S.; Gao, L.; Lowrance, A.; et al. Mutations in SUFU predispose to medulloblastoma. *Nat. Genet.* **2002**, *31*, 306–310. [CrossRef]

44. Taylor, M.D.; Northcott, P.A.; Korshunov, A.; Remke, M.; Cho, Y.J.; Clifford, S.C.; Eberhart, C.G.; Parsons, D.W.; Rutkowski, S.; Gajjar, A.; et al. Molecular subgroups of medulloblastoma: The current consensus. *Acta Neuropathol.* **2012**, *123*, 465–472. [CrossRef]

45. Eberhart, C.G. Medulloblastoma in mice lacking p53 and PARP: All roads lead to Gli. *Am. J. Pathol.* **2003**, *162*, 7–10. [CrossRef]

46. Hatton, B.A.; Villavicencio, E.H.; Tsuchiya, K.D.; Pritchard, J.I.; Ditzler, S.; Pullar, B.; Hansen, S.; Knoblaugh, S.E.; Lee, D.; Eberhart, C.G.; et al. The Smo/Smo model: Hedgehog-induced medulloblastoma with 90% incidence and leptomeningeal spread. *Cancer Res.* **2008**, *68*, 1768–1776. [CrossRef]

47. Kieran, M.W. Targeted treatment for sonic hedgehog-dependent medulloblastoma. *Neuro-Oncol.* **2014**, *16*, 1037–1047. [CrossRef]

48. Li, Y.; Song, Q.; Day, B.W. Phase I and phase II sonidegib and vismodegib clinical trials for the treatment of paediatric and adult MB patients: A systemic review and meta-analysis. *Acta Neuropathol. Commun.* **2019**, *7*, 123. [CrossRef]

49. Tamayo-Orrego, L.; Charron, F. Recent advances in SHH medulloblastoma progression: Tumor suppressor mechanisms and the tumor microenvironment. *F1000Research* **2019**, *8*. [CrossRef]

50. Margol, A.S.; Robison, N.J.; Gnanachandran, J.; Hung, L.T.; Kennedy, R.J.; Vali, M.; Dhall, G.; Finlay, J.L.; Erdreich-Epstein, A.; Krieger, M.D.; et al. Tumor-associated macrophages in SHH subgroup of medulloblastomas. *Clin. Cancer Res.* **2015**, *21*, 1457–1465. [CrossRef]

51. Maximov, V.; Chen, Z.; Wei, Y.; Robinson, M.H.; Herting, C.J.; Shanmugam, N.S.; Rudneva, V.A.; Goldsmith, K.C.; MacDonald, T.J.; Northcott, P.A.; et al. Tumour-associated macrophages exhibit anti-tumoural properties in sonic hedgehog medulloblastoma. *Nat. Commun.* **2019**, *10*, 2410. [CrossRef]

52. Cai, H.; Li, H.; Li, J.; Li, X.; Li, Y.; Shi, Y.; Wang, D. Sonic hedgehog signaling pathway mediates development of hepatocellular carcinoma. *Tumor Biol.* **2016**. [CrossRef]

53. Ferlay, J.; Soerjomataram, I.; Dikshit, R.; Eser, S.; Mathers, C.; Rebelo, M.; Parkin, D.M.; Forman, D.; Bray, F. Cancer incidence and mortality worldwide: Sources, methods and major patterns in GLOBOCAN 2012. *Int. J. Cancer* **2015**, *136*, E359–E386. [CrossRef]

54. Swiderska-Syn, M.; Xie, G.; Michelotti, G.A.; Jewell, M.L.; Premont, R.T.; Syn, W.K.; Diehl, A.M. Hedgehog regulates yes-associated protein 1 in regenerating mouse liver. *Hepatology* **2016**, *64*, 232–244. [CrossRef]

55. Michelotti, G.A.; Xie, G.; Swiderska, M.; Choi, S.S.; Karaca, G.; Krüger, L.; Premont, R.; Yang, L.; Syn, W.-K.; Metzger, D. Smoothened is a master regulator of adult liver repair. *J. Clin. Investig.* **2013**, *123*, 2380–2394.

56. Sicklick, J.K.; Li, Y.-X.; Jayaraman, A.; Kannangai, R.; Qi, Y.; Vivekanandan, P.; Ludlow, J.W.; Owzar, K.; Chen, W.; Torbenson, M.S. Dysregulation of the hedgehog pathway in human hepatocarcinogenesis. *Carcinogenesis* **2005**, *27*, 748–757.

57. Ding, X.; Yang, Y.; Han, B.; Du, C.; Xu, N.; Huang, H.; Cai, T.; Zhang, A.; Han, Z.-G.; Zhou, W. Transcriptomic characterization of hepatocellular carcinoma with CTNNB1 mutation. *PLoS ONE* **2014**, *9*, e95307.

58. Liu, Z.; Tu, K.; Wang, Y.; Yao, B.; Li, Q.; Wang, L.; Dou, C.; Liu, Q.; Zheng, X. Hypoxia accelerates aggressiveness of hepatocellular carcinoma cells involving oxidative stress, epithelial-mesenchymal transition and non-canonical hedgehog signaling. *Cell. Physiol. Biochem.* **2017**, *44*, 1856–1868.

59. Chen, J.-S.; Huang, X.-H.; Wang, Q.; Huang, J.-Q.; Zhang, L.-J.; Chen, X.-L.; Lei, J.; Cheng, Z.-X. Sonic hedgehog signaling pathway induces cell migration and invasion through focal adhesion kinase/AKT signaling-mediated activation of matrix metalloproteinase (MMP)-2 and MMP-9 in liver cancer. *Carcinogenesis* **2012**, *34*, 10–19.

60. Arzumanyan, A.; Sambandam, V.; Clayton, M.M.; Choi, S.S.; Xie, G.; Diehl, A.M.; Yu, D.-Y.; Feitelson, M.A. Hedgehog signaling blockade delays hepatocarcinogenesis induced by hepatitis B virus X protein. *Cancer Res.* **2012**, *72*, 5912–5920.

61. Chen, X.-L.; Cheng, Q.-Y.; She, M.-R.; Wang, Q.; Huang, X.-H.; Cao, L.-Q.; Fu, X.-H.; Chen, J.-S. Expression of sonic hedgehog signaling components in hepatocellular carcinoma and cyclopamine-induced apoptosis through Bcl-2 downregulation in vitro. *Arch. Med. Res.* **2010**, *41*, 315–323.

62. Jeng, K.S.; Sheen, I.S.; Jeng, W.J.; Lin, C.C.; Lin, C.K.; Su, J.C.; Yu, M.C.; Fang, H.Y. High expression of patched homolog-1 messenger RNA and glioma-associated oncogene-1 messenger RNA of sonic hedgehog signaling pathway indicates a risk of postresection recurrence of hepatocellular carcinoma. *Ann. Surg. Oncol.* **2013**, *20*, 464–473. [CrossRef]

63. Wang, P.; Song, W.; Li, H.; Wang, C.; Shi, B.; Guo, W.; Zhong, L. Association between donor and recipient smoothened gene polymorphisms and the risk of hepatocellular carcinoma recurrence following orthotopic liver transplantation in a Han Chinese population. *Tumor Biol.* **2015**, *36*, 7807–7815.

64. Czauderna, C.; Castven, D.; Mahn, F.L.; Marquardt, J.U. Context-dependent role of NF-kappaB signaling in primary liver cancer-from tumor development to therapeutic implications. *Cancers (Basel)* **2019**, 11. [CrossRef]

65. Li, H.; Zhang, L. Liver regeneration microenvironment of hepatocellular carcinoma for prevention and therapy. *Oncotarget* **2017**, *8*, 1805–1813. [CrossRef]

66. Petty, A.J.; Li, A.; Wang, X.; Dai, R.; Heyman, B.; Hsu, D.; Huang, X.; Yang, Y. Hedgehog signaling promotes tumor-associated macrophage polarization to suppress intratumoral CD8+ T cell recruitment. *J. Clin. Investig.* **2019**, *129*, 5151–5162. [CrossRef]

67. Razumilava, N.; Gradilone, S.A.; Smoot, R.L.; Mertens, J.C.; Bronk, S.F.; Sirica, A.E.; Gores, G.J. Non-canonical hedgehog signaling contributes to chemotaxis in cholangiocarcinoma. *J. Hepatol.* **2014**, *60*, 599–605.

68. Fingas, C.D.; Bronk, S.F.; Werneburg, N.W.; Mott, J.L.; Guicciardi, M.E.; Cazanave, S.C.; Mertens, J.C.; Sirica, A.E.; Gores, G.J. Myofibroblast-derived PDGF-BB promotes hedgehog survival signaling in cholangiocarcinoma cells. *Hepatology* **2011**, *54*, 2076–2088.

69. El Khatib, M.; Kalnytska, A.; Palagani, V.; Kossatz, U.; Manns, M.P.; Malek, N.P.; Wilkens, L.; Plentz, R.R. Inhibition of hedgehog signaling attenuates carcinogenesis in vitro and increases necrosis of cholangiocellular carcinoma. *Hepatology* **2013**, *57*, 1035–1045.

70. Higashiyama, H.; Ozawa, A.; Sumitomo, H.; Uemura, M.; Fujino, K.; Igarashi, H.; Imaimatsu, K.; Tsunekawa, N.; Hirate, Y.; Kurohmaru, M.; et al. Embryonic cholecystitis and defective gallbladder contraction in the Sox17-haploinsufficient mouse model of biliary atresia. *Development* **2017**, *144*, 1906–1917. [CrossRef]

71. Matsushita, S.; Onishi, H.; Nakano, K.; Nagamatsu, I.; Imaizumi, A.; Hattori, M.; Oda, Y.; Tanaka, M.; Katano, M. Hedgehog signaling pathway is a potential therapeutic target for gallbladder cancer. *Cancer Sci.* **2014**, *105*, 272–280.

72. Xie, F.; Xu, X.; Xu, A.; Liu, C.; Liang, F.; Xue, M.; Bai, L. Aberrant activation of Sonic hedgehog signaling in chronic cholecystitis and gallbladder carcinoma. *Hum. Pathol.* **2014**, *45*, 513–521. [CrossRef]

73. Li, J.; Wu, T.; Lu, J.; Cao, Y.; Song, N.; Yang, T.; Dong, R.; Yang, Y.; Zang, L.; Du, X.; et al. Immunohistochemical evidence of the prognostic value of hedgehog pathway components in primary gallbladder carcinoma. *Surg. Today* **2012**, *42*, 770–775. [CrossRef]

74. Dixit, R.; Pandey, M.; Tripathi, S.K.; Dwivedi, A.N.; Shukla, V.K. Comparative analysis of mutational profile of sonic hedgehog gene in gallbladder cancer. *Dig. Dis. Sci.* **2017**, *62*, 708–714. [CrossRef]

75. Hanna, A.; Shevde, L.A. Hedgehog signaling: Modulation of cancer properies and tumor mircroenvironment. *Mol. Cancer* **2016**, *15*, 24. [CrossRef]

76. Nielsen, S.K.; Mollgard, K.; Clement, C.A.; Veland, I.R.; Awan, A.; Yoder, B.K.; Novak, I.; Christensen, S.T. Characterization of primary cilia and Hedgehog signaling during development of the human pancreas and in human pancreatic duct cancer cell lines. *Dev. Dyn.* **2008**, *237*, 2039–2052. [CrossRef]

77. Lodh, S.; O'Hare, E.A.; Zaghloul, N.A. Primary cilia in pancreatic development and disease. *Birth Defects Res. Part C Embryo Today* **2014**, *102*, 139–158. [CrossRef]

78. Bailey, J.M.; Mohr, A.M.; Hollingsworth, M.A. Sonic hedgehog paracrine signaling regulates metastasis and lymphangiogenesis in pancreatic cancer. *Oncogene* **2009**, *28*, 3513–3525. [CrossRef]

79. Li, X.; Ma, Q.; Duan, W.; Liu, H.; Xu, H.; Wu, E. Paracrine sonic hedgehog signaling derived from tumor epithelial cells: A key regulator in the pancreatic tumor microenvironment. *Crit. Rev. Eukaryot. Gene Expr.* **2012**, *22*, 97–108.

80. Kumar, V.; Chaudhary, A.K.; Dong, Y.; Zhong, H.A.; Mondal, G.; Lin, F.; Kumar, V.; Mahato, R.I. Design, synthesis and biological evaluation of novel hedgehog inhibitors for treating pancreatic cancer. *Sci. Rep.* **2017**, *7*, 1665.

81. Niyaz, M.; Khan, M.S.; Wani, R.A.; Shah, O.J.; Besina, S.; Mudassar, S. Nuclear localization and overexpression of smoothened in pancreatic and colorectal cancers. *J. Cell. Biochem.* **2019**. [CrossRef]

82. Tian, H.; Callahan, C.A.; DuPree, K.J.; Darbonne, W.C.; Ahn, C.P.; Scales, S.J.; de Sauvage, F.J. Hedgehog signaling is restricted to the stromal compartment during pancreatic carcinogenesis. *Proc. Natl. Acad. Sci. USA* **2009**, *106*, 4254–4259. [CrossRef]

83. Saini, F.; Argent, R.H.; Grabowska, A.M. Sonic hedgehog ligand: A role in formation of a mesenchymal niche in human pancreatic ductal adenocarcinoma. *Cells* **2019**, *8*. [CrossRef]

84. Wang, Y.; Jin, G.; Li, Q.; Wang, Z.; Hu, W.; Li, P.; Li, S.; Wu, H.; Kong, X.; Gao, J. Hedgehog signaling non-canonical activated by pro-inflammatory cytokines in pancreatic ductal adenocarcinoma. *J. Cancer* **2016**, *7*, 2067.

85. Rucki, A.A.; Foley, K.; Zhang, P.; Xiao, Q.; Kleponis, J.; Wu, A.A.; Sharma, R.; Mo, G.; Liu, A.; Van Eyk, J.; et al. Heterogeneous stromal signaling within the tumor microenvironment controls the metastasis of pancreatic cancer. *Cancer Res.* **2017**, *77*, 41–52. [CrossRef]

86. Feldmann, G.; Dhara, S.; Fendrich, V.; Bedja, D.; Beaty, R.; Mullendore, M.; Karikari, C.; Alvarez, H.; Iacobuzio-Donahue, C.; Jimeno, A.; et al. Blockade of hedgehog signaling inhibits pancreatic cancer invasion and metastases: A new paradigm for combination therapy in solid cancers. *Cancer Res.* **2007**, *67*, 2187–2196. [CrossRef]

87. Pitarresi, J.R.; Liu, X.; Avendano, A.; Thies, K.A.; Sizemore, G.M.; Hammer, A.M.; Hildreth, B.E., III; Wang, D.J.; Steck, S.A.; Donohue, S.; et al. Disruption of stromal hedgehog signaling initiates RNF5-mediated proteasomal degradation of PTEN and accelerates pancreatic tumor growth. *Life Sci. Alliance* **2018**, *1*, e201800190. [CrossRef]

88. Yauch, R.L.; Gould, S.E.; Scales, S.J.; Tang, T.; Tian, H.; Ahn, C.P.; Marshall, D.; Fu, L.; Januario, T.; Kallop, D.; et al. A paracrine requirement for hedgehog signalling in cancer. *Nature* **2008**, *455*, 406–410. [CrossRef]

89. Rhim, A.D.; Oberstein, P.E.; Thomas, D.H.; Mirek, E.T.; Palermo, C.F.; Sastra, S.A.; Dekleva, E.N.; Saunders, T.; Becerra, C.P.; Tattersall, I.W.; et al. Stromal elements act to restrain, rather than support, pancreatic ductal adenocarcinoma. *Cancer Cell* **2014**, *25*, 735–747. [CrossRef]

90. Liu, X.; Pitarresi, J.R.; Cuitiño, M.C.; Kladney, R.D.; Woelke, S.A.; Sizemore, G.M.; Nayak, S.G.; Egriboz, O.; Schweickert, P.G.; Yu, L. Genetic ablation of Smoothened in pancreatic fibroblasts increases acinar–ductal metaplasia. *Genes Dev.* **2016**, *30*, 1943–1955.

91. Zhou, Q.; Zhou, Y.; Liu, X.; Shen, Y. GDC-0449 improves the antitumor activity of nano-doxorubicin in pancreatic cancer in a fibroblast-enriched microenvironment. *Sci. Rep.* **2017**, *7*, 13379.

92. Olive, K.P.; Jacobetz, M.A.; Davidson, C.J.; Gopinathan, A.; McIntyre, D.; Honess, D.; Madhu, B.; Goldgraben, M.A.; Caldwell, M.E.; Allard, D. Inhibition of hedgehog signaling enhances delivery of chemotherapy in a mouse model of pancreatic cancer. *Science* **2009**, *324*, 1457–1461.

93. Von Ahrens, D.; Bhagat, T.D.; Nagrath, D.; Maitra, A.; Verma, A. The role of stromal cancer-associated fibroblasts in pancreatic cancer. *J. Hematol. Oncol.* **2017**, *10*, 76. [CrossRef]

94. Wang, F.; Ma, L.; Zhang, Z.; Liu, X.; Gao, H.; Zhuang, Y.; Yang, P.; Kornmann, M.; Tian, X.; Yang, Y. Hedgehog signaling regulates epithelial-mesenchymal transition in pancreatic cancer stem-like cells. *J. Cancer* **2016**, *7*, 408.

95. Li, S.H.; Fu, J.; Watkins, D.N.; Srivastava, R.K.; Shankar, S. Sulforaphane regulates self-renewal of pancreatic cancer stem cells through the modulation of sonic hedgehog-GLI pathway. *Mol. Cell. Biochem.* **2013**, *373*, 217–227. [CrossRef]

96. Merchant, J.L.; Ding, L. Hedgehog signaling links chronic inflammation to gastric cancer precursor lesions. *Cell. Mol. Gastroenterol. Hepatol.* **2017**, *3*, 201–210. [CrossRef]

97. Rawadi, G.; Vayssiere, B.; Dunn, F.; Baron, R.; Roman-Roman, S. BMP-2 controls alkaline phosphatase expression and osteoblast mineralization by a wnt autocrine loop. *J. Bone Min. Res.* **2003**, *18*, 1842–1853. [CrossRef]

98. Van den Brink, G.R.; Hardwick, J.C.; Nielsen, C.; Xu, C.; ten Kate, F.J.; Glickman, J.; van Deventer, S.J.; Roberts, D.J.; Peppelenbosch, M.P. Sonic hedgehog expression correlates with fundic gland differentiation in the adult gastrointestinal tract. *Gut* **2002**, *51*, 628–633. [CrossRef]

99. Litingtung, Y.; Lei, L.; Westphal, H.; Chiang, C. Sonic hedgehog is essential to foregut development. *Nat. Genet.* **1998**, *20*, 58–61. [CrossRef]

100. Ramalho-Santos, M.; Melton, D.A.; McMahon, A.P. Hedgehog signals regulate multiple aspects of gastrointestinal development. *Development* **2000**, *127*, 2763–2772.

101. Konstantinou, D.; Bertaux-Skeirik, N.; Zavros, Y. Hedgehog signaling in the stomach. *Curr. Opin. Pharm.* **2016**, *31*, 76–82. [CrossRef]

102. Xiao, C.; Feng, R.; Engevik, A.C.; Martin, J.R.; Tritschler, J.A.; Schumacher, M.; Koncar, R.; Roland, J.; Nam, K.T.; Goldenring, J.R.; et al. Sonic hedgehog contributes to gastric mucosal restitution after injury. *Lab. Investig.* **2013**, *93*, 96–111. [CrossRef]

103. Shiotani, A.; Iishi, H.; Uedo, N.; Ishiguro, S.; Tatsuta, M.; Nakae, Y.; Kumamoto, M.; Merchant, J.L. Evidence that loss of sonic hedgehog is an indicator of Helicobater pylori-induced atrophic gastritis progressing to gastric cancer. *Am. J. Gastroenterol.* **2005**, *100*, 581–587. [CrossRef]

104. Yang, Z.; Lv, Y.; Wang, L.; Chen, Y.; Han, J.; Zhao, S.; Liu, W. Inhibition of hedgehog pathway reveals the regulatory role of SMO in gastric cancer cells. *Tumor Biol.* **2017**, *39*, 1010428317715546.

105. Chong, Y.; Tang, D.; Gao, J.; Jiang, X.; Xu, C.; Xiong, Q.; Huang, Y.; Wang, J.; Zhou, H.; Shi, Y. Galectin-1 induces invasion and the epithelial-mesenchymal transition in human gastric cancer cells via non-canonical activation of the hedgehog signaling pathway. *Oncotarget* **2016**, *7*, 83611–83626.

106. Wu, C.; Cheng, J.; Hu, S.; Deng, R.; Muangu, Y.W.; Shi, L.; Wu, K.; Zhang, P.; Chang, W.; Wang, G. Reduced proliferation and increased apoptosis of the SGC-7901 gastric cancer cell line on exposure to GDC-0449. *Mol. Med. Rep.* **2016**, *13*, 1434–1440.

107. Ma, H.; Tian, Y.; Yu, X. Targeting smoothened sensitizes gastric cancer to chemotherapy in experimental models. *Med Sci. Monit.* **2017**, *23*, 1493–1500.

108. Zhang, X.; Zhang, S.-S.; Wei, G.-J.; Deng, Z.-M.; Hu, Y. Dysregulation of hedgehog signaling pathway related components in the evolution of colonic carcinogenesis. *Int. J. Clin. Exp. Med.* **2015**, *8*, 21379–21385.

109. Li, T.; Liao, X.; Lochhead, P.; Morikawa, T.; Yamauchi, M.; Nishihara, R.; Inamura, K.; Kim, S.A.; Mima, K.; Sukawa, Y. SMO expression in colorectal cancer: Associations with clinical, pathological, and molecular features. *Ann. Surg. Oncol.* **2014**, *21*, 4164–4173.

110. Ding, Y.-L.; Wang, Q.-S.; Zhao, W.-M.; Xiang, L. Expression of smoothened protein in colon cancer and its prognostic value for postoperative liver metastasis. *Asian Pac. J. Cancer Prev.* **2012**, *13*, 4001–4005.

111. Regan, J.L.; Schumacher, D.; Staudte, S.; Steffen, A.; Haybaeck, J.; Keilholz, U.; Schweiger, C.; Golob-Schwarzl, N.; Mumberg, D.; Henderson, D.; et al. Non-canonical hedgehog signaling is a positive regulator of the WNT pathway and is required for the survival of colon cancer stem cells. *Cell Rep.* **2017**, *21*, 2813–2828. [CrossRef]

112. Wu, C.; Hu, S.; Cheng, J.; Wang, G.; Tao, K. Smoothened antagonist GDC-0449 (Vismodegib) inhibits proliferation and triggers apoptosis in colon cancer cell lines. *Exp. Ther. Med.* **2017**, *13*, 2529–2536.

113. Magistri, P.; Battistelli, C.; Strippoli, R.; Petrucciani, N.; Pellinen, T.; Rossi, L.; Mangogna, L.; Aurello, P.; D'Angelo, F.; Tripodi, M. SMO inhibition modulates cellular plasticity and invasiveness in colorectal cancer. *Front. Pharmacol.* **2018**, *8*, 956.

114. Gallego, M.I.; Beachy, P.A.; Hennighausen, L.; Robinson, G.W. Differential requirements for shh in mammary tissue and hair follicle morphogenesis. *Dev. Biol.* **2002**, *249*, 131–139. [CrossRef]

115. Kouros-Mehr, H.; Werb, Z. Candidate regulators of mammary branching morphogenesis identified by genome-wide transcript analysis. *Dev. Dyn.* **2006**, *235*, 3404–3412. [CrossRef]

116. Lewis, M.T.; Ross, S.; Strickland, P.A.; Sugnet, C.W.; Jimenez, E.; Hui, C.; Daniel, C.W. The Gli2 transcription factor is required for normal mouse mammary gland development. *Dev. Biol.* **2001**, *238*, 133–144. [CrossRef]

117. Michno, K.; Boras-Granic, K.; Mill, P.; Hui, C.C.; Hamel, P.A. Shh expression is required for embryonic hair follicle but not mammary gland development. *Dev. Biol.* **2003**, *264*, 153–165. [CrossRef]

118. Velanovich, V.; Yood, M.U.; Bawle, U.; Nathanson, S.D.; Strand, V.F.; Talpos, G.B.; Szymanski, W.; Lewis, F.R., Jr. Racial differences in the presentation and surgical management of breast cancer. *Surgery* **1999**, *125*, 375–379.

119. Hatsell, S.J.; Cowin, P. Gli3-mediated repression of hedgehog targets is required for normal mammary development. *Development* **2006**, *133*, 3661–3670. [CrossRef]

120. Chang, H.; Balenci, L.; Okolowsky, N.; Muller, W.J.; Hamel, P.A. Mammary epithelial-restricted expression of activated c-src rescues the block to mammary gland morphogenesis due to the deletion of the C-terminus of Patched-1. *Dev. Biol.* **2012**, *370*, 187–197. [CrossRef]

121. Okolowsky, N.; Furth, P.A.; Hamel, P.A. Oestrogen receptor-alpha regulates non-canonical hedgehog-signalling in the mammary gland. *Dev. Biol.* **2014**, *391*, 219–229. [CrossRef]

122. Harvey, M.C.; Fleet, A.; Okolowsky, N.; Hamel, P.A. Distinct effects of the mesenchymal dysplasia gene variant of murine Patched-1 protein on canonical and non-canonical Hedgehog signaling pathways. *J. Biol. Chem.* **2014**, *289*, 10939–10949. [CrossRef]

123. McDermott, K.M.; Liu, B.Y.; Tlsty, T.D.; Pazour, G.J. Primary cilia regulate branching morphogenesis during mammary gland development. *Curr. Biol.* **2010**, *20*, 731–737. [CrossRef]

124. Moraes, R.C.; Zhang, X.; Harrington, N.; Fung, J.Y.; Wu, M.F.; Hilsenbeck, S.G.; Allred, D.C.; Lewis, M.T. Constitutive activation of smoothened (SMO) in mammary glands of transgenic mice leads to increased proliferation, altered differentiation and ductal dysplasia. *Development* **2007**, *134*, 1231–1242. [CrossRef]

125. Visbal, A.P.; LaMarca, H.L.; Villanueva, H.; Toneff, M.J.; Li, Y.; Rosen, J.M.; Lewis, M.T. Altered differentiation and paracrine stimulation of mammary epithelial cell proliferation by conditionally activated Smoothened. *Dev. Biol.* **2011**, *352*, 116–127. [CrossRef]

126. Guerrini, G.; Criscuoli, M.; Filippi, I.; Naldini, A.; Carraro, F. Inhibition of smoothened in breast cancer cells reduces CAXII expression and cell migration. *J. Cell. Physiol.* **2018**, *233*, 9799–9811. [CrossRef]

127. Riaz, S.K.; Khan, J.S.; Shah, S.T.A.; Wang, F.; Ye, L.; Jiang, W.G.; Malik, M.F.A. Involvement of hedgehog pathway in early onset, aggressive molecular subtypes and metastatic potential of breast cancer. *Cell Commun. Signal.* **2018**, *16*, 3. [CrossRef]

128. Colavito, S.A.; Zou, M.R.; Yan, Q.; Nguyen, D.X.; Stern, D.F. Significance of glioma-associated oncogene homolog 1 (GLI1) expression in claudin-low breast cancer and crosstalk with the nuclear factor kappa-light-chain-enhancer of activated B cells (NFkappaB) pathway. *Breast Cancer Res.* **2014**, *16*, 444. [CrossRef]

129. Goel, H.L.; Pursell, B.; Chang, C.; Shaw, L.M.; Mao, J.; Simin, K.; Kumar, P.; Vander Kooi, C.W.; Shultz, L.D.; Greiner, D.L.; et al. GLI1 regulates a novel neuropilin-2/alpha6beta1 integrin based autocrine pathway that contributes to breast cancer initiation. *EMBO Mol. Med.* **2013**, *5*, 488–508. [CrossRef]

130. O'Toole, S.A.; Machalek, D.A.; Shearer, R.F.; Millar, E.K.; Nair, R.; Schofield, P.; McLeod, D.; Cooper, C.L.; McNeil, C.M.; McFarland, A.; et al. Hedgehog overexpression is associated with stromal interactions and predicts for poor outcome in breast cancer. *Cancer Res.* **2011**, *71*, 4002–4014. [CrossRef]

131. Harris, L.G.; Pannell, L.K.; Singh, S.; Samant, R.S.; Shevde, L.A. Increased vascularity and spontaneous metastasis of breast cancer by hedgehog signaling mediated upregulation of cyr61. *Oncogene* **2012**, *31*, 3370–3380. [CrossRef]

132. Ramaswamy, B.; Lu, Y.; Teng, K.Y.; Nuovo, G.; Li, X.; Shapiro, C.L.; Majumder, S. Hedgehog signaling is a novel therapeutic target in tamoxifen-resistant breast cancer aberrantly activated by PI3K/AKT pathway. *Cancer Res.* **2012**, *72*, 5048–5059. [CrossRef]

133. Sun, Y.; Wang, Y.; Fan, C.; Gao, P.; Wang, X.; Wei, G.; Wei, J. Estrogen promotes stemness and invasiveness of ER-positive breast cancer cells through Gli1 activation. *Mol. Cancer* **2014**, *13*, 137. [CrossRef]

134. Thomas, Z.I.; Gibson, W.; Sexton, J.Z.; Aird, K.M.; Ingram, S.M.; Aldrich, A.; Lyerly, H.K.; Devi, G.R.; Williams, K.P. Targeting GLI1 expression in human inflammatory breast cancer cells enhances apoptosis and attenuates migration. *Br. J. Cancer* **2011**, *104*, 1575–1586. [CrossRef]

135. Johnson, R.W.; Nguyen, M.P.; Padalecki, S.S.; Grubbs, B.G.; Merkel, A.R.; Oyajobi, B.O.; Matrisian, L.M.; Mundy, G.R.; Sterling, J.A. TGF-beta promotion of Gli2-induced expression of parathyroid hormone-related protein, an important osteolytic factor in bone metastasis, is independent of canonical hedgehog signaling. *Cancer Res.* **2011**, *71*, 822–831. [CrossRef]

136. Benvenuto, M.; Masuelli, L.; De Smaele, E.; Fantini, M.; Mattera, R.; Cucchi, D.; Bonanno, E.; Di Stefano, E.; Frajese, G.V.; Orlandi, A. In vitro and in vivo inhibition of breast cancer cell growth by targeting the Hedgehog/GLI pathway with SMO (GDC-0449) or GLI (GANT-61) inhibitors. *Oncotarget* **2016**, *7*, 9250–9270.

137. Han, B.; Qu, Y.; Jin, Y.; Yu, Y.; Deng, N.; Wawrowsky, K.; Zhang, X.; Li, N.; Bose, S.; Wang, Q. FOXC1 activates smoothened-independent hedgehog signaling in basal-like breast cancer. *Cell Rep.* **2015**, *13*, 1046–1058.

138. Di Mauro, C.; Rosa, R.; D'Amato, V.; Ciciola, P.; Servetto, A.; Marciano, R.; Orsini, R.C.; Formisano, L.; De Falco, S.; Cicatiello, V.; et al. Hedgehog signalling pathway orchestrates angiogenesis in triple-negative breast cancers. *Br. J. Cancer* **2017**, *116*, 1425–1435. [CrossRef]

139. O'Brien, C.S.; Farnie, G.; Howell, S.J.; Clarke, R.B. Breast cancer stem cells and their role in resistance to endocrine therapy. *Horm. Cancer* **2011**, *2*, 91–103. [CrossRef]

140. Souzaki, M.; Kubo, M.; Kai, M.; Kameda, C.; Tanaka, H.; Taguchi, T.; Tanaka, M.; Onishi, H.; Katano, M. Hedgehog signaling pathway mediates the progression of non-invasive breast cancer to invasive breast cancer. *Cancer Sci.* **2011**, *102*, 373–381. [CrossRef]

141. Cao, X.; Geradts, J.; Dewhirst, M.W.; Lo, H.W. Upregulation of VEGF-A and CD24 gene expression by the tGLI1 transcription factor contributes to the aggressive behavior of breast cancer cells. *Oncogene* **2012**, *31*, 104–115. [CrossRef]

142. Menzl, I.; Lebeau, L.; Pandey, R.; Hassounah, N.B.; Li, F.W.; Nagle, R.; Weihs, K.; McDermott, K.M. Loss of primary cilia occurs early in breast cancer development. *Cilia* **2014**, *3*, 7. [CrossRef]

143. Kwon, Y.J.; Hurst, D.R.; Steg, A.D.; Yuan, K.; Vaidya, K.S.; Welch, D.R.; Frost, A.R. Gli1 enhances migration and invasion via up-regulation of MMP-11 and promotes metastasis in ERalpha negative breast cancer cell lines. *Clin. Exp. Metastasis* **2011**, *28*, 437–449. [CrossRef]

144. Das, S.; Samant, R.S.; Shevde, L.A. Nonclassical activation of Hedgehog signaling enhances multidrug resistance and makes cancer cells refractory to Smoothened-targeting hedgehog inhibition. *J. Biol. Chem.* **2013**, *288*, 11824–11833. [CrossRef]

145. Ruiz-Borrego, M.; Jimenez, B.; Antolín, S.; García-Saenz, J.A.; Corral, J.; Jerez, Y.; Trigo, J.; Urruticoechea, A.; Colom, H.; Gonzalo, N. A phase Ib study of sonidegib (LDE225), an oral small molecule inhibitor of smoothened or Hedgehog pathway, in combination with docetaxel in triple negative advanced breast cancer patients: GEICAM/2012–12 (EDALINE) study. *Investig. N. Drugs* **2019**, *37*, 98–108.

146. Nwabo Kamdje, A.H.; Seke Etet, P.F.; Vecchio, L.; Tagne, R.S.; Amvene, J.M.; Muller, J.M.; Krampera, M.; Lukong, K.E. New targeted therapies for breast cancer: A focus on tumor microenvironmental signals and chemoresistant breast cancers. *World J. Clin. Cases* **2014**, *2*, 769–786. [CrossRef]

147. Peiris-Pages, M.; Sotgia, F.; Lisanti, M.P. Chemotherapy induces the cancer-associated fibroblast phenotype, activating paracrine Hedgehog-GLI signalling in breast cancer cells. *Oncotarget* **2015**, *6*, 10728–10745. [CrossRef]

148. Spivak-Kroizman, T.R.; Hostetter, G.; Posner, R.; Aziz, M.; Hu, C.; Demeure, M.J.; Von Hoff, D.; Hingorani, S.R.; Palculict, T.B.; Izzo, J.; et al. Hypoxia triggers hedgehog-mediated tumor-stromal interactions in pancreatic cancer. *Cancer Res.* **2013**, *73*, 3235–3247. [CrossRef]

149. Fernandes-Silva, H.; Correia-Pinto, J.; Moura, R.S. Canonical sonic hedgehog signaling in early lung development. *J. Dev. Biol.* **2017**, 5. [CrossRef]

150. Kugler, M.C.; Joyner, A.L.; Loomis, C.A.; Munger, J.S. Sonic hedgehog signaling in the lung. From development to disease. *Am. J. Respir. Cell Mol. Biol.* **2015**, *52*, 1–13. [CrossRef]

151. Xu, L.; Xiong, H.; Shi, W.; Zhou, F.; Zhang, M.; Hu, G.; Mei, J.; Luo, S.; Chen, L. Differential expression of sonic hedgehog in lung adenocarcinoma and lung squamous cell carcinoma. *Neoplasma* **2019**, *66*, 839–846. [CrossRef]

152. Lim, S.; Lim, S.M.; Kim, M.J.; Park, S.Y.; Kim, J.H. Sonic hedgehog pathway as the prognostic marker in patients with extensive stage small cell lung cancer. *Yonsei Med. J.* **2019**, *60*, 898–904. [CrossRef]

153. Szczepny, A.; Rogers, S.; Jayasekara, W.S.N.; Park, K.; McCloy, R.A.; Cochrane, C.R.; Ganju, V.; Cooper, W.A.; Sage, J.; Peacock, C.D. The role of canonical and non-canonical Hedgehog signaling in tumor progression in a mouse model of small cell lung cancer. *Oncogene* **2017**, *36*, 5544–5550.

154. Sun, X.; Sun, B.; Cui, M.; Zhou, Z. HERC4 exerts an anti-tumor role through destabilizing the oncoprotein Smo. *Biochem. Biophys. Res. Commun.* **2019**, *513*, 1013–1018.

155. Khatra, H.; Bose, C.; Sinha, S. Discovery of hedgehog antagonists for cancer therapy. *Curr. Med. Chem.* **2017**, *24*, 2033–2058.

156. Peer, E.; Tesanovic, S.; Aberger, F. Next-generation hedgehog/GLI pathway inhibitors for cancer therapy. *Cancers* **2019**, *11*, 538.

157. Mille, F.; Thibert, C.; Fombonne, J.; Rama, N.; Guix, C.; Hayashi, H.; Corset, V.; Reed, J.C.; Mehlen, P. The patched dependence receptor triggers apoptosis through a DRAL-caspase-9 complex. *Nat. Cell Biol.* **2009**, *11*, 739–746. [CrossRef]

158. Petrova, E.; Matevossian, A.; Resh, M.D. Hedgehog acyltransferase as a target in pancreatic ductal adenocarcinoma. *Oncogene* **2015**, *34*, 263–268. [CrossRef]

159. Matevossian, A.; Resh, M.D. Hedgehog acyltransferase as a target in estrogen receptor positive, HER2 amplified, and tamoxifen resistant breast cancer cells. *Mol. Cancer* **2015**, *14*, 72. [CrossRef]

International Journal of
Molecular Sciences

Review

Hedgehog Signaling for Urogenital Organogenesis and Prostate Cancer: An Implication for the Epithelial–Mesenchyme Interaction (EMI)

Taiju Hyuga [1,†], Mellissa Alcantara [1,†], Daiki Kajioka [1], Ryuma Haraguchi [2], Kentaro Suzuki [1], Shinichi Miyagawa [3], Yoshiyuki Kojima [4], Yutaro Hayashi [5] and Gen Yamada [1,*]

[1]   Department of Developmental Genetics, Institute of Advanced Medicine, Wakayama Medical University, Kimiidera 811-1, Wakayama 641-8509, Japan; hyuga520@wakayama-med.ac.jp (T.H.); mcalcant@wakayama-med.ac.jp (M.A.); 09p851kd@wakayama-med.ac.jp (D.K.); k-suzuki@wakayama-med.ac.jp (K.S.)

[2]   Department of Molecular Pathology, Ehime University Graduate School of Medicine, Shitsukawa, Toon City, Ehime 791-0295, Japan; ryumaha@m.ehime-u.ac.jp

[3]   Department of Biological Science and Technology, Faculty of Industrial Science and Technology, Tokyo University of Science, Tokyo 125-8585, Japan; miyagawa@rs.tus.ac.jp

[4]   Department of Urology, Fukushima Medical University School of Medicine, 1 Hikarigaoka, Fukushima 960-1295, Japan; ykojima@fmu.ac.jp

[5]   Department of Pediatric Urology, Nagoya City University, Graduate School of Medical Sciences, 1 Kawasumi, Mizuho-cho, Mizuho-ku, Nagoya 467-8601, Japan; yutaro@med.nagoya-cu.ac.jp

*   Correspondence: gensan7@wakayama-med.ac.jp; Tel.: +81-73-441-0849; Fax: +81-73-499-5026

†   These authors contributed equally to this work.

Received: 19 November 2019; Accepted: 18 December 2019; Published: 20 December 2019

Abstract: Hedgehog (Hh) signaling is an essential growth factor signaling pathway especially in the regulation of epithelial–mesenchymal interactions (EMI) during the development of the urogenital organs such as the bladder and the external genitalia (EXG). The Hh ligands are often expressed in the epithelia, affecting the surrounding mesenchyme, and thus constituting a form of paracrine signaling. The development of the urogenital organ, therefore, provides an intriguing opportunity to study EMI and its relationship with other pathways, such as hormonal signaling. Cellular interactions of prostate cancer (PCa) with its neighboring tissue is also noteworthy. The local microenvironment, including the bone metastatic site, can release cellular signals which can affect the malignant tumors, and vice versa. Thus, it is necessary to compare possible similarities and divergences in Hh signaling functions and its interaction with other local growth factors, such as BMP (bone morphogenetic protein) between organogenesis and tumorigenesis. Additionally, this review will discuss two pertinent research aspects of Hh signaling: (1) the potential signaling crosstalk between Hh and androgen signaling; and (2) the effect of signaling between the epithelia and the mesenchyme on the status of the basement membrane with extracellular matrix structures located on the epithelial–mesenchymal interface.

Keywords: hedgehog; epithelial–mesenchymal interaction (EMI); prostate cancer; external genitalia; androgen; basement membrane; bone morphogenetic protein

---

1. The Basic Architecture of Hedgehog Signaling

Hedgehog (Hh) signaling is an effective and critical regulatory growth factor signal for organogenesis. There are three ligands found in mammals: Sonic hedgehog (Shh), Indian hedgehog (Ihh), and Desert hedgehog (Dhh). Among these three, Shh is the best studied Hh ligand. During development, the expression of Shh is found often in the epithelia [1–3]. In addition to its ligands, various genes involved in the Hh signaling pathway, such as its receptors and transcription factors, have been identified [4]. Ptch (Patched) is a transmembrane receptor that functions as the primary regulator of Hh signaling [5–8]. In the absence of a Hh ligand, Ptch localizes to the base of the primary cilium, the center of Hh signaling in mammals, and prevents the movement of Smo (Smoothened), another membrane signaling component, into the primary cilium. The accumulation of Smo within the primary cilium results in its activation and the processing of downstream mediators. Thus, in the absence of a Hh ligand, Ptch blocks the Hh signaling pathway activation by inhibiting Smo. Therefore, it is the critical step to initiate the signaling cascade.

The downstream effects of the Hh signaling pathway are mediated by the Gli family of transcription factors [9–12]. This family consists of three major regulators: Gli1, Gli2, and Gli3. Gli1 is the transcriptional activator (Gli-A), while Gli2 and Gli3 can be processed into both activator and repressor forms (Gli-R). The accumulation of activated Smo suppresses the generation of Gli-R forms and allows Gli-A proteins to translocate into the nucleus and bind to their target genes. Thus, the canonical Hh signaling pathway is noted with Hh ligand binding to the membrane receptor, which activates Smo regulating the Gli transcription factors.

The Hh ligands are secreted molecules, often from the epithelia. These signals can be received by the neighboring mesenchymal layer, and the presence of both negative and positive regulators of the signaling pathway has been reported in these signal-receiving cells. This establishes a signaling crosstalk between the epithelia and mesenchyme, known as epithelial–mesenchymal interactions (EMI), which drives cell proliferation, differentiation, migration, and other developmental processes. The role of Hh signaling in EMI in both development and cancer will be discussed below.

2. Hedgehog Signaling and EMI in Urogenital Organogenesis

Loss of Hh signaling during mammalian development results in several defects such as polydactyly, cyclopia, and limb malformation, indicating the essential functions of Hh signaling pathway in organogenesis. In several cases, other growth factor signals are also involved in Hh signaling regulation during development. These factors are released from either the epithelia or the mesenchyme, depending on the developmental context. Urogenital organ development is a dynamic process, involving several signaling pathways which changes depending on tissue contexts. The urogenital system includes the urinary tract: composed of the kidney, the ureter, the bladder, and the urethra; and the reproductive tract: composed of the testes, the accessory glands, and the external genitalia (EXG). Recently, an increasing number of regulatory genes for EXG development has been reported [13–17]. Thus, possible insights gleaned from comparisons between urogenital organogenesis with other medical topics, such as prostate cancer, is valuable. This review offers such perspectives based on the recent findings for these topics.

During urogenital tissue formation, especially in the EXG, Hh signaling and its role during EMI is essential. One of its ligands, Shh, is expressed in the endodermal epithelia of the embryonic urethra [18,19]. The Shh released from the epithelia can regulate the differentiation of the bilateral mesenchyme, the tissue located immediately adjacent to the urethra [15,20,21]. Such regulation of the mesenchyme by epithelial signals is a key event for EXG organogenesis, as loss of this signal leads to the agenesis of the EXG anlage, the genital tubercle (GT) [19]. The interactions between this epithelial signal and the mesenchyme has been investigated in the development of several embryonic regions and stages – from the cloaca in the early stages of development (~E10.5) and later (from ~E13.5) in the bladder, the urethra, and the EXG [22]. The role of Hh signaling in EMI in late-stage, androgen-dependent formation of the EXG, including the formation of the male-type urethra, will be discussed below.

Several works report on the role of Shh signaling and EMI during urogenital tract development. In these organs, the Hh ligands are typically expressed in the epithelia; and the expressed growth factor ligands emanate signals towards the mesenchyme, which, in turn, relays signals to the epithelia. This can induce cellular proliferation or differentiation and, in some cases, may also result in the expression of other growth factor signaling genes. Such signaling between the two tissue layers constitutes an essential crosstalk during organogenesis. The growth factor ligands that have been reported to be responsive to Hh include BMP (bone morphogenetic protein), fibroblast growth factor, and Wnt signaling [18,23–26]. To fully understand the essential functions of Hh signaling, investigation of interacting signaling pathways is also necessary. Among them, some BMP ligands are expressed in the developing mesenchyme of the urogenital tract [26–28]. Several signaling studies have reported on the vital role of BMP signaling in mesenchymal differentiation during ureter and bladder smooth muscle development [27,29,30]. In the ureter, this BMP-expressing mesenchyme will form ordered layers with different type of cells, including the lamina propria and the surrounding smooth muscle cells [31]. One of the effective strategies to study the role of BMP signaling is to analyze the corresponding mutant mouse models for such signaling genes. Introduction of loss of function mutations of the Bmp signal-component genes in both ureteric and bladder mesenchyme led to impaired development such as hypoplasia of the smooth muscle [31,32]. Thus, proper BMP signaling has been shown to be essential in urogenital tract development.

Because Shh is expressed in the developing cloacal and ureteric epithelia, the interaction between epithelia-derived Hh signal and corresponding BMP signaling from the nearby mesenchyme was investigated [22]. To analyze this interaction, studies were performed on Gli1-expressing mesenchyme in response to Hh ligand simulation. $CreER^{T2}$ is a tamoxifen inducible Cre recombinase gene, and it is knocked into the Gli1 gene locus by homologous recombination. Thus, taking advantage of the inducible nature of Gli1 gene expression, gene modulation in the target mesenchyme cells was performed using a Gli1-$CreER^{T2}$ modifier mouse strain [22]. Following activation of Hh signaling, BMP signaling increased in the bladder, suggesting that BMP signaling is a downstream event to Hh [27,33]. BmprIa is the major Bmp type I receptor. It is expressed on the cell surface and interacts with the type II receptor. Gli1-mediated gene knock-out of BmprIa led to hypoplastic ureteral smooth muscle formation [27]. As a result, the mutant mouse develops hydronephrosis, a severe urogenital phenotype [27]. Thus, signaling relays from Hh, particularly Shh, toward BMP signaling has been suggested. However, contradictory reports have also been published about such interactions. Addition of Shh to cultured neonatal prostate glands increased Bmp4 expression in the adjacent mesenchyme; however, addition of Noggin, the Bmp antagonist, did not override the Shh-induced growth inhibition [34]. This suggests that the increase of Bmp4 does not mediate the effect of Shh signaling. Further investigations of the downstream cellular functions using conditional mutant mice and other in vivo systems are necessary. The urogenital developmental process, therefore, offers an intriguing research opportunity for understanding EMI between Hh and its interacting signal cascades (Table 1).

Table 1. Representative references showing various mechanisms and functions of epithelial-mesenchymal interactions (EMI) in organogenesis and tumorigenesis (shown by blue color). Hormonal signal has been also reported as locally produced and their actions are cited in the table. Representative EMI processes include hedgehog ligands, interacting growth factor (BMP: bone morphogenetic protein)-Bmp receptor, local hormones (steroidogenesis) and basement membrane (BM). Yellow colored columns represent works showing the crosstalks for hedgehog (Hh)/Bmp and Hh-steroidogenesis. Two papers are marked with parentheses for the roles of BM in EMI.

		Prostate Cancer Progression	Exg Organogenesis
Epithelial–Mesenchymal Interaction (EMI)	Ligands	Hedgehog Signaling *Shh (Sonic hedgehog)* Fan et al., 2004, *Endocrinology* [35] Wilkinson et al., 2013, *Prostate* [36] Nishimori et al., 2012, *J Biol Chem* [37]	*Shh* Haraguchi et al., 2001, *Development* [19] Haraguchi et al., 2007, *Development* [22]
	Mediators	*Gli* Sanchez et al., 2004, *PNAS* [38] Chen et al., 2010, *Mol cancer* [39]	*Gli* Miyagawa et al., 2011, *Endocrinology* [40] He et al., 2016, *PLoS One* [16]
	Hh/Bmp Crosstalk	Shaw et al., 2010, *Differentiation* [41]	Haraguchi et al., 2012, *PLoS One* [27]
	Bmp Ligands	Bmp Signaling *BMP2* Horvath et al., 2004, *Prostate* [42] *BMP4* Lee at al., 2011, *Cancer Res* [43] *BMP7* Masuda et al., 2003, *Prostate* [44]	*BMP4* Kajioka et al., 2019, *Congenit Anom* [24] Ching et al., 2018, *Hum Mol Genet* [28] *BMP7* Suzuki et al., 2008, *Eur J Hum Genet* [26]
	Bmp Receptor	Ide et al., 1997, *Oncogene* [45] Kim et al., 2000, *Cancer Res* [46]	Suzuki et al., 2003, *Development* [23]
	Paracrine Action of Local Steroidogenesis	Lubik et al., 2016, *Int J Cancer* [47] Levina et al., 2012, *prostate* [48]	Suzuki et al., 2017, *Andrology* [20]
		Backdoor production of dihydrotestosterone (DHT) Chang et al., 2011, *PNAS* [49]	*Backdoor production of DHT* O'Shaughnessy et al., 2019, *PLoS Biol* [50]
	Role of the BM	Davis et al., 2001, *Prostate* [51] Hao et al., 1996, *Am J Pathol* [52] Hao et al., 2001, *Am J Pathol* [53] Nagle, 2003, *J Biol Chem* [54]	Lin et al., 2016, Mech Dev [55] Gao et al., 2008, *Genes Dev* [56] Pickering et al., 2017, *Matrix Biol* [57]

3. Hedgehog Signaling and EMI in Prostate Cancer Tumorigenesis

Hh signaling has also been often implicated in tumorigenesis. Over the recent years, prostate cancer (PCa) has been recognized as one of the life-threatening cancers in men. PCa arises in the prostatic epithelia, and it has been reported to interact with the nearby mesenchyme [35,36,41]. In the adult tissue, the expression of the Hh signaling gene is not prominently detected, but it has been found to be present in the regenerating prostatic epithelium [58]. Considering this, the involvement of Shh in PCa tumorigenesis has been suggested. Shh expression has been frequently correlated with a higher Gleason score, the indicator of malignant type PCa [59–61]. In fact, Shh overexpression induced the growth of the tumor, the LNCaP cells, in a mouse xenograft model [35]. Furthermore, Shh has been reported to be highly correlated with the recurrence of the prostate specific antigen (PSA), a protein that is highly expressed in PCa [61]. The expression of Hh signaling components, such as Gli, is also correlated with a higher Gleason score [59–61]. During prostate development, Gli signaling is involved in the differentiation of prostate progenitor cells [62], and Gli mediates the oncogenic transformation of prostate basal cells [63]. Furthermore, Gli1 confers basal-like characteristics onto LNCaP cells, leading to the acquisition of PCa hormone independence [64]. Lastly, it has been reported that Hh signaling through Gli can support androgen signaling in both androgen-deprived and androgen-independent PCa [39,65]. Thus, experimental manipulation of Hh signaling leads to the oncogenic transformation of PCa.

Several studies have been published on the autocrine and paracrine modes of Hh signaling. As Hh signaling proteins can either be expressed solely in the epithelia, or in both epithelia and mesenchyme, there is still no strong consensus on the primary type of Hh signaling in PCa [61,66]. There is a growing concern on the possibility of EMI in PCa progression, as it will provide new information for the development of its treatments [66]. It is possible that Shh released from the prostatic epithelia can affect the nearby mesenchyme and modulate proliferation and differentiation [35]. This interaction may eventually promote PCa tumorigenesis. In addition to paracrine signaling, increasing attention has recently been given to autocrine Hh signaling in PCa [38,66,67]. It has been suggested that some PCa cells change from paracrine type of Hh to autocrine Hh, therefore giving them the ability to produce and respond to the ligand.

Comparing paracrine Hh signaling in organogenesis and tumorigenesis may provide new insights for future research directions [68,69]. BMP has also been investigated as a candidate responsive signaling pathway in PCa. PCa cells are highly invasive towards bone tissue, a phenomenon referred to as bone metastasis [70]. This occurs most prominently in the spine, particularly the lumbar vertebrae [71,72]. The BMP protein was originally identified from bone extracts, suggesting it as a candidate growth factor signal in the local environment. In such PCa metastatic sites, bone-derived signals such as BMP may play a role in modulating PCa malignancy. To investigate bone metastasis, bone-derived cell lines, such as osteoblast cells, were co-incubated with PCa cell lines, such as LNCaP, in experimental systems. Such in vitro systems enabled researchers to investigate the effects of soluble factors for cellular interactions, mimicking the metastatic site of PCa in bones [37]. However, aside from osteoblasts, there are also osteoclasts, fibroblasts, and immune cells present at the metastatic bone sites in vivo, reflecting the complex conditions of the invasive nature of PCa. This merits the use of other experimental assays for investigating the mechanism of bone metastases in PCa. This has led to the development of xenograft models. In this system, the xenograft, which contains a set of cancer cells, can be grafted into a host mouse and analyzed in vivo, simulating the cancer microenvironment. The expression or inhibitory function of the BMP antagonist, Noggin, has been implicated in such xenograft model experiments [41]. In this study, the authors overexpressed Shh in an LNCaP xenograft and found increased BMP7 expression in the neighboring environment, even in the presence of Noggin [41]. The role of Noggin in PCa has yet to be elucidated.

In addition to the above investigations, application of recent analytical techniques may be required. Many PCa tumors contain various types of cancer cells. The mesenchymal cells adjacent to the PCa cells are often termed as cancer associated fibroblasts (CAF). Interactions between PCa and CAFs has been noted to be essential for PCa pathogenesis [36]. This interaction can be regarded as a form of EMI, with the PCa taking the role of the epithelium. It is also noteworthy to mention the heterogeneity of CAFs in PCa [73]. Furthermore, some types of PCa cells are also known for circulating tumor cells [74]. Hence, single-cell level analyses could be a suitable approach to analyze the interaction between PCa and CAF.

In addition to such experimental models, information on the dysregulation of BMP signaling in some metastatic sites is available. Changes in the expression levels of several BMP ligands and BMP receptors have been implicated in bone metastasis [44]. BMPRs are typically expressed in the epithelia in normal tissues, but it has been reported that BMPRIA, BMPRIB, and BMPRII are all down-regulated with increasing prostate malignancy [46]. A separate group also reported that there is a lower expression level of BMPRIB after androgen withdrawal in both cancer-bearing prostate glands and PCa cell lines [45]. As for the mode of release of BMPs in the tumor environment, some BMP ligands have been reported to be released from PCa. In bone metastasis experimental systems, bone-derived BMP7 and PCa-derived BMP4 functions have been reported [43,44,75,76]. Furthermore, loss of BMP2 tends to be associated with an increase in Gleason score [42].

Locally released BMPs and its relationship with Hh signals can also be noted in other reproductive organ tumors, such as ovarian cancer. BMP signals have been reported to function in cell differentiation in the ovary, particularly in the maturation of ovarian follicles [77] and has been reported to play a role in ovarian cancer pathogenesis [78,79]. Similarly, Hh has also been implicated in ovarian cancer progression [80]. All three type of Hh ligands—Dhh, Shh, Ihh—are expressed in ovarian cancers. Dhh expression, in particular, has been suggested to be correlated with its poor prognosis [81]. Interestingly, the presence of ovarian cancer-derived Hh signals and BMP4 from the surrounding mesenchyme has been reported [82]. Furthermore, they report that both Shh and Ihh form a signaling loop with BMP4, leading to chemotherapy resistance [82]. Thus, the importance of Shh has been reported in this system.

Overall, various Hh ligands and its signals have been detected in the epithelia in both urogenital organ development and reproductive organ cancers, such as PCa and ovarian cancer. BMP is often expressed in the mesenchyme of these systems, and signals between these two tissue layers are important for both development and pathogenesis. This putative Hh-Bmp interaction and its role in cancer can be an interesting research topic, based on the transduction by the epithelial Hh signal interpreted at the level of the adjacent mesenchyme during urogenital organ development (Table 1, Figure 1).

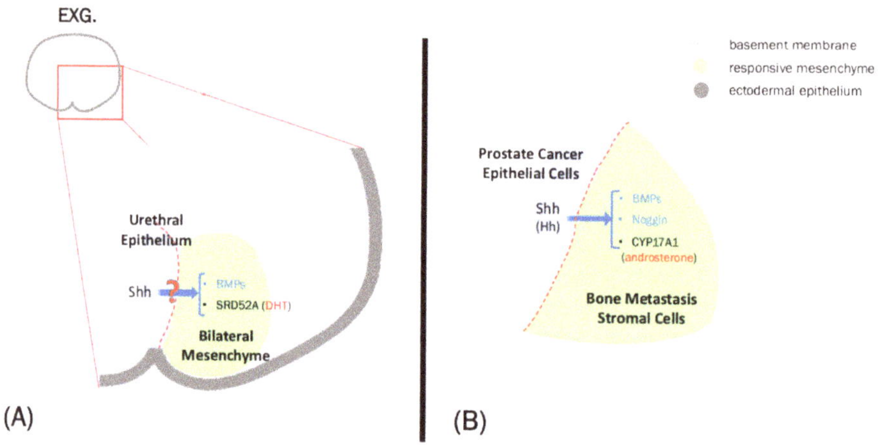

Figure 1. Overview of the role of hedgehog (Hh) in external genitalia (EXG) organogenesis and prostate cancer (PCa). (**A**) Sonic hedgehog (Shh) is expressed in the urethral epithelium and possibly regulates genes expressed in the mesenchyme. The result of this interaction may contribute to the masculinization of the EXG. (**B**) Similarly, Hedgehog ligands, particularly Shh, is expressed in the PCa epithelium and can promote PCa progression by regulating the expression of genes, such as bone morphogenetic protein (BMP), in the stroma. Shh can also promote the backdoor androgen pathway through CYP17A1 up-regulation.

What kind of consequences are expected as a result of such signaling crosstalk? The promotion and inhibition of PCa cell proliferation has been investigated in the field of cancer biology, but whether such environment-derived BMP render positive or negative effects for PCa cell proliferation requires further investigation. Recent studies indicate that the addition of recombinant BMP protein inhibits PCa growth in vitro [83]. Additionally, the effect of this crosstalk in cellular growth has also been analyzed. In general, both positive and negative regulation of proliferation are incorporated during developmental programs [24].

4. Modulation of Hedgehog Signaling by Androgen, the Male Hormone

Recently, progress on the mechanism of sexual differentiation has been described [84], and sexual differentiation with hormonal regulation is given much attention. The hormonal system has been classically described as a remotely acting biological system, as the signal emanates from the central nervous system and affects the target organs through the bloodstream. Testosterone, a gonadal hormone, is produced in the embryonic testes under the control of the central nervous system and regulates the masculinization of the reproductive tract, including the EXG anlage (GT) and other organs [85,86]. The definition of hormonal effects has been recently modified and extended to include their local action and production. These local actions have been suggested to participate in EMI [87], and recent evidence also suggests local modulation of hormone production. Hence, the possible significance of the interaction between local hormone signaling and growth factors is a growing concern.

In this section, the relationship of the male hormonal system with the local growth factor, Hh signal, is discussed in the context of urogenital organ development and PCa.

In mice, EXG development occurs independently of androgen during the early stages. In the male, this is subsequently followed by an androgen-dependent masculinization process [15,88,89]. Robust mouse embryonic GT outgrowth is promoted by testicular androgens, starting from mid-gestation (~E14.0–15.0) until after birth [90]. The development of the male GT continues with the formation of a tubular urethra, a well-developed prepuce, and the condensation of a bilateral prospective corporal body. The male urethra is incorporated into the glans. This entire process is mediated by androgen, and these effects are generally described as consequences of "positive" androgen actions. Proper development of the male urethra enables efficient ejaculation during copulation and physiological reproductive male functions. In contrast to male development, the female GT does not form a tubular urethra. Thus, urethral formation is a useful landmark for investigating the mechanisms of masculinization and androgen-dependent signaling cascades.

Androgens, such as testosterone and 5α-dihydrotestosterone (DHT), are produced locally in both embryonic and adult tissues. Both androgens possess essential roles for GT organogenesis. In fact, androgen administration is used during treatment of several conditions in human patients. As administration of androgen can promote penis elongation, it is used as a treatment for micropenis [91]. Similarly, urologists administer androgen to boost the outgrowth of a patient's penis before operating on conditions such as hypospadias [92,93]. A recent study has suggested that DHT can negatively regulate cell proliferation in the ventral side of the GT during urethral formation [20]. The enzyme which converts testosterone into DHT, type II 5α-reductase, is expressed in the bilateral mesenchyme prior to male-type urethral formation. This enzyme is encoded by SRD5A2 [20]. Direct measurement analysis using mass spectrometry revealed the differential distribution of DHT. Data showed that there is a higher production of DHT in the ventral side of male GT [94], which is the same region of the bilateral mesenchyme showing reduced rate of cellular proliferation. Thus, higher levels of locally converted DHT can reduce cell proliferation, an event necessary for male-type urethral formation. Although DHT-target regulators for organogenesis have yet to be identified, MafB has been reported to be a DHT-responsive gene for urethral formation [95,96]. Therefore, it is possible that the regulated production of DHT leads to reduced cell proliferation in the bilateral mesenchyme under the control of transcription factor MafB, which regulates male-type urethral formation.

The regulation of cell proliferation in the different stages of PC by androgens has been reported. In contrast to the developmental process, PCa progression is initially dependent on androgens but may switch into androgen-independent during the advanced stages. It has long been established that the proliferation and tumor growth of PCa is driven by androgen [97,98]. LNCaP cells proliferated in response to the concentration of DHT in vitro [99]. Furthermore, after injecting LNCaP cells in nude mice, the frequency of tumor development was significantly higher in the male versus female mice [99], indicating that the presence of androgens is necessary for tumor growth. Several other studies also report on the regulation of cell-cycle genes by androgen in other PCa models [100,101] and in normal conditions [102]. In both cases, androgen increased the expression of cyclins and their kinases while decreasing the expression of its inhibitors. This is further evidence of the proliferative effect of the androgens towards PCa. Similar to the developmental context, however, there are reports of inhibitory activities on PCa cell proliferation by androgen. In human castration-resistant prostate cancer (CRPC), addition of DHT did not usually exhibit a positive effect on cellular proliferation but instead inhibited it [103].

Due to the dependence of tumor growth on androgens, castration and administration of androgen signaling inhibitors are frequent treatments for PCa [104]. Although most cases respond positively to androgen deprivation, some cases proceed to a more lethal form of PCa. This stage, known as CRPC, is often associated with bone metastasis and has been reported to progress despite approximately castration levels of androgens. The formation of CRPC is reported to occur either through the androgen receptor (AR) or independently of AR. The AR is a nuclear receptor, which, upon binding its ligand, mediates downstream signaling. AR is reported to be present in CRPC tumors. In fact, it has been reported that AR is often overexpressed or mutated in CRPC [105,106]. AR levels has been reported to be higher in CRPC xenografts compared to primary tumors or benign prostatic hyperplasia [106]. This increase in AR has been suggested to increase sensitivity of the PCa to low levels of androgens, thus allowing the cancer to progress further.

Changes in steroid levels due to local steroidogenesis has been reported in CRPC using mass spectrometry [107], and this has been suggested as another mechanism by which CRPC progresses in a low-androgen environment. In the backdoor pathway, DHT is derived from androstenedione, an intermediate in the classical androgen biosynthesis pathway [49]. Gene expression analysis has confirmed the presence these enzymes and intermediates in PCa sites [49], and addition of androstenedione to androgen-independent PCa lines significantly enhanced tumor growth [49]. Also, the conversion of androgens via the backdoor pathway has been reported to induce cell proliferation [108]. Furthermore, androgen deprivation has been reported to promote DHT production via the backdoor pathway in PCa [109]. The backdoor pathway was first reported in the development of the prostate and testes in marsupials [110,111]. Since then, it has also been reported in mice [112,113] and humans [50,113,114]. It has recently been suggested that backdoor androgens play a role during masculinization. During human reproductive development, it has been reported that DHT levels in the fetal circulation and fetal testes are low. Instead, androsterone, an intermediate in the backdoor androgen synthesis pathway, is the major androgen within the circulation. It is, however, not expressed in the fetal testes but was instead detected in non-gonadal tissues such as the placenta. Furthermore, androsterone and testosterone are the only androgens that is higher in the male than in the female fetal circulation [50]. Taken together, this suggests that masculinization is dependent not only on the fetal testes, but also on other tissues capable of locally producing androgens. As DHT is low in the fetal testes, it is also suggested that androsterone may be converted into DHT within the GT [50]. Further studies related to the production of DHT and the GT are necessary.

Of note, it has also been suggested that paracrine type Shh signaling is involved in the local steroidogenesis in PCa [47]. Although it is generally believed that local steroidogenesis occurs within the tumor, there have been several reports stating that the stroma is also capable of steroidogenesis [47,48,115]. Steroidogenesis related with Hh signal in the mesenchyme of bones is also discussed in such works. This may be recognized as a form of EMI between Hh signaling and androgens in the cancer environment (Table 1, Figure 1). Whether this paracrine interaction between growth factor signals, such as Shh, and hormonal signals occurs during organogenesis is still unclear. Modes of paracrine action may differ from stage to stage in organogenesis, as well as in PCa status. Hence, the role of Hh and androgen signaling in development and cancer may be discussed not only in the context of EMI but also in the chronological context.

Androgen signaling is mediated by the nuclear androgen receptor AR. Gli1, the major Hh downstream transcriptional regulator, may interact with AR in some types of PCa cell lines [39]. BMP signal is transduced in the responding cell by pSmad signaling. There are some studies indicating the presence of pSmad–AR interactions in the responding cell nucleus [116]. Several other cofactors have been reported to interact with AR and pSmad, respectively [116]. The involvement of male hormones and its potential crosstalk with Hh and BMP signals in PCa progression and organogenesis requires further analysis.

5. The Status of the Basement Membrane (BM) As a New Aspect of EMI Regulation

Previously, the description of EMI has been limited to the cellular layers, but recently, the participation of the basement membrane (BM) in EMI has been suggested [56]. When the BM was first described, its primary function was stated as a structural protein that acts as a barrier or as an adhesion site. Since then, it has been revealed that the BM possesses a wide range of functions that is critical in development and cancer [54,117,118]. For example, one of the main BM proteins, laminin, can regulate various biological processes by directly activating signaling pathways through its interactions with integrin receptors [119–121]. More recently, the status of the BM, including BM protein deposition and the BM integrity, has been described as an emerging regulator of signaling related with the function of EMI.

The proper development of the BM is essential during organogenesis: its formation and subsequent removal is an essential step in the development of several organs. This removal is particularly essential in cases wherein there is tissue fusion in the midline region, such as in the formation of the male-type urethra and the palate. Morphological and loss of function studies on palatal fusion revealed the necessity of regulating the status of the BM during this process [122]. It has been suggested that soluble growth factor signals, such as TGF-β, may participate in regulating this process [123]. Moreover, the failure of the midline to fuse properly has been suggested to result in palatal cleft phenotypes. Hence, the attention to defects in the regulation of the BM components, which results in developmental abnormalities, has been increasing.

Shh is one of the important soluble factors involved in the maintenance of the BM during organogenesis. The BM is composed of several extracellular matrix proteins, including laminin and collagen. During urogenital organ development, loss of the α5 chain of laminin has been reported to result in the formation of an abnormal urethra [55]. As mentioned above, growth factor signals, including Hh, are essential in the regulation of urethral formation during early mouse development (E11.5–E14) [18,19,22]. Shh is prominently expressed in the embryonic urethral epithelium in the mouse, and functional disruption of Shh during early development (~E10.5) leads to GT agenesis [19]. Subsequent conditional gene knock-out studies for Shh in later stages (E11–13.5) revealed that Shh is essential in urethral formation, as such knock-out models showed a hypospadias-like phenotype [40]. Detailed gene expression analysis revealed that absence of Shh expression during later stages of urethral development (E15.5–16.5) might be associated with the reduced level of laminin in the BM [Alcantara, unpublished]. In fact, Shh has been shown to be necessary for laminin deposition and BM formation in the myotome [124]. Thus, the loss of laminin may indicate that Shh signaling is required for the maintenance of BM structure in the prospective fusion site in the murine embryonic urethra. The effect of Shh on laminin regulation, however, during male-type urethral fusion should be analyzed further.

How the BM participates in EMI is a growing research field. As the disappearance of the epithelia is associated with the breakdown of the BM, it is likely that the changes in the BM are mediated, at least in part, by the epithelia. As mentioned, epithelial Shh can induce the deposition of laminin in the BM. Loss of Hh signaling results in defective myotome in both mice and zebrafish [57,124]. However, the possible role of the mesenchyme should not be discounted. In the zebrafish myotome, epithelial Shh-induced laminin deposition inhibited BMP signaling in the mesoderm leading to myotomal cell fate specification [57]. Thus, Hh from the epithelia may participate in paracrine signaling via the BM for the above organogenesis. In contrast, laminin can induce Shh in the hair follicle epithelia by triggering Noggin expression in the mesenchyme [56]. This evidence shows that the BM should be taken into consideration during EMI studies.

Another emerging possibility for the role of BM integrity in EMI is its potential to regulate protein diffusion between the epithelia and the mesenchyme. A study performed on MCF10A acini showed that the status of the BM is positively correlated with molecular permeability [125]. As the breast acini matures, the BM develops and thickens with it. Therefore, they compared the rate of permeation by dextran in low-matured and semi-matured acini. They revealed that in low-matured acini, the dextran penetrated the thin BM efficiently. On the other hand, semi-matured acini with thicker BM could significantly slow down permeation of the dextran [125]. Hence, it is possible that the changes in the BM integrity is responsible for allowing the diffusion of proteins, consequently influencing organogenesis. A similar case might occur in the urethra, where there is loss of BM integrity specifically at the fusion site of the embryonic urethral epithelia [Alcantara, unpublished]. Further investigation is necessary to establish such a role of BM in development.

Cell migration is a key event during metastasis, and it is important to consider the role of the BM during this event. The proper regulation and dysregulation of the BM can be related with PCa progression, as there is differential expression of BM proteins in normal and malignant prostate cells [54,126]. In fact, loss of BM continuity has been correlated with PCa malignancy [127]. It has also been reported that malignant PCa cells lack the $\gamma2$ sub-chain of laminin-5 (laminin-332) [51,52,54]. This loss of laminin-5 function may occur only in PCa, despite being highly expressed in other cancers, such as colon and breast carcinomas [51]. This has been suggested to contribute to accelerated migratory behavior in PCa [53], suggesting that the status of the BM is important for metastasis. It is also interesting to note that this loss in laminin-5 is concurrent with the activation of Shh signaling during progression of PCa. It has long been established that Shh is detected in metastasizing PCa, regulating the signaling pathways for proliferation and invasion [128]. In fact, overexpressing Shh can transform normal prostate cells into metastatic pCa cells [129]. There appears to be a possibility that Shh promotes PCa metastasis by affecting the BM condition via laminin-5. Whether these observations can be applied to the varying conditions of PCa in vivo should be studied further. Thus, the involvement of Hh signaling in the regulation of the BM during PCa formation or progression has yet to be established. Further studies are necessary to investigate the possible correlations of abnormal cell migration to the status of the BM in mutant mice. Alteration of actions of growth factors and hormones and cellular behaviors across the affected BM region should be investigated further (Table 1, Figure 1).

Modulating Hh signal may lead to altered laminin production in the BM. Matrix metalloproteinases (MMPs), which can cleave extracellular matrix proteins, have been suggested to be involved in tumorigenesis in several cancers, including PCa [130,131]. Whether the correlation of such enzymes with abnormal distribution of proteins is possible and requires further analysis. It has been reported that Shh signaling can regulate the expression of MMPs, leading to accelerated migration in ovarian, liver, and gastric cancers [132–134]. Hence, Hh signaling may also indirectly affect the status of the BM via MMP regulation.

6. Conclusions

EMI (epithelial–mesenchymal interaction) is a fundamental concept which governs both organogenesis and cancer progression. In this text, we use the development of the EXG and prostate cancer as representatives of these events. Hh signaling has been shown to play a vital role in both processes, and it is worth noting that the effect of Hh signaling might either be positive or detrimental depending on the tissue and developmental context. This review offers a summary of the possible and known roles of Hh signaling during EMI. Although there are several other signaling pathways by which Hh can interact, we show here the importance of the Hh/Bmp signaling interactions in the two systems. This crosstalk is necessary for the proper development of the urogenital system and is also responsible for driving part of PCa malignancy. Furthermore, Shh can induce local steroidogenesis of androgens, a hormone that is critical for both events of interest. This local action is currently a concern as it offers possible therapeutic targets for PCa. Lastly, we discussed the potential role of the BM during EMI. The BM has long been believed to function as a barrier between the epithelia and the mesenchyme, but we describe that Hh can participate in EMI via the BM. Insights from the role of Hh in each biological event may contribute further understanding of this complex signaling pathway. Studying this pathway using comparative and interdisciplinary viewpoints would be necessary to understand even better the Hh signaling function and its implications in EMI.

Author Contributions: Conceptualization: G.Y.; Data curation: G.Y., K.S.; Funding acquisition: G.Y., S.M.; Project administration: G.Y., K.S.; Visualization: R.H., Y.K., Y.H.; Writing—original draft: G.Y., T.H., M.A., D.K. All authors have read and agreed to the published version of the manuscript.

Funding: This research was supported by the Japan Society for the Promotion of Science grants 18H02474, 18K06938, 18K06837, 17H06432, 17K18024, 15H04300, 15K15403, 15K10647, 15K19013, and 15J11033.

Acknowledgments: The authors thank Sohei Kitazawa, Hideo Nakai, Alvin R. Acebedo, Gail S. Prins, Gerald R. Cunha, Xue Li, and Marilyn B. Renfree for their encouragements. The authors also thank Y. Rim and all laboratory colleagues for their assistance.

Conflicts of Interest: The authors declare no conflict of interest.

Abbreviations

Hh	hedgehog
EMI	epithelial–mesenchymal interactions
EXG	external genitalia
PCa	prostate cancer
BMP	bone morphogenetic protein
Shh	Sonic hedgehog
Ihh	Indian hedgehog
Dhh	Desert hedgehog
Ptch	Patched
Smo	Smoothened
GT	genital tubercle
PSA	prostate specific antigen
CAF	cancer associated fibroblasts
DHT	dihydrotestosterone
CRPC	castration-resistant prostate cancer
AR	androgen receptor
BM	basement membrane
MMP	matrix metalloproteinases

References

1. Varjosalo, M.; Taipale, J. Hedgehog: Functions and mechanisms. *Genes Dev.* **2008**, *22*, 2454–2472. [CrossRef]
2. Prins, G.S.; Putz, O. Molecular signaling pathways that regulate prostate gland development. *Differentiation* **2008**, *76*, 641–659. [CrossRef] [PubMed]
3. Guo, C.; Sun, Y.; MacDonald, B.T.; Borer, J.G.; Li, X. Dkk1 in the peri-cloaca mesenchyme regulates formation of anorectal and genitourinary tracts. *Dev. Biol.* **2014**, *385*, 41–51. [CrossRef] [PubMed]
4. Jiang, J.; Hui, C.C. Hedgehog signaling in development and cancer. *Dev. Cell* **2008**, *15*, 801–812. [CrossRef] [PubMed]
5. Stone, D.M.; Hynes, M.; Armanini, M.; Swanson, T.A.; Gu, Q.; Johnson, R.L.; Scott, M.P.; Pennica, D.; Goddard, A.; Phillips, H.; et al. The tumour-suppressor gene patched encodes a candidate receptor for Sonic hedgehog. *Nature* **1996**, *384*, 129–134. [CrossRef]
6. Shima, Y.; Morohashi, K.I. Leydig progenitor cells in fetal testis. *Mol. Cell Endocrinol.* **2017**, *445*, 55–64. [CrossRef]
7. Boonen, S.E.; Stahl, D.; Kreiborg, S.; Rosenberg, T.; Kalscheuer, V.; Larsen, L.A.; Tommerup, N.; Brøndum-Nielsen, K.; Tümer, Z. Delineation of an interstitial 9q22 deletion in basal cell nevus syndrome. *Am. J. Med. Genet. A* **2005**, *132*, 324–328. [CrossRef]
8. Chen, C.P.; Lin, S.P.; Wang, T.H.; Chen, Y.J.; Chen, M.; Wang, W. Perinatal findings and molecular cytogenetic analyses of de novo interstitial deletion of 9q (9q22.3→q31.3) associated with Gorlin syndrome. *Prenat. Diagn.* **2006**, *26*, 725–729. [CrossRef]
9. Démurger, F.; Ichkou, A.; Mougou-Zerelli, S.; Le Merrer, M.; Goudefroye, G.; Delezoide, A.L.; Quélin, C.; Manouvrier, S.; Baujat, G.; Fradin, M.; et al. New insights into genotype-phenotype correlation for GLI3 mutations. *Eur. J. Hum. Genet.* **2015**, *23*, 92–102. [CrossRef]
10. Gustavsson, P.; Schoumans, J.; Staaf, J.; Jönsson, G.; Carlsson, F.; Kristoffersson, U.; Borg, A.; Nordenskjöld, M.; Dahl, N. Hemizygosity for chromosome 2q14.2-q22.1 spanning the GLI2 and PROC genes associated with growth hormone deficiency, polydactyly, deep vein thrombosis and urogenital abnormalities. *Clin. Genet.* **2006**, *69*, 441–443. [CrossRef]
11. Greally, M.T.; Robinson, E.; Allen, N.M.; O'Donovan, D.; Crolla, J.A. De novo interstitial deletion 2q14.1q22.1: Is there a recognizable phenotype? *Am. J. Med. Genet. A* **2014**, *164*, 3194–3202. [CrossRef] [PubMed]

12. Ma, D.; Marion, R.; Punjabi, N.P.; Pereira, E.; Samanich, J.; Agarwal, C.; Li, J.; Huang, C.K.; Ramesh, K.H.; Cannizzaro, L.A.; et al. A de novo 10.79 Mb interstitial deletion at 2q13q14.2 involving PAX8 causing hypothyroidism and mullerian agenesis: A novel case report and literature review. *Mol. Cytogenet.* **2014**, *7*, 85. [CrossRef] [PubMed]

13. Yamada, G.; Satoh, Y.; Baskin, L.S.; Cunha, G.R. Cellular and molecular mechanisms of development of the external genitalia. *Differentiation* **2003**, *71*, 445–460. [CrossRef] [PubMed]

14. Hashimoto, D.; Hyuga, T.; Acebedo, A.R.; Alcantara, M.C.; Suzuki, K.; Yamada, G. Developmental mutant mouse models for external genitalia formation. *Congenit. Anom.* **2019**, *59*, 74–80. [CrossRef] [PubMed]

15. Hyuga, T.; Suzuki, K.; Acebedo, A.R.; Hashimoto, D.; Kajimoto, M.; Miyagawa, S.; Enmi, J.I.; Yoshioka, Y.; Yamada, G. Regulatory roles of epithelial-mesenchymal interaction (EMI) during early and androgen dependent external genitalia development. *Differentiation* **2019**, *110*, 29–35. [CrossRef]

16. He, F.; Akbari, P.; Mo, R.; Zhang, J.J.; Hui, C.C.; Kim, P.C.; Farhat, W.A. Adult Gli2+/−;Gli3Δ699/+ Male and Female Mice Display a Spectrum of Genital Malformation. *PLoS ONE* **2016**, *11*, e0165958. [CrossRef]

17. Seifert, A.W.; Zheng, Z.; Ormerod, B.K.; Cohn, M.J. Sonic hedgehog controls growth of external genitalia by regulating cell cycle kinetics. *Nat. Commun.* **2010**, *1*, 23. [CrossRef]

18. Haraguchi, R.; Suzuki, K.; Murakami, R.; Sakai, M.; Kamikawa, M.; Kengaku, M.; Sekine, K.; Kawano, H.; Kato, S.; Ueno, N.; et al. Molecular analysis of external genitalia formation: The role of fibroblast growth factor (Fgf) genes during genital tubercle formation. *Development* **2000**, *127*, 2471–2479.

19. Haraguchi, R.; Mo, R.; Hui, C.; Motoyama, J.; Makino, S.; Shiroishi, T.; Gaffield, W.; Yamada, G. Unique functions of Sonic hedgehog signaling during external genitalia development. *Development* **2001**, *128*, 4241–4250.

20. Suzuki, H.; Matsushita, S.; Suzuki, K.; Yamada, G. 5α-Dihydrotestosterone negatively regulates cell proliferation of the periurethral ventral mesenchyme during urethral tube formation in the murine male genital tubercle. *Andrology* **2017**, *5*, 146–152. [CrossRef]

21. Matsushita, S.; Suzuki, K.; Murashima, A.; Kajioka, D.; Acebedo, A.R.; Miyagawa, S.; Haraguchi, R.; Ogino, Y.; Yamada, G. Regulation of masculinization: Androgen signalling for external genitalia development. *Nat. Rev. Urol.* **2018**, *15*, 358–368. [CrossRef] [PubMed]

22. Haraguchi, R.; Motoyama, J.; Sasaki, H.; Satoh, Y.; Miyagawa, S.; Nakagata, N.; Moon, A.; Yamada, G. Molecular analysis of coordinated bladder and urogenital organ formation by Hedgehog signaling. *Development* **2007**, *134*, 525–533. [CrossRef] [PubMed]

23. Suzuki, K.; Bachiller, D.; Chen, Y.P.; Kamikawa, M.; Ogi, H.; Haraguchi, R.; Ogino, Y.; Minami, Y.; Mishina, Y.; Ahn, K.; et al. Regulation of outgrowth and apoptosis for the terminal appendage: External genitalia development by concerted actions of BMP signaling [corrected]. *Development* **2003**, *130*, 6209–6220. [CrossRef] [PubMed]

24. Kajioka, D.; Suzuki, K.; Nakada, S.; Matsushita, S.; Miyagawa, S.; Takeo, T.; Nakagata, N.; Yamada, G. Bmp4 is an essential growth factor for the initiation of genital tubercle (GT) outgrowth. *Congenit. Anom.* **2019**. [CrossRef]

25. Harada, M.; Omori, A.; Nakahara, C.; Nakagata, N.; Akita, K.; Yamada, G. Tissue-specific roles of FGF signaling in external genitalia development. *Dev. Dyn.* **2015**, *244*, 759–773. [CrossRef]

26. Suzuki, K.; Haraguchi, R.; Ogata, T.; Barbieri, O.; Alegria, O.; Vieux-Rochas, M.; Nakagata, N.; Ito, M.; Mills, A.A.; Kurita, T.; et al. Abnormal urethra formation in mouse models of split-hand/split-foot malformation type 1 and type 4. *Eur. J. Hum. Genet.* **2008**, *16*, 36–44. [CrossRef]

27. Haraguchi, R.; Matsumaru, D.; Nakagata, N.; Miyagawa, S.; Suzuki, K.; Kitazawa, S.; Yamada, G. The hedgehog signal induced modulation of bone morphogenetic protein signaling: An essential signaling relay for urinary tract morphogenesis. *PLoS ONE* **2012**, *7*, e42245. [CrossRef]

28. Ching, S.T.; Infante, C.R.; Du, W.; Sharir, A.; Park, S.; Menke, D.B.; Klein, O.D. Isl1 mediates mesenchymal expansion in the developing external genitalia via regulation of Bmp4, Fgf10 and Wnt5a. *Hum. Mol. Genet.* **2018**, *27*, 107–119. [CrossRef]

29. Brenner-Anantharam, A.; Cebrian, C.; Guillaume, R.; Hurtado, R.; Sun, T.T.; Herzlinger, D. Tailbud-derived mesenchyme promotes urinary tract segmentation via BMP4 signaling. *Development* **2007**, *134*, 1967–1975. [CrossRef]

30. Miyazaki, Y.; Oshima, K.; Fogo, A.; Ichikawa, I. Evidence that bone morphogenetic protein 4 has multiple biological functions during kidney and urinary tract development. *Kidney Int.* **2003**, *63*, 835–844. [CrossRef]

31. Bohnenpoll, T.; Kispert, A. Ureter growth and differentiation. *Semin. Cell Dev. Biol.* **2014**, *36*, 21–30. [CrossRef] [PubMed]

32. Miyazaki, Y.; Oshima, K.; Fogo, A.; Hogan, B.L.; Ichikawa, I. Bone morphogenetic protein 4 regulates the budding site and elongation of the mouse ureter. *J. Clin. Investig.* **2000**, *105*, 863–873. [CrossRef] [PubMed]

33. Pan, A.; Chang, L.; Nguyen, A.; James, A.W. A review of hedgehog signaling in cranial bone development. *Front. Physiol.* **2013**, *4*, 61. [CrossRef] [PubMed]

34. Pu, Y.; Huang, L.; Prins, G.S. Sonic hedgehog-patched Gli signaling in the developing rat prostate gland: Lobe-specific suppression by neonatal estrogens reduces ductal growth and branching. *Dev. Biol.* **2004**, *273*, 257–275. [CrossRef]

35. Fan, L.; Pepicelli, C.V.; Dibble, C.C.; Catbagan, W.; Zarycki, J.L.; Laciak, R.; Gipp, J.; Shaw, A.; Lamm, M.L.; Munoz, A.; et al. Hedgehog signaling promotes prostate xenograft tumor growth. *Endocrinology* **2004**, *145*, 3961–3970. [CrossRef]

36. Wilkinson, S.E.; Furic, L.; Buchanan, G.; Larsson, O.; Pedersen, J.; Frydenberg, M.; Risbridger, G.P.; Taylor, R.A. Hedgehog signaling is active in human prostate cancer stroma and regulates proliferation and differentiation of adjacent epithelium. *Prostate* **2013**, *73*, 1810–1823. [CrossRef]

37. Nishimori, H.; Ehata, S.; Suzuki, H.I.; Katsuno, Y.; Miyazono, K. Prostate cancer cells and bone stromal cells mutually interact with each other through bone morphogenetic protein-mediated signals. *J. Biol. Chem.* **2012**, *287*, 20037–20046. [CrossRef]

38. Sanchez, P.; Hernández, A.M.; Stecca, B.; Kahler, A.J.; DeGueme, A.M.; Barrett, A.; Beyna, M.; Datta, M.W.; Datta, S.; Ruiz i Altaba, A. Inhibition of prostate cancer proliferation by interference with SONIC HEDGEHOG-GLI1 signaling. *Proc. Natl. Acad. Sci. USA* **2004**, *101*, 12561–12566. [CrossRef]

39. Chen, M.; Feuerstein, M.A.; Levina, E.; Baghel, P.S.; Carkner, R.D.; Tanner, M.J.; Shtutman, M.; Vacherot, F.; Terry, S.; de la Taille, A.; et al. Hedgehog/Gli supports androgen signaling in androgen deprived and androgen independent prostate cancer cells. *Mol. Cancer* **2010**, *9*, 89. [CrossRef]

40. Miyagawa, S.; Matsumaru, D.; Murashima, A.; Omori, A.; Satoh, Y.; Haraguchi, R.; Motoyama, J.; Iguchi, T.; Nakagata, N.; Hui, C.C.; et al. The role of sonic hedgehog-Gli2 pathway in the masculinization of external genitalia. *Endocrinology* **2011**, *152*, 2894–2903. [CrossRef]

41. Shaw, A.; Gipp, J.; Bushman, W. Exploration of Shh and BMP paracrine signaling in a prostate cancer xenograft. *Differentiation* **2010**, *79*, 41–47. [CrossRef] [PubMed]

42. Horvath, L.G.; Henshall, S.M.; Kench, J.G.; Turner, J.J.; Golovsky, D.; Brenner, P.C.; O'Neill, G.F.; Kooner, R.; Stricker, P.D.; Grygiel, J.J.; et al. Loss of BMP2, Smad8, and Smad4 expression in prostate cancer progression. *Prostate* **2004**, *59*, 234–242. [CrossRef] [PubMed]

43. Lee, Y.C.; Cheng, C.J.; Bilen, M.A.; Lu, J.F.; Satcher, R.L.; Yu-Lee, L.Y.; Gallick, G.E.; Maity, S.N.; Lin, S.H. BMP4 promotes prostate tumor growth in bone through osteogenesis. *Cancer Res.* **2011**, *71*, 5194–5203. [CrossRef] [PubMed]

44. Masuda, H.; Fukabori, Y.; Nakano, K.; Takezawa, Y.; CSuzuki, T.; Yamanaka, H. Increased expression of bone morphogenetic protein-7 in bone metastatic prostate cancer. *Prostate* **2003**, *54*, 268–274. [CrossRef]

45. Ide, H.; Katoh, M.; Sasaki, H.; Yoshida, T.; Aoki, K.; Nawa, Y.; Osada, Y.; Sugimura, T.; Terada, M. Cloning of human bone morphogenetic protein type IB receptor (BMPR-IB) and its expression in prostate cancer in comparison with other BMPRs. *Oncogene* **1997**, *14*, 1377–1382. [CrossRef]

46. Kim, I.Y.; Lee, D.H.; Ahn, H.J.; Tokunaga, H.; Song, W.; Devereaux, L.M.; Jin, D.; Sampath, T.K.; Morton, R.A. Expression of bone morphogenetic protein receptors type-IA, -IB and -II correlates with tumor grade in human prostate cancer tissues. *Cancer Res.* **2000**, *60*, 2840–2844.

47. Lubik, A.A.; Nouri, M.; Truong, S.; Ghaffari, M.; Adomat, H.H.; Corey, E.; Cox, M.E.; Li, N.; Guns, E.S.; Yenki, P.; et al. Paracrine sonic hedgehog signaling contributes significantly to acquired steroidogenesis in the prostate tumor microenvironment. *Int. J. Cancer* **2017**, *140*, 358–369. [CrossRef]

48. Levina, E.; Chen, M.; Carkner, R.; Shtutman, M.; Buttyan, R. Paracrine Hedgehog increases the steroidogenic potential of prostate stromal cells in a Gli-dependent manner. *Prostate* **2012**, *72*, 817–824. [CrossRef]

49. Chang, K.H.; Li, R.; Papari-Zareei, M.; Watumull, L.; Zhao, Y.D.; Auchus, R.J.; Sharifi, N. Dihydrotestosterone synthesis bypasses testosterone to drive castration-resistant prostate cancer. *Proc. Natl. Acad. Sci. USA* **2011**, *108*, 13728–13733. [CrossRef]

50. O'Shaughnessy, P.J.; Antignac, J.P.; Le Bizec, B.; Morvan, M.L.; Svechnikov, K.; Söder, O.; Savchuk, I.; Monteiro, A.; Soffientini, U.; Johnston, Z.C.; et al. Alternative (backdoor) androgen production and masculinization in the human fetus. *PLoS Biol.* **2019**, *17*, e3000002. [CrossRef]

51. Davis, T.L.; Cress, A.E.; Dalkin, B.L.; Nagle, R.B. Unique expression pattern of the alpha6beta4 integrin and laminin-5 in human prostate carcinoma. *Prostate* **2001**, *46*, 240–248. [CrossRef]

52. Hao, J.; Yang, Y.; McDaniel, K.M.; Dalkin, B.L.; Cress, A.E.; Nagle, R.B. Differential expression of laminin 5 (alpha 3 beta 3 gamma 2) by human malignant and normal prostate. *Am. J. Pathol.* **1996**, *149*, 1341–1349. [PubMed]

53. Hao, J.; Jackson, L.; Calaluce, R.; McDaniel, K.; Dalkin, B.L.; Nagle, R.B. Investigation into the mechanism of the loss of laminin 5 (alpha3beta3gamma2) expression in prostate cancer. *Am. J. Pathol.* **2001**, *158*, 1129–1135. [CrossRef]

54. Nagle, R.B. Role of the extracellular matrix in prostate carcinogenesis. *J. Cell Biochem.* **2004**, *91*, 36–40. [CrossRef]

55. Lin, C.; Werner, R.; Ma, L.; Miner, J.H. Requirement for basement membrane laminin α5 during urethral and external genital development. *Mech. Dev.* **2016**, *141*, 62–69. [CrossRef]

56. Gao, J.; DeRouen, M.C.; Chen, C.H.; Nguyen, M.; Nguyen, N.T.; Ido, H.; Harada, K.; Sekiguchi, K.; Morgan, B.A.; Miner, J.H.; et al. Laminin-511 is an epithelial message promoting dermal papilla development and function during early hair morphogenesis. *Genes Dev.* **2008**, *22*, 2111–2124. [CrossRef]

57. Pickering, J.; Cunliffe, V.T.; Van Eeden, F.; Borycki, A.G. Hedgehog signalling acts upstream of Laminin alpha1 transcription in the zebrafish paraxial mesoderm. *Matrix Biol.* **2017**, *62*, 58–74. [CrossRef]

58. Freestone, S.H.; Marker, P.; Grace, O.C.; Tomlinson, D.C.; Cunha, G.R.; Harnden, P.; Thomson, A.A. Sonic hedgehog regulates prostatic growth and epithelial differentiation. *Dev. Biol.* **2003**, *264*, 352–362. [CrossRef]

59. Sheng, T.; Li, C.; Zhang, X.; Chi, S.; He, N.; Chen, K.; McCormick, F.; Gatalica, Z.; Xie, J. Activation of the hedgehog pathway in advanced prostate cancer. *Mol. Cancer* **2004**, *3*, 29. [CrossRef]

60. Azoulay, S.; Terry, S.; Chimingqi, M.; Sirab, N.; Faucon, H.; Gil Diez de Medina, S.; Moutereau, S.; Maillé, P.; Soyeux, P.; Abbou, C.; et al. Comparative expression of Hedgehog ligands at different stages of prostate carcinoma progression. *J. Pathol.* **2008**, *216*, 460–470. [CrossRef]

61. Kim, T.J.; Lee, J.Y.; Hwang, T.K.; Kang, C.S.; Choi, Y.J. Hedgehog signaling protein expression and its association with prognostic parameters in prostate cancer: A retrospective study from the view point of new 2010 anatomic stage/prognostic groups. *J. Surg. Oncol.* **2011**, *104*, 472–479. [CrossRef] [PubMed]

62. Li, Q.; Alsaidan, O.A.; Rai, S.; Wu, M.; Shen, H.; Beharry, Z.; Almada, L.L.; Fernandez-Zapico, M.E.; Wang, L.; Cai, H. Stromal Gli signaling regulates the activity and differentiation of prostate stem and progenitor cells. *J. Biol. Chem.* **2018**, *293*, 10547–10560. [CrossRef] [PubMed]

63. Wu, M.; Ingram, L.; Tolosa, E.J.; Vera, R.E.; Li, Q.; Kim, S.; Ma, Y.; Spyropoulos, D.D.; Beharry, Z.; Huang, J.; et al. Gli Transcription Factors Mediate the Oncogenic Transformation of Prostate Basal Cells Induced by a Kras-Androgen Receptor Axis. *J. Biol. Chem.* **2016**, *291*, 25749–25760. [CrossRef] [PubMed]

64. Nadendla, S.K.; Hazan, A.; Ward, M.; Harper, L.J.; Moutasim, K.; Bianchi, L.S.; Naase, M.; Ghali, L.; Thomas, G.J.; Prowse, D.M.; et al. GLI1 confers profound phenotypic changes upon LNCaP prostate cancer cells that include the acquisition of a hormone independent state. *PLoS ONE* **2011**, *6*, e20271. [CrossRef] [PubMed]

65. Sirab, N.; Terry, S.; Giton, F.; Caradec, J.; Chimingqi, M.; Moutereau, S.; Vacherot, F.; de la Taille, A.; Kouyoumdjian, J.C.; Loric, S. Androgens regulate Hedgehog signalling and proliferation in androgen-dependent prostate cells. *Int. J. Cancer* **2012**, *131*, 1297–1306. [CrossRef] [PubMed]

66. Bushman, W. Hedgehog Signaling in Prostate Development, Regeneration and Cancer. *J. Dev. Biol.* **2016**, *4*, 30. [CrossRef] [PubMed]

67. Karhadkar, S.S.; Bova, G.S.; Abdallah, N.; Dhara, S.; Gardner, D.; Maitra, A.; Isaacs, J.T.; Berman, D.M.; Beachy, P.A. Hedgehog signalling in prostate regeneration, neoplasia and metastasis. *Nature* **2004**, *431*, 707–712. [CrossRef]

68. Podlasek, C.A.; Barnett, D.H.; Clemens, J.Q.; Bak, P.M.; Bushman, W. Prostate development requires Sonic hedgehog expressed by the urogenital sinus epithelium. *Dev. Biol.* **1999**, *209*, 28–39. [CrossRef]

69. Cunha, G.R.; Ricke, W.; Thomson, A.; Marker, P.C.; Risbridger, G.; Hayward, S.W.; Wang, Y.Z.; Donjacour, A.A.; Kurita, T. Hormonal, cellular, and molecular regulation of normal and neoplastic prostatic development. *J. Steroid Biochem. Mol. Biol.* **2004**, *92*, 221–236. [CrossRef]

70. Wong, S.K.; Mohamad, N.V.; Giaze, T.R.; Chin, K.Y.; Mohamed, N.; Ima-Nirwana, S. Prostate Cancer and Bone Metastases: The Underlying Mechanisms. *Int. J. Mol. Sci.* **2019**, *20*, 2587. [CrossRef]

71. Wang, C.Y.; Wu, G.Y.; Shen, M.J.; Cui, K.W.; Shen, Y. Comparison of distribution characteristics of metastatic bone lesions between breast and prostate carcinomas. *Oncol. Lett.* **2013**, *5*, 391–397. [CrossRef] [PubMed]

72. Onken, J.S.; Fekonja, L.S.; Wehowsky, R.; Hubertus, V.; Vajkoczy, P. Metastatic dissemination patterns of different primary tumors to the spine and other bones. *Clin. Exp. Metastasis* **2019**, *36*, 493–498. [CrossRef] [PubMed]

73. Kato, M.; Placencio-Hickok, V.R.; Madhav, A.; Haldar, S.; Tripathi, M.; Billet, S.; Mishra, R.; Smith, B.; Rohena-Rivera, K.; Agarwal, P.; et al. Heterogeneous cancer-associated fibroblast population potentiates neuroendocrine differentiation and castrate resistance in a CD105-dependent manner. *Oncogene* **2019**, *38*, 716–730. [CrossRef] [PubMed]

74. Pantel, K.; Hille, C.; Scher, H.I. Circulating Tumor Cells in Prostate Cancer: From Discovery to Clinical Utility. *Clin. Chem.* **2019**, *65*, 87–99. [CrossRef]

75. Yang, S.; Pham, L.K.; Liao, C.P.; Frenkel, B.; Reddi, A.H.; Roy-Burman, P. A novel bone morphogenetic protein signaling in heterotypic cell interactions in prostate cancer. *Cancer Res.* **2008**, *68*, 198–205. [CrossRef]

76. Nohno, T.; Ishikawa, T.; Saito, T.; Hosokawa, K.; Noji, S.; Wolsing, D.H.; Rosenbaum, J.S. Identification of a human type II receptor for bone morphogenetic protein-4 that forms differential heteromeric complexes with bone morphogenetic protein type I receptors. *J. Biol. Chem.* **1995**, *270*, 22522–22526. [CrossRef]

77. Moore, R.K.; Shimasaki, S. Molecular biology and physiological role of the oocyte factor, BMP-15. *Mol. Cell Endocrinol.* **2005**, *234*, 67–73. [CrossRef]

78. Peng, J.; Yoshioka, Y.; Mandai, M.; Matsumura, N.; Baba, T.; Yamaguchi, K.; Hamanishi, J.; Kharma, B.; Murakami, R.; Abiko, K.; et al. The BMP signaling pathway leads to enhanced proliferation in serous ovarian cancer-A potential therapeutic target. *Mol. Carcinog.* **2016**, *55*, 335–345. [CrossRef]

79. Le Page, C.; Puiffe, M.L.; Meunier, L.; Zietarska, M.; de Ladurantaye, M.; Tonin, P.N.; Provencher, D.; Mes-Masson, A.M. BMP-2 signaling in ovarian cancer and its association with poor prognosis. *J. Ovarian Res.* **2009**, *2*, 4. [CrossRef]

80. Li, H.; Li, J.; Feng, L. Hedgehog signaling pathway as a therapeutic target for ovarian cancer. *Cancer Epidemiol.* **2016**, *40*, 152–157. [CrossRef]

81. Chen, X.; Horiuchi, A.; Kikuchi, N.; Osada, R.; Yoshida, J.; Shiozawa, T.; Konishi, I. Hedgehog signal pathway is activated in ovarian carcinomas, correlating with cell proliferation: it's inhibition leads to growth suppression and apoptosis. *Cancer Sci.* **2007**, *98*, 68–76. [CrossRef] [PubMed]

82. Coffman, L.G.; Choi, Y.J.; McLean, K.; Allen, B.L.; di Magliano, M.P.; Buckanovich, R.J. Human carcinoma-associated mesenchymal stem cells promote ovarian cancer chemotherapy resistance via a BMP4/HH signaling loop. *Oncotarget* **2016**, *7*, 6916–6932. [CrossRef] [PubMed]

83. Miyazaki, H.; Watabe, T.; Kitamura, T.; Miyazono, K. BMP signals inhibit proliferation and in vivo tumor growth of androgen-insensitive prostate carcinoma cells. *Oncogene* **2004**, *23*, 9326–9335. [CrossRef] [PubMed]

84. Kashimada, K.; Koopman, P. Sry: The master switch in mammalian sex determination. *Development* **2010**, *137*, 3921–3930. [CrossRef]

85. Chen, Y.; Yu, H.; Pask, A.J.; Fujiyama, A.; Suzuki, Y.; Sugano, S.; Shaw, G.; Renfree, M.B. Hormone-responsive genes in the SHH and WNT/β-catenin signaling pathways influence urethral closure and phallus growth. *Biol. Reprod.* **2018**, *99*, 806–816. [CrossRef]

86. Zheng, Z.; Armfield, B.A.; Cohn, M.J. Timing of androgen receptor disruption and estrogen exposure underlies a spectrum of congenital penile anomalies. *Proc. Natl. Acad. Sci. USA* **2015**, *112*, E7194–E7203. [CrossRef]

87. Matsumoto, H.; Zhao, X.; Das, S.K.; Hogan, B.L.; Dey, S.K. Indian hedgehog as a progesterone-responsive factor mediating epithelial-mesenchymal interactions in the mouse uterus. *Dev. Biol.* **2002**, *245*, 280–290. [CrossRef]

88. Murashima, A.; Kishigami, S.; Thomson, A.; Yamada, G. Androgens and mammalian male reproductive tract development. *Biochim. Biophys. Acta* **2015**, *1849*, 163–170. [CrossRef]

89. Ipulan, L.A.; Raga, D.; Suzuki, K.; Murashima, A.; Matsumaru, D.; Cunha, G.; Yamada, G. Investigation of sexual dimorphisms through mouse models and hormone/hormone-disruptor treatments. *Differentiation* **2016**, *91*, 78–89. [CrossRef]

90. Welsh, M.; Saunders, P.T.; Fisken, M.; Scott, H.M.; Hutchison, G.R.; Smith, L.B.; Sharpe, R.M. Identification in rats of a programming window for reproductive tract masculinization, disruption of which leads to hypospadias and cryptorchidism. *J. Clin. Investig.* **2008**, *118*, 1479–1490. [CrossRef]

91. Bin-Abbas, B.; Conte, F.A.; Grumbach, M.M.; Kaplan, S.L. Congenital hypogonadotropic hypogonadism and micropenis: Effect of testosterone treatment on adult penile size why sex reversal is not indicated. *J. Pediatr.* **1999**, *134*, 579–583. [CrossRef]

92. Gearhart, J.P.; Jeffs, R.D. The use of parenteral testosterone therapy in genital reconstructive surgery. *J. Urol.* **1987**, *138*, 1077–1078. [CrossRef]

93. Luo, C.C.; Lin, J.N.; Chiu, C.H.; Lo, F.S. Use of parenteral testosterone prior to hypospadias surgery. *Pediatr. Surg. Int.* **2003**, *19*, 82–84. [CrossRef] [PubMed]

94. Suzuki, H.; Suzuki, K.; Yamada, G. Systematic analyses of murine masculinization processes based on genital sex differentiation parameters. *Dev. Growth Differ.* **2015**, *57*, 639–647. [CrossRef] [PubMed]

95. Suzuki, K.; Numata, T.; Suzuki, H.; Raga, D.D.; Ipulan, L.A.; Yokoyama, C.; Matsushita, S.; Hamada, M.; Nakagata, N.; Nishinakamura, R.; et al. Sexually dimorphic expression of Mafb regulates masculinization of the embryonic urethral formation. *Proc. Natl. Acad. Sci. USA* **2014**, *111*, 16407–16412. [CrossRef]

96. Matsushita, S.; Suzuki, K.; Ogino, Y.; Hino, S.; Sato, T.; Suyama, M.; Matsumoto, T.; Omori, A.; Inoue, S.; Yamada, G. Androgen Regulates Mafb Expression Through its 3'UTR During Mouse Urethral Masculinization. *Endocrinology* **2016**, *157*, 844–857. [CrossRef]

97. Huggins, C.; Hodges, C.V. Studies on Prostate Cancer. I. The effect of castration, of estrogen and of androgen injection on serum phosphates in metastatic carcinoma of the prostate. *Cancer Res.* **1941**, *1*, 293–297.

98. He, Y.; Hooker, E.; Yu, E.J.; Cunha, G.R.; Liao, L.; Xu, J.; Earl, A.; Wu, H.; Gonzalgo, M.L.; Sun, Z. Androgen signaling is essential for development of prostate cancer initiated from prostatic basal cells. *Oncogene* **2019**, *38*, 2337–2350. [CrossRef]

99. Horoszewicz, J.S.; Leong, S.S.; Kawinski, E.; Karr, J.P.; Rosenthal, H.; Chu, T.M.; Mirand, E.A.; Murphy, G.P. LNCaP model of human prostatic carcinoma. *Cancer Res.* **1983**, *43*, 1809–1818.

100. Gregory, C.W.; Johnson, R.T.; Presnell, S.C.; Mohler, J.L.; French, F.S. Androgen receptor regulation of G1 cyclin and cyclin-dependent kinase function in the CWR22 human prostate cancer xenograft. *J. Androl.* **2001**, *22*, 537–548.

101. Ye, D.; Mendelsohn, J.; Fan, Z. Androgen and epidermal growth factor down-regulate cyclin-dependent kinase inhibitor p27Kip1 and costimulate proliferation of MDA PCa 2a and MDA PCa 2b prostate cancer cells. *Clin. Cancer Res.* **1999**, *5*, 2171–2177. [PubMed]

102. Chen, Y.; Robles, A.I.; Martinez, L.A.; Liu, F.; Gimenez-Conti, I.B.; Conti, C.J. Expression of G1 cyclins, cyclin-dependent kinases, and cyclin-dependent kinase inhibitors in androgen-induced prostate proliferation in castrated rats. *Cell Growth Differ.* **1996**, *7*, 1571–1578. [PubMed]

103. Kosaka, T.; Miyajima, A.; Nagata, H.; Maeda, T.; Kikuchi, E.; Oya, M. Human castration resistant prostate cancer rather prefer to decreased 5α-reductase activity. *Sci. Rep.* **2013**, *3*, 1268. [CrossRef] [PubMed]

104. Labrie, F.; Dupont, A.; Belanger, A.; Cusan, L.; Lacourciere, Y.; Monfette, G.; Laberge, J.G.; Emond, J.P.; Fazekas, A.T.; Raynaud, J.P.; et al. New hormonal therapy in prostatic carcinoma: Combined treatment with an LHRH agonist and an antiandrogen. *Clin. Investig. Med.* **1982**, *5*, 267–275.

105. Linja, M.J.; Savinainen, K.J.; Saramäki, O.R.; Tammela, T.L.; Vessella, R.L.; Visakorpi, T. Amplification and overexpression of androgen receptor gene in hormone-refractory prostate cancer. *Cancer Res.* **2001**, *61*, 3550–3555.

106. Yuan, X.; Balk, S.P. Mechanisms mediating androgen receptor reactivation after castration. *Urol. Oncol.* **2009**, *27*, 36–41. [CrossRef]

107. Arai, S.; Miyashiro, Y.; Shibata, Y.; Tomaru, Y.; Kobayashi, M.; Honma, S.; Suzuki, K. Effect of castration monotherapy on the levels of adrenal androgens in cancerous prostatic tissues. *Steroids* **2011**, *76*, 301–308. [CrossRef]

108. Kumagai, J.; Hofland, J.; Erkens-Schulze, S.; Dits, N.F.; Steenbergen, J.; Jenster, G.; Homma, Y.; de Jong, F.H.; van Weerden, W.M. Intratumoral conversion of adrenal androgen precursors drives androgen receptor-activated cell growth in prostate cancer more potently than de novo steroidogenesis. *Prostate* **2013**, *73*, 1636–1650. [CrossRef]

109. Ishizaki, F.; Nishiyama, T.; Kawasaki, T.; Miyashiro, Y.; Hara, N.; Takizawa, I.; Naito, M.; Takahashi, K. Androgen deprivation promotes intratumoral synthesis of dihydrotestosterone from androgen metabolites in prostate cancer. *Sci. Rep.* **2013**, *3*, 1528. [CrossRef]

110. Shaw, G.; Renfree, M.B.; Leihy, M.W.; Shackleton, C.H.; Roitman, E.; Wilson, J.D. Prostate formation in a marsupial is mediated by the testicular androgen 5 alpha-androstane-3 alpha,17 beta-diol. *Proc. Natl. Acad. Sci. USA* **2000**, *97*, 12256–12259. [CrossRef]

111. Wilson, J.D.; Auchus, R.J.; Leihy, M.W.; Guryev, O.L.; Estabrook, R.W.; Osborn, S.M.; Shaw, G.; Renfree, M.B. 5alpha-androstane-3alpha,17beta-diol is formed in tammar wallaby pouch young testes by a pathway involving 5alpha-pregnane-3alpha,17alpha-diol-20-one as a key intermediate. *Endocrinology* **2003**, *144*, 575–580. [CrossRef] [PubMed]

112. Mahendroo, M.; Wilson, J.D.; Richardson, J.A.; Auchus, R.J. Steroid 5alpha-reductase 1 promotes 5alpha-androstane-3alpha,17beta-diol synthesis in immature mouse testes by two pathways. *Mol. Cell Endocrinol.* **2004**, *222*, 113–120. [CrossRef] [PubMed]

113. Fukami, M.; Homma, K.; Hasegawa, T.; Ogata, T. Backdoor pathway for dihydrotestosterone biosynthesis: Implications for normal and abnormal human sex development. *Dev. Dyn.* **2013**, *242*, 320–329. [CrossRef] [PubMed]

114. Flück, C.E.; Meyer-Böni, M.; Pandey, A.V.; Kempná, P.; Miller, W.L.; Schoenle, E.J.; Biason-Lauber, A. Why boys will be boys: Two pathways of fetal testicular androgen biosynthesis are needed for male sexual differentiation. *Am. J. Hum. Genet.* **2011**, *89*, 201–218. [CrossRef] [PubMed]

115. Arnold, J.T.; Gray, N.E.; Jacobowitz, K.; Viswanathan, L.; Cheung, P.W.; McFann, K.K.; Le, H.; Blackman, M.R. Human prostate stromal cells stimulate increased PSA production in DHEA-treated prostate cancer epithelial cells. *J. Steroid Biochem. Mol. Biol.* **2008**, *111*, 240–246. [CrossRef]

116. Chipuk, J.E.; Cornelius, S.C.; Pultz, N.J.; Jorgensen, J.S.; Bonham, M.J.; Kim, S.J.; Danielpour, D. The androgen receptor represses transforming growth factor-beta signaling through interaction with Smad3. *J. Biol. Chem.* **2002**, *277*, 1240–1248. [CrossRef]

117. De Arcangelis, A.; Mark, M.; Kreidberg, J.; Sorokin, L.; Georges-Labouesse, E. Synergistic activities of alpha3 and alpha6 integrins are required during apical ectodermal ridge formation and organogenesis in the mouse. *Development* **1999**, *126*, 3957–3968.

118. Chang, J.; Chaudhuri, O. Beyond proteases: Basement membrane mechanics and cancer invasion. *J. Cell Biol.* **2019**, *218*, 2456–2469. [CrossRef]

119. Estrach, S.; Cailleteau, L.; Franco, C.A.; Gerhardt, H.; Stefani, C.; Lemichez, E.; Gagnoux-Palacios, L.; Meneguzzi, G.; Mettouchi, A. Laminin-binding integrins induce Dll4 expression and Notch signaling in endothelial cells. *Circ. Res.* **2011**, *109*, 172–182. [CrossRef]

120. Sekiguchi, R.; Yamada, K.M. Basement Membranes in Development and Disease. *Curr. Top. Dev. Biol.* **2018**, *130*, 143–191. [CrossRef]

121. Lei, W.L.; Xing, S.G.; Deng, C.Y.; Ju, X.C.; Jiang, X.Y.; Luo, Z.G. Laminin/β1 integrin signal triggers axon formation by promoting microtubule assembly and stabilization. *Cell Res.* **2012**, *22*, 954–972. [CrossRef] [PubMed]

122. Shuler, C.F.; Guo, Y.; Majumder, A.; Luo, R.Y. Molecular and morphologic changes during the epithelial-mesenchymal transformation of palatal shelf medial edge epithelium in vitro. *Int. J. Dev. Biol.* **1991**, *35*, 463–472. [PubMed]

123. Proetzel, G.; Pawlowski, S.A.; Wiles, M.V.; Yin, M.; Boivin, G.P.; Howles, P.N.; Ding, J.; Ferguson, M.W.; Doetschman, T. Transforming growth factor-beta 3 is required for secondary palate fusion. *Nat. Genet.* **1995**, *11*, 409–414. [CrossRef] [PubMed]

124. Anderson, C.; Thorsteinsdóttir, S.; Borycki, A.G. Sonic hedgehog-dependent synthesis of laminin alpha1 controls basement membrane assembly in the myotome. *Development* **2009**, *136*, 3495–3504. [CrossRef] [PubMed]

125. Gaiko-Shcherbak, A.; Fabris, G.; Dreissen, G.; Merkel, R.; Hoffmann, B.; Noetzel, E. The Acinar Cage: Basement Membranes Determine Molecule Exchange and Mechanical Stability of Human Breast Cell Acini. *PLoS ONE* **2015**, *10*, e0145174. [CrossRef] [PubMed]

126. Cress, A.E.; Rabinovitz, I.; Zhu, W.; Nagle, R.B. The alpha 6 beta 1 and alpha 6 beta 4 integrins in human prostate cancer progression. *Cancer Metastasis Rev.* **1995**, *14*, 219–228. [CrossRef]

127. Fuchs, M.E.; Brawer, M.K.; Rennels, M.A.; Nagle, R.B. The relationship of basement membrane to histologic grade of human prostatic carcinoma. *Mod. Pathol.* **1989**, *2*, 105–111.

128. Datta, S.; Datta, M.W. Sonic Hedgehog signaling in advanced prostate cancer. *Cell Mol. Life Sci.* **2006**, *63*, 435–448. [CrossRef]

129. Chang, H.H.; Chen, B.Y.; Wu, C.Y.; Tsao, Z.J.; Chen, Y.Y.; Chang, C.P.; Yang, C.R.; Lin, D.P. Hedgehog overexpression leads to the formation of prostate cancer stem cells with metastatic property irrespective of androgen receptor expression in the mouse model. *J. Biomed. Sci.* **2011**, *18*, 6. [CrossRef]

130. Anand-Apte, B.; Bao, L.; Smith, R.; Iwata, K.; Olsen, B.R.; Zetter, B.; Apte, S.S. A review of tissue inhibitor of metalloproteinases-3 (TIMP-3) and experimental analysis of its effect on primary tumor growth. *Biochem. Cell Biol.* **1996**, *74*, 853–862. [CrossRef]

131. Bair, E.L.; Chen, M.L.; McDaniel, K.; Sekiguchi, K.; Cress, A.E.; Nagle, R.B.; Bowden, G.T. Membrane type 1 matrix metalloprotease cleaves laminin-10 and promotes prostate cancer cell migration. *Neoplasia* **2005**, *7*, 380–389. [CrossRef] [PubMed]

132. Liao, X.; Siu, M.K.; Au, C.W.; Wong, E.S.; Chan, H.Y.; Ip, P.P.; Ngan, H.Y.; Cheung, A.N. Aberrant activation of hedgehog signaling pathway in ovarian cancers: Effect on prognosis, cell invasion and differentiation. *Carcinogenesis* **2009**, *30*, 131–140. [CrossRef] [PubMed]

133. Chen, J.S.; Huang, X.H.; Wang, Q.; Huang, J.Q.; Zhang, L.J.; Chen, X.L.; Lei, J.; Cheng, Z.X. Sonic hedgehog signaling pathway induces cell migration and invasion through focal adhesion kinase/AKT signaling-mediated activation of matrix metalloproteinase (MMP)-2 and MMP-9 in liver cancer. *Carcinogenesis* **2013**, *34*, 10–19. [CrossRef] [PubMed]

134. Yoo, Y.A.; Kang, M.H.; Kim, J.S.; Oh, S.C. Sonic hedgehog signaling promotes motility and invasiveness of gastric cancer cells through TGF-beta-mediated activation of the ALK5-Smad 3 pathway. *Carcinogenesis* **2008**, *29*, 480–490. [CrossRef] [PubMed]

International Journal of
Molecular Sciences

Article

The Role of Sonic Hedgehog Signaling in the Tumor Microenvironment of Oral Squamous Cell Carcinoma

Kiyofumi Takabatake [1,*], Tsuyoshi Shimo [2], Jun Murakami [3], Chang Anqi [1,4], Hotaka Kawai [1], Saori Yoshida [1], May Wathone Oo [1], Omori Haruka [1], Shintaro Sukegawa [1,5], Hidetsugu Tsujigiwa [1,6], Keisuke Nakano [1] and Hitoshi Nagatsuka [1]

[1] Department of Oral Pathology and Medicine, Graduate School of Medicine, Dentistry and Pharmaceutical Sciences, Okayama University, Okayama 7008525, Japan; pjvr21b0@s.okayama-u.ac.jp (C.A.); de18018@s.okayama-u.ac.jp (H.K.); de20052@s.okayama-u.ac.jp (S.Y.); p1qq7mbu@s.okayama-u.ac.jp (M.W.O.); p4628fuz@s.okayama-u.ac.jp (O.H.); gouwan19@gmail.com (S.S.); tsuji@dls.ous.ac.jp (H.T.); pir19btp@okayama-u.ac.jp (K.N.); jin@okayama-u.ac.jp (H.N.)
[2] Division of Reconstructive Surgery for Oral and Maxillofacial Region, Department of Human Biology and Pathophysiology, School of Dentistry, Health Sciences University of Hokkaido, Hokkaido 0610293, Japan; shimotsu@hoku-iryo-u.ac.jp
[3] Department of Oral and Maxillofacial Radiology, Graduate School of Medicine, Dentistry and Pharmaceutical Sciences, Okayama University, Okayama 7008525, Japan; jun-m@md.okayama-u.ac.jp
[4] Department of Anatomy, Basic Medical Science College, Harbin Medical University, Harbin 150081, China
[5] Department of Oral and Maxillofacial Surgery, Kagawa Prefectural Central Hospital, Kagawa 7608557, Japan
[6] Department of Life Science, Faculty of Science, Okayama University of Science, Okayama 7000005, Japan
* Correspondence: gmd422094@s.okayama-u.ac.jp; Tel.: +81-086-235-6651

Received: 30 October 2019; Accepted: 15 November 2019; Published: 17 November 2019

Abstract: Sonic hedgehog (SHH) and its signaling have been identified in several human cancers, and increased levels of SHH expression appear to correlate with cancer progression. However, the role of SHH in the tumor microenvironment (TME) of oral squamous cell carcinoma (OSCC) is still unclear. No studies have compared the expression of SHH in different subtypes of OSCC and focused on the relationship between the tumor parenchyma and stroma. In this study, we analyzed SHH and expression of its receptor, Patched-1 (PTCH), in the TME of different subtypes of OSCC. Fifteen endophytic-type cases (ED type) and 15 exophytic-type cases (EX type) of OSCC were used. H&E staining, immunohistochemistry (IHC), double IHC, and double-fluorescent IHC were performed on these samples. ED-type parenchyma more strongly expressed both SHH and PTCH than EX-type parenchyma. In OSCC stroma, CD31-positive cancer blood vessels, CD68- and CD11b-positive macrophages, and α-smooth muscle actin-positive cancer-associated fibroblasts partially expressed PTCH. On the other hand, in EX-type stroma, almost no double-positive cells were observed. These results suggest that autocrine effects of SHH induce cancer invasion, and paracrine effects of SHH govern parenchyma-stromal interactions of OSCC. The role of the SHH pathway is to promote growth and invasion.

Keywords: sonic hedgehog (SHH); oral squamous cell carcinoma (OSCC); tumor microenvironment (TME); tumor-associated macrophages (TAMs); cancer-associated fibroblasts (CAFs); tumor-associated angiogenesis

1. Introduction

Oral squamous cell carcinoma (OSCC) is a malignant tumor that comprises up to 90% of tumors in the head and neck region [1] and is a heterogeneous group of tumors arising from the mucosal surfaces of the oral cavity.

OSCC is classified into various subtypes as described in the World Health Organization (WHO) Classification of Head and Neck Tumors 4th Ed [2]. In addition, macroscopic subtypes have also been identified, based on the clinical invasion pattern. These subtypes have important differences in prognosis due to differences in invasive ability. Clinically, tumors primarily show exophytic growth (exophytic type—EX type) or endophytic growth (endophytic type—ED type).

OSCC is a malignant epithelial tumor. Similar to many other solid tumors, the tumor microenvironment (TME) of OSCC consists of the tumor parenchyma and stroma. The stroma of the TME is composed of multiple different cell types, such as macrophages, endothelial cells, cancer-associated fibroblasts (CAFs), and immune cells. These subpopulations of cells interact with each other as well as with cancer cells via complex communication networks through various secreted cytokines, chemokines, growth factors, and proteins of the extracellular matrix.

Macrophages, especially tumor-associated macrophages (TAMs), largely contribute to proliferation, invasion, and metastasis of the tumor. Recently, some studies have suggested that a relationship exists between the level of infiltration of TAMs and a poor outcome of OSCC and that the relationship could be used as a potential prognostic marker [3–5]. CD11b is monocytes macrophage lineage-specific marker. Numerous studies have reported that CD11b plays a role in invasion and metastasis and CD11b is a marker of TAMs in tumors [6].

Blood vessels in the tumor stroma play an important role. Tumor blood vessels nourish not only cancer cells but also stromal cells. Tumor blood vessels are positive for CD31, which was first characterized as a protein that is expressed by human hematopoietic progenitor cells and has been considered a definitive marker of angiogenesis in neoplastic lesions [7]. Recently, it has been reported that CXCR4 plays a crucial role in tumor angiogenesis, which is required for OSCC progression [8]. Moreover, stromal cell-derived factor 1 (SDF-1), which is the ligand for CXCR4, is also expressed in various types of cancers [9,10].

Additionally, CAFs are the predominant cell type within the tumor stroma, and their main function is to maintain a favorable microenvironment for tumor cell growth and proliferation. The most common marker used to detect CAFs in the tumor stroma is α-smooth muscle actin (α-SMA), a specific marker of myofibroblasts [11–14]. This myofibroblast phenotype of CAFs is frequently observed in OSCC, and upregulation of α-SMA is correlated with poor prognosis [15].

Thus, the biological properties of the tumor stroma are closely related to the growth and spread of cancer. However, the details of how the TME varies depending on the proliferation type (such as ED or EX type) of OSCC have not been reported.

Hedgehog signaling is essential for proper pattern formation and morphogenesis during embryogenesis [16–19]. The interaction of the Hedgehog protein with its receptor, Patched-1 (PTCH), leads to activation of the transcription factor, Gli, which induces downstream target genes including PTCH and Gli themselves [20]. Among the three types of Hedgehog genes, sonic hedgehog (SHH), Indian hedgehog, and desert hedgehog, the role of the SHH pathway in blood vessel formation has been well studied in several animal models [21,22]. SDF-1 recruits endothelial progenitors, and SHH increases the expression of SDF-1 in a concentration-dependent manner [23].

Hedgehog signaling contributes to the development and progression of many cancers [24–32]. In OSCC, reactivation and overexpression of the Hedgehog pathway plays a key role in development and progression [33–36], and SHH signal activation is associated with worse prognosis in OSCC [37,38].

However, how Hedgehog signaling is involved in the TME of OSCC is unclear, and no studies have focused on the expression of SHH signaling according to the subtype (such as ED or EX type) of OSCC.

In the present study, we analyzed the relationship between SHH signaling and TEM in OSCC comparing resected ED-type or EX-type OSCC samples.

2. Results

2.1. SHH and PTCH Expression in OSCC

To investigate how the expression of SHH and its receptor PTCH influence on the invasion of OSCC, we examined the expression of SHH and PTCH in the parenchymal and stromal regions comparing ED-type and EX-type specimens.

As seen with H&E staining, the existing basal membrane in ED type was destroyed and infiltrated by forming cancer nests in the subepithelial connective tissue (Figure 1a). On the other hand, the cancer cells did not invade over the existing basal membrane, and cancer cells growth was observed as extrovert in the EX type (Figure 1b).

In the cancer parenchyma, SHH was strongly expressed in the cytoplasm of cancer cells in ED type: Moderate (+2), five patients; marked (+3), 10 patients, and more strongly in the cancer nests at the front line of invasion (Figure 1c). On the other hand, in EX type, strong expression was observed in the cytoplasm of part of the basal layer, and expression was observed from the spinous layer to the surface layer. However, the expression was weaker than that in ED type: Weak (+1), four patients; moderate (+2), 11 patients (Figure 1d).

In the cancer parenchyma, PTCH was strongly expressed in the cytoplasm of cancer cells in ED type: Moderate (+2), five patients; marked (+3), 10 patients, particularly in the cancer nests at the invasion front (Figure 1e). On the other hand, the cytoplasm of cancer cells from the basal layer to the surface layer was weakly positive in EX type: No reactivity (0), one patient; weak (+1), 12 patients; moderate (+2), two patients, (Figure 1f).

SHH in the cancer stroma was strongly expressed in blood vessels, spindle cells, and round cells in ED type (Figure 1g). SHH in EX-type stroma was also strongly expressed in blood vessels and round cells (Figure 1h). The expression of PTCH in the cancer stroma was stronger in EX type than in ED type, and blood vessels, round cells, and spindle-shaped cells were positive for PTCH (Figure 1i,j).

2.2. Involvement of SHH Signaling in Tumor Progression

2.2.1. Tumor Angiogenesis

To clarify whether the SHH pathway is involved in angiogenesis in OSCC, we first determined the effect of SHH/PTCH in neovascularization using CD31 IHC staining. CD31 is a marker associated with angiogenesis and labels vascular endothelial cells of blood vessels in tumors. Vascular endothelial cells of blood vessels were positive for CD31 in both ED-type and EX-type cancer stroma. Especially in ED-type stroma, CD31-positive expression was observed not only in blood vessels but also round cells (Figure 2a). When comparing the area of blood vessels in the cancer stroma between ED type and EX type, ED type showed superior angiogenesis compared to EX type (Figure 2b).

Many tumor blood vessels, which expressed both SHH and PTCH, were observed in the cancer stroma, and the accumulation of these blood vessels was adjacent to the site where SHH was strongly expressed in the cancer parenchyma (Figure 2c). However, connective tissue adjacent to epithelial tissue in the non-cancerous region had few blood vessels that expressed SHH (Figure 2d). The expression of PTCH in tumor blood vessels also showed the same tendency as the expression of SHH, and PTCH was not expressed in blood vessels in the connective tissue adjacent to the non-cancerous region epithelium (Figure 2e,f).

In double-fluorescent IHC staining, both PTCH and CD31 were positive in blood vessels in the cancer stroma in ED type and EX type. In addition, accumulation of CD31-positive round cells was observed around blood vessels in the stroma of ED type, however these cells did not express PTCH (Figure 2g). EX type also showed PTCH expression in almost all CD31-positive blood vessels in the cancer stroma (Figure 2h).

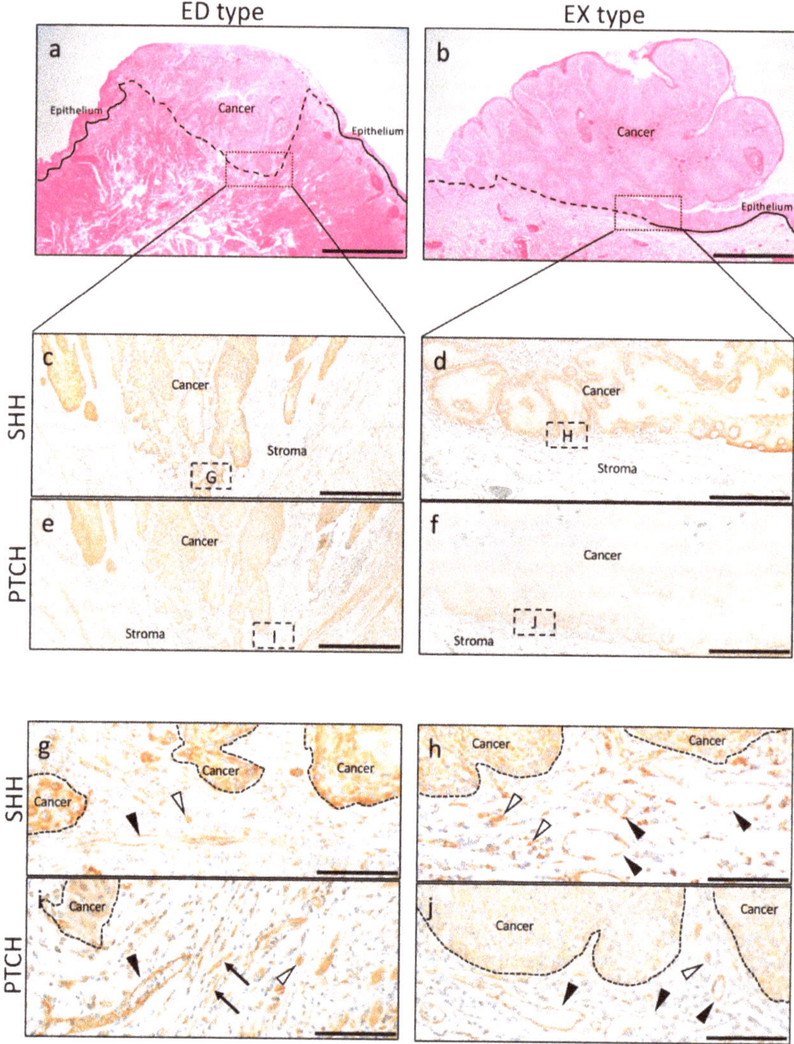

Figure 1. Representative pictures of immunohistochemical staining of sonic hedgehog (SHH) and Patched-1 (PTCH) in endophytic (ED)-type and exophytic (EX)-type oral squamous cell carcinoma (OSCC). (**a**,**b**) H&E staining of ED type and EX type. Scale bars: 2 mm. (**c**,**d**) Low-power magnification of SHH immunohistochemistry (IHC) staining. Scar bar: 1 mm. (**e**,**f**) Low-power magnification of PTCH IHC staining. Scale bar: 1 mm. (**g**,**h**) High-power magnification of SHH IHC staining. The expression of SHH in the cancer stroma was expressed in blood vessels (block arrowheads) and round cells (white arrowheads). Scale bar: 100 μm. (**i**,**j**) High-power magnification of PTCH IHC staining. The expression of PTCH in the cancer stroma was stronger in EX type than in ED type and was positive for blood vessels (block arrowheads), round cells (white arrowheads), and spindle-shaped cells (arrows). Scale bar: 100 μm.

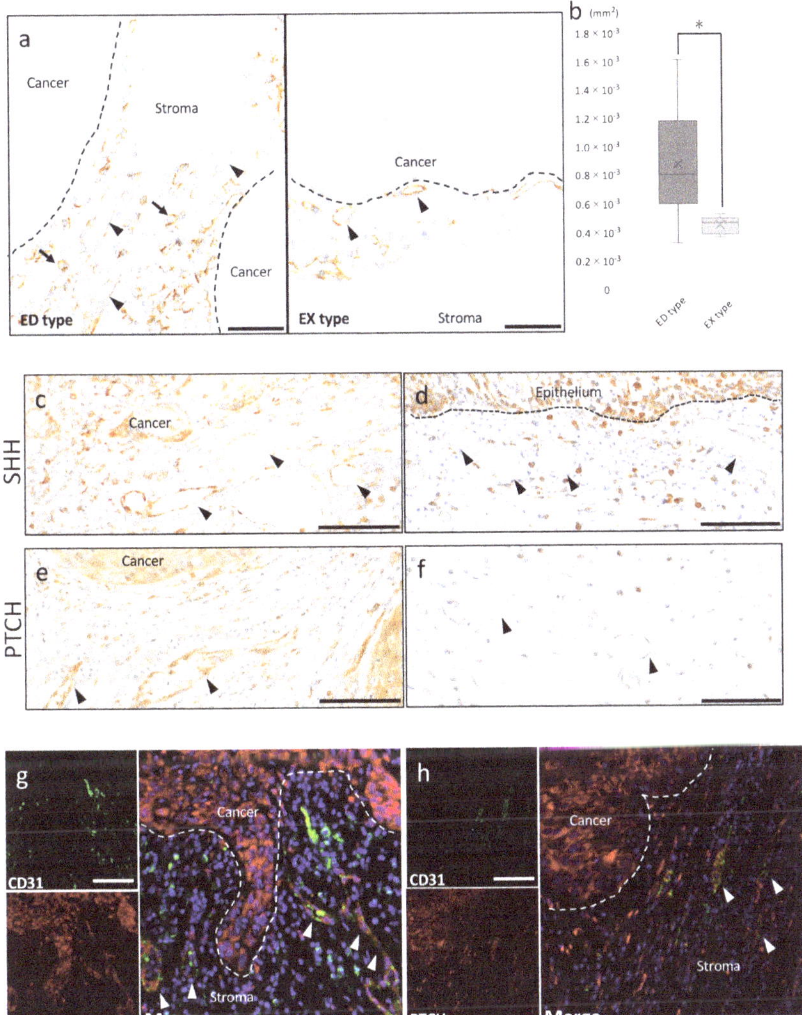

Figure 2. IHC and immunofluorescence of CD31. (**a**) IHC staining for CD31 in ED type and EX type. CD31 positive expression was observed in blood vessels (arrowheads). CD31 positive round-shape cells were observed (arrows). Scale bar: 50 μm. (**b**) Quantification of the angiogenesis area in ED type and EX type. ED type showed superior angiogenesis compared to EX type. * $p < 0.05$ as indicated. (**c**) IHC feature of SHH in ED-type stroma. The accumulation of these blood vessels (arrowheads) was shown adjacent to the site where the SHH was strongly expressed in the cancer parenchyma. Scale bar: 100 μm. (**d**) IHC feature of SHH in non-cancerous area. The connective tissue adjacent to normal epithelial tissue had few blood vessels that expressed SHH weekly (arrowheads). Scale bar: 100 μm. (**e**) IHC feature of PTCH in ED-type stroma. The PTCH positive blood vessels (arrowheads) was shown adjacent to the cancer parenchyma. Scale bar: 100 μm. (**f**) IHC feature of PTCH in non-cancerous area. The PTCH positive blood vessels were not observed (arrowheads). Scale bar: 100 μm. (**g**,**h**) Double-fluorescent IHC in ED type and EX type. Double-fluorescent IHC for PTCH-CD31 demonstrated that both PTCH and CD31 were positive in blood vessels (arrowheads) in the cancer stroma in ED type and EX type. Scale bar: 100 μm.

SDF-1/CXCR4 plays multiple roles in tumor pathogenesis. CXCR4 promotes tumor growth and malignancy, enhances tumor angiogenesis, and participates in tumor metastasis [39–42]. SDF-1 was expressed in both ED-type and EX-type cancer parenchyma, and the expression was stronger in ED type than in EX type. SDF-1 expression was observed in the cancer parenchyma; however, SDF-1 was not expressed in the non-cancerous epithelium adjacent to the cancer parenchyma, similar to SHH expression (Figure 3a). Importantly, more intense signals of SDF-1 were detected in the microvascular cells at the cancer's invasive front (Figure 3b). In EX type, SDF-1 expression was detected in blood vessels of stroma (Figure 3c). CXCR4 was also expressed in both ED-type and EX-type cancer parenchyma; however, CXCR4 was not expressed in the non-cancerous epithelium adjacent to the cancer parenchyma, similar to PTCH expression (Figure 3d). In ED type, CXCR4 were detected in the blood vessels at the cancer stroma (Figure 3e); however, in EX type, CXCR4 expression was not detected in blood vessels of stroma (Figure 3f). Double-fluorescent IHC staining showed that these CXCR4-positive structures expressed PTCH in ED type (Figure 3g).

2.2.2. Tumor-Associated Macrophages

To clarify whether the SHH pathway is involved in OSCC invasion via TAMs, we first determined the effect of SHH/PTCH in OSCC invasion using CD68 IHC staining. CD68 is a pan-macrophage marker and is considered a marker of TAMs. Recently, CD68-positive cells have been reported to be a poor prognostic factor [5] and TAMs recruitment in SHH expression tumor recruit TAMs significantly [43].

In both ED and EX type, accumulation of CD68-positive round or dendritic-shaped cells was observed in the cancer stroma (Figure 4a). The number of CD68-positive cells in the cancer stroma was significantly higher in ED type than in EX type (Figure 4b). Double immunostaining for SHH and CD68 showed that the expression of these molecules did not merge. However, a large concentration of CD68 was observed at the site close to the cancer parenchyma where SHH expression was strong (Figure 4c). Double-fluorescent IHC staining for CD68 and PTCH showed abundant CD68-positive cells that overlapped with PTCH in ED type (Figure 4d), whereas almost all CD68-positive cells did not merge with PTCH in EX type.

CD11b is a specific monocyte macrophage marker, and recently it was reported that CD11b is a marker of macrophages, some of which are TAMs in tumors [44]. Thus, we next determined the effect of SHH/PTCH in OSCC invasion using CD11b IHC staining. In both ED type and EX type, aggregation of CD11b-positive cells was observed in the cancer stroma. Many round CD11b-positive cells were observed in the ED-type cancer stroma, and CD11b-positive round cells were also observed in the EX-type cancer stroma (Figure 4e). The number of CD11b-positive cells was higher in ED type than EX type (Figure 4f). Double immunostaining for SHH and CD11b showed that the expression of these molecules did not merge. However, a large concentration of CD11b was present at the site adjacent to the cancer parenchyma where SHH expression was strong (Figure 4g). When the expression of PTCH and CD11b was examined, CD11b-positive round cells merged with PTCH expression (Figure 4h). On the other hand, almost no expression of PTCH merged with CD11b-positive round cells in the EX-type stroma.

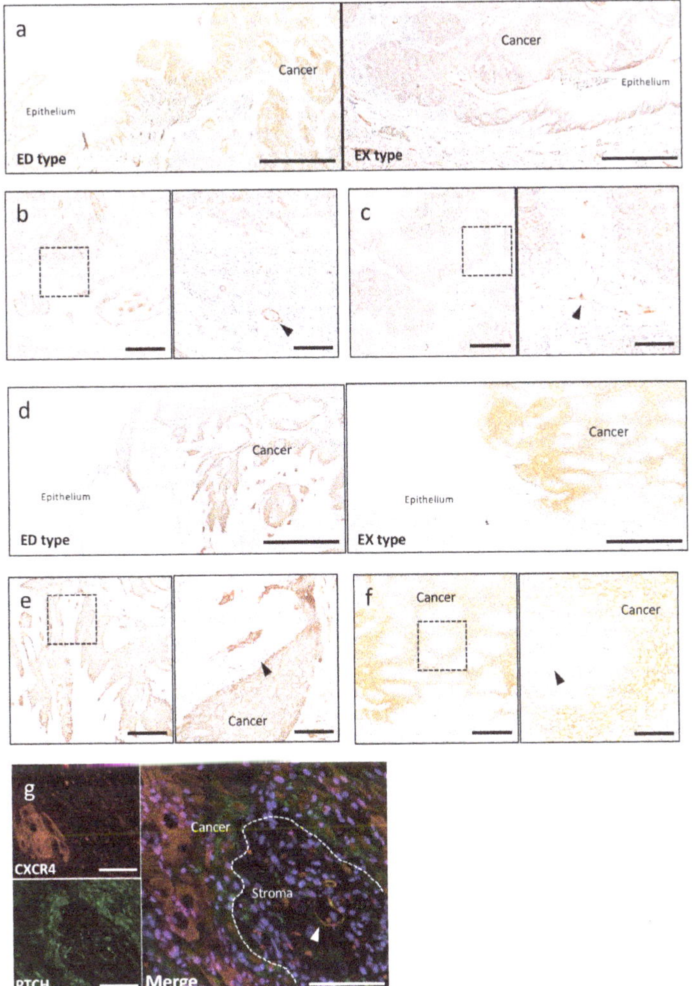

Figure 3. IHC and immunofluorescence of SDF-1 and CXCR4. (**a**) IHC staining for SDF-1 in ED type and EX type. SDF-1 was expressed in both ED-type and EX-type cancer parenchyma; however, SDF-1 was not expressed in the non-cancerous epithelium adjacent to the cancer parenchyma. Scale bar: 1 mm. (**b**) IHC feature of SDF-1 in ED type. Blood vessels (arrowhead) were positive for SDF-1 in invasive front of cancer. Scale bar, left: 500 μm, right: 100 μm. (**c**) IHC feature of SDF-1 in EX type. Blood vessels (arrowhead) were positive for SDF-1 weakly in cancer stroma. Scale bar, left: 500 μm, right: 100 μm. (**d**) IHC staining for CXCR4 in ED type and EX type. CXCR4 was expressed in both ED-type and EX-type cancer parenchyma; however, CXCR4 was not expressed in the non-cancerous epithelium adjacent to the cancer parenchyma. Scale bar: 1 mm. (**e**) IHC feature of CXCR4 in ED type. Blood vessels (arrowhead) were positive for CXCR4 in cancer stroma. Scale bar, left: 500 μm, right: 100 μm. (**f**) IHC feature of CXCR4 in EX type. Blood vessels (arrowhead) were not positive for CXCR4 in cancer stroma. Scale bar, left: 500 μm, right: 100 μm. (**g**) Double-fluorescent IHC in ED type. Double-fluorescent IHC staining of CXCR4 and PTCH demonstrated that the CXCR4-positive blood vessels merged with PTCH in ED type (arrowhead). Scale bar: 100 μm.

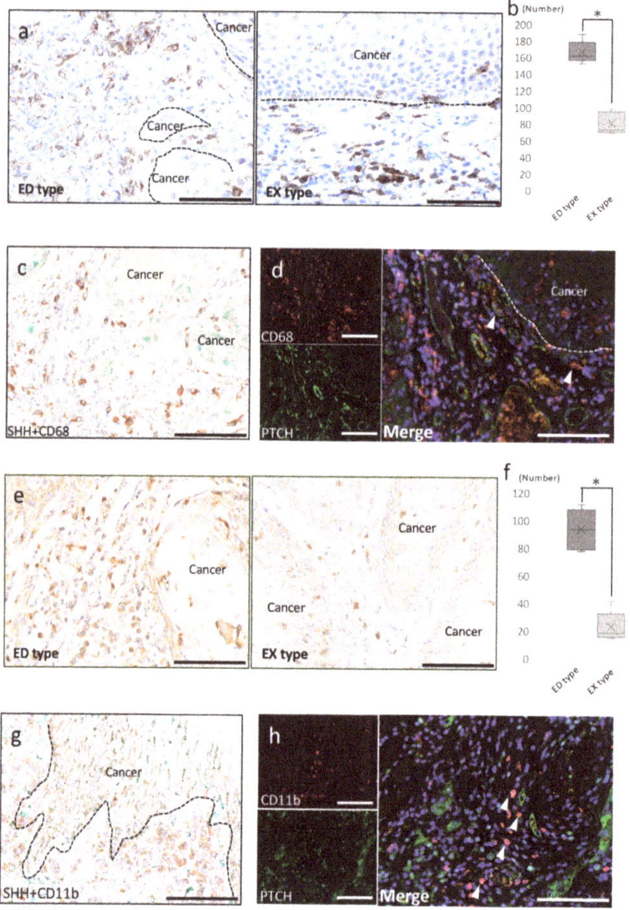

Figure 4. IHC, double IHC, and immunofluorescence of CD68 and CD11b. (**a**) IHC staining for CD68 in ED type and EX type. The accumulation of CD68-positive round- or dendritic-shape cells was observed in the cancer stroma. Scale bar: 100 μm. (**b**) Quantification of the number of CD68-positive cells in ED type and EX type. The number of CD68-positive cells in the cancer stroma was significantly higher in ED type than in EX type. * $p < 0.05$ as indicated. (**c**) Double IHC staining for SHH (green)-CD68 (brown) in ED type. The expression of SHH and CD68 did not merge. However, there was a large concentration of CD68 at the site close to the cancer parenchyma where SHH expression was strong. Scale bar: 100 μm. (**d**) Double-fluorescent IHC in ED type. Double-fluorescent IHC staining of CD68 and PTCH demonstrated that the abundant CD68-positive cells merged (arrowheads) with PTCH in ED type. Scale bar: 100 μm. (**e**) IHC staining for CD11b in ED type and EX type. The accumulation of CD11b-positive round- or dendritic-shape cells was observed in the cancer stroma. Scale bar: 100 μm. (**f**) Quantification of the number of CD11b-positive cells in ED type and EX type. The number of CD11b-positive cells in the cancer stroma was significantly higher in ED type than in EX type. * $p < 0.05$ as indicated. (**g**) Double IHC staining for SHH (green)-CD11b (brown) in ED type The expression of SHH and CD11b did not merge. However, there was a large concentration of CD68 at the site close to the cancer parenchyma where SHH expression was strong. Scale bar: 100 μm. (**h**) Double-fluorescent IHC in ED type. Double-fluorescent IHC staining of CD11b and PTCH demonstrated that the abundant CD11b-positive cells merged (arrowheads) with PTCH in ED type. Scale bar: 100 μm.

2.2.3. Cancer-Associated Fibroblasts

To clarify whether the SHH pathway is involved in OSCC invasion via CAFs, we investigated the effect of SHH/PTCH in OSCC invasion using α-SMA IHC staining. α-SMA is a common marker of myoepithelial cells, especially CAFs, in many human tumors. In ED-type stroma, abundant spindle-shaped cells were observed around the cancer nests, and these cells were positive for α-SMA (Figure 5a). In EX-type stroma, α-SMA-positive blood vessels were observed in the cancer stroma. However, almost no spindle-shaped α-SMA-positive cells were observed (Figure 5b). Double immunostaining for SHH and α-SMA showed that the expression of these molecules did not overlap. However, a large concentration of α-SMA was seen at the site close to the cancer parenchyma where SHH expression was strong (Figure 5c).

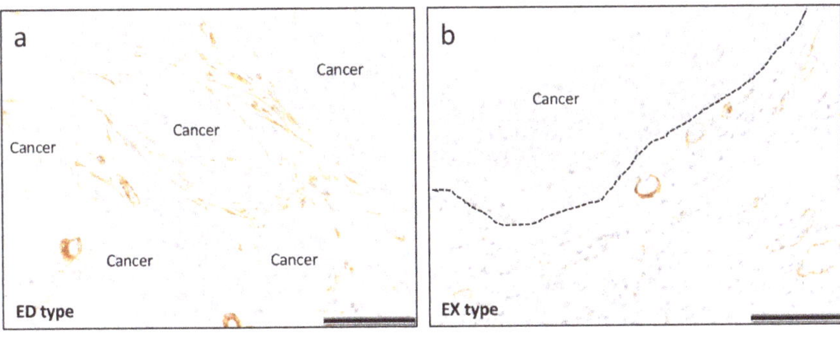

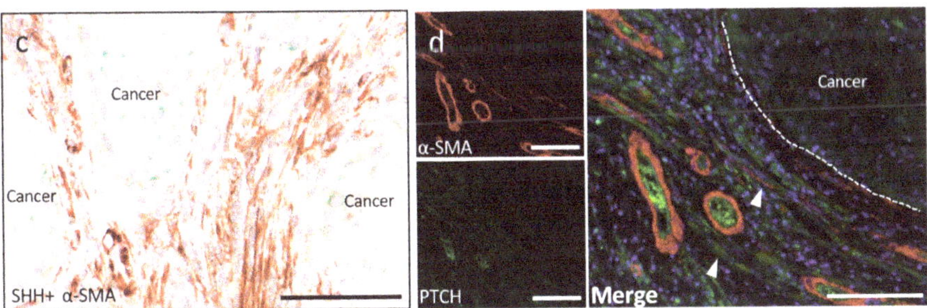

Figure 5. IHC, double IHC and immunofluorescence of α-smooth muscle actin (α-SMA). (**a**) IHC staining for α-SMA in ED type. The abundant α-SMA-positive spindle-shaped cells were observed around the cancer nests. Scale bar: 100 μm. (**b**) IHC staining for α-SMA in EX type. Almost all spindle-shaped α-SMA-positive cells were not observed. Scale bar: 100 μm. (**c**) Double IHC staining for α-SMA (green)-CD11b (brown) in ED type. There was a large concentration of α-SMA at the site close to the cancer parenchyma where SHH expression was strong. Scale bar: 100 μm. (**d**) Double-fluorescent IHC in ED type. Double-fluorescent IHC staining of α-SMA and PTCH demonstrated that the abundant α-SMA-positive cells merged (arrowheads) with PTCH partially in ED type. Scale bar: 100 μm.

Double-fluorescent IHC staining showed that some α-SMA-positive spindle-shaped cells expressed PTCH in ED-type stroma (Figure 5d). PTCH-positive cells were observed in EX-type cancer stroma, but no positive cells that overlapped with α-SMA were observed.

3. Discussion

In this study, the role of SHH in OSCC TME was clarified by comparing the expression of SHH by ED type and EX type by immunostaining.

First, we demonstrated autocrine expression of SHH in OSCC cancer parenchyma. Our data showed that both SHH and PTCH were expressed in the cancer parenchyma, thus suggesting that SHH signaling may operate in an autocrine manner in the cancer parenchyma. The expression of SHH and PTCH was higher in ED-type cancer parenchyma than in EX-type parenchyma, suggesting that SHH signaling may be involved in not only cancer growth but also cancer invasion. Autocrine SHH signaling depends on secretion of SHH ligands by the cancer parenchyma. SHH acts on itself in a positive feedback loop. Active Hedgehog signaling has been reported in various cancers such as prostate, lung, liver, thyroid, bladder, ovarian, and colon cancers [45–51]. SHH overexpression promotes tumor growth and metastasis, and higher levels of SHH are associated with poor survival and poor prognosis. Some studies reported that autocrine activation of SHH is present in colorectal cancer and breast cancer [52–55]. In addition, the SHH pathway induces epithelial-to-mesenchymal transition in gastric tumors, pancreatic cancer, and breast cancer [56–58]. Our data also showed that SHH expression was different in ED-type parenchyma compared to EX-type parenchyma. Therefore, SHH signaling in the cancer parenchyma of OSCC may be involved in causing epithelial-to-mesenchymal transition and invasion, similar to other tumors.

Second, we demonstrated paracrine signaling in which the cancer parenchyma secretes SHH ligands that bind receptors on the surrounding stroma, thus activating stromal SHH signaling.

The role of SHH during tumor-associated angiogenesis has not been clarified in OSCC. Here, we demonstrated a novel role for SHH in the development of the tumor vasculature. Our results imply that SHH secreted from cancer cells facilitates cancer invasion not only by stimulating proliferation of cancer cells in an autocrine manner but also by promoting angiogenesis in a paracrine manner. Angiogenesis of tumors has critical effects on development of the tumor. Some studies have reported the details of the contribution of SHH to tumor angiogenesis [59] and also reported the contribution of SHH to angiogenesis of OSCC [33,34]. PTCH expression was not expressed in blood vessels observed in the connective tissue adjacent to the normal epithelium. However, PTCH-positive tumor blood vessels were observed abundantly in sites close to the cancer parenchyma where SHH was strongly expressed. Thus, these results suggest that SHH acts in a paracrine manner on tumor blood vessels in a concentration-dependent manner. Additionally, our studies also showed that SHH was expressed in tumor blood vessels. Thus, SHH signaling had both autocrine and paracrine effects on tumor angiogenesis in OSCC.

Both CXCR4 and SDF-1 were expressed in the cancer cells themselves. SDF-1 expression is related to ovarian tumorigenesis and malignant transformation [60]. Pancreatic cancer [61], neuroblastoma cells [62], and glioblastoma [63] also express both CXCR4 and SDF-1 proteins. In addition, SDF-1 recruits endothelial progenitors, and SHH increases the expression of SDF-1 in a concentration-dependent manner [23]. Recently, some studies have indicated that CXCR4 is essential for the formation of large blood vessels that feed the gastrointestinal tract in the fetal stage [64,65]. PTCH-positive blood vessels have the potential to be a therapeutic target because CXCR4 is expressed in specific tumor blood vessels [8].

The SHH signaling pathway is not only involved in regulation of tumor angiogenesis but is also important for tumor migration and invasion. The number of macrophages that infiltrated the cancer stroma and expressed CD68 was significantly higher in ED-type stroma than in EX-type stroma. In other tumors, the high accumulation of CD68-positive cells may be associated with poor prognosis [3,66], suggesting that CD68 may also be correlated with prognosis of OSCC. In addition, CD68 overlapped with PTCH expression in ED-type stroma, suggesting that SHH signaling may be involved in macrophage aggregation in invasive cancer. Shimo et al. reported that the progression of oral cancer to the bone marrow of the jaw is correlated with prognosis, and that SHH produced from cancer cells directly affects CD68-positive osteoclast precursor cells and mature osteoclasts [67].

CD11b is generally known as a marker of monocytes, macrophages, and TAMs [5]. TAMs are involved in tumor invasion and metastasis [68]. In our study, CD11b-positive cells that contacted the cancer parenchyma were observed in ED-type stroma and EX-type stroma, and the number of

CD11b-positive cells was higher in the former than the latter type. In ED type, some CD11b-positive cells were PTCH positive. On the other hand, in EX type, almost no CD11b-positive cells were PTCH positive. ED type that expressed SHH strongly induced CD11b/PTCH-double positive cells, which participate in cancer invasion around the front of cancer nests. Considering the characteristics of their distribution and shape, CD11b-positive cells in OSCC may represent TAMs, and some CD11b-positive cells may have been recruited by SHH signaling in the cancer parenchyma.

α-SMA is a popular marker of myoepithelial cells and CAFs in tumors. In our study, we found many spindle-shaped α-SMA-positive cells surrounding the cancer parenchyma in ED-type stroma. In ED-type stroma, some of these spindle-shaped α-SMA-positive cells expressed PTCH. On the other hand, we did not observe α-SMA-positive cells surrounding the cancer parenchyma in EX-type stroma. Thus, the SHH pathway has been identified as activated in CAFs in OSCC, especially in ED type. Previous work identified SHH as a mediator of the desmoplastic response in pancreatic cancer and suggested that the stroma may serve as a barrier to delivery of therapeutic compounds [69]. In a previous in vivo study, human pancreatic CAFs overexpressed the Hedgehog receptor Smoothened (SMO). Increased SMO expression indicates increased Hedgehog pathway activity in these cells, suggesting evidence for Hedgehog pathway activity in pancreatic cancer–associated stromal cells [70]. Our study is consistent with these previous reports and indicates that the SHH pathway in OSCC may play an important role in inducing CAFs through a paracrine mechanism.

The cancer stroma consists of various types of cells. Some cancer stromal cells are derived from BMDCs [71,72]. These stromal cells derived from BMDCs are recruited by SDF-1 and induced by SHH. Stromal cells differentiate from endothelial cells, TAMs, and CAFs. Therefore, SHH signaling affects not only tumor growth but also the cells that make up the cancer stroma and is an important pathway in the TME.

The limitations of this study are due to using the patients tissue and the lack of understanding of the cell biology and genetics approaches. Future research should include the cell biology and genetics approaches.

In conclusion, given all the findings we observed from OSCC of ED type and EX type, the SHH pathway may participate in the processes of tumorigenesis and cancer development. The histological comparison between ED type and EX type provides authentic evidence for the involvement of the SHH pathway in the development of cancer via differentiation into various kinds of cells in the cancer stroma, such as macrophages, fibroblasts, and endothelial cells. Our results suggest roles for these cells in tumorigenesis due to their multilineage differentiation potential. This study is, to the best our knowledge, the first to show the direct and indirect influence of the SHH pathway on the TME. Our findings clearly show that OSCC-derived SHH is involved in progression and invasion in the OSCC TME.

4. Materials and Methods

4.1. Patients and Samples

Patient samples were obtained from the Oral Pathology Department of Okayama University. This study was approved by the Ethics Committee of Okayama University Graduate School of Medicine, Dentistry and Pharmaceutical Sciences (the project identification code: 1608–018; date of approval: 10 March 2017; and name of the ethics committee: Analysis of biological property of oral cancer). A total of 30 cases of tongue OSCC classified as T2 according to the Union for International Cancer Control (UICC) 8th Ed criteria were enrolled in the retrospective study, including 15 ED-type cases and 15 EX-type cases of OSCC. These tissue samples from 30 patients were collected during excision. None of the patients received chemotherapy, radiotherapy, or immunotherapy before sampling.

Tissues were processed and embedded in paraffin wax according to routine histological preparation methods and sectioned at 3 μm thickness. The sections were used for hematoxylin-eosin (H&E) staining

(Wako Pure Chemical Industries, Ltd., Osaka, Japan), immunohistochemistry (IHC), double IHC, and double-fluorescent IHC.

4.2. Immunohistochemistry and Evaluation

IHC was carried out using the antibodies detailed in Table 1. Following antigen retrieval, sections were treated with 10% normal serum for 30 min, and then incubated with primary antibodies at 4 °C overnight. Tagging of primary antibody was achieved by the subsequent application of anti-rabbit, anti-goat, or anti-mouse IgG and avidin–biotin complexes (Rabbit/Goat/Mouse ABC kit; Vector Laboratories, Inc., Burlingame, CA, USA). Immunoreactivity was visualized using diaminobenzidine (DAB)/H_2O_2 solution (Histofine DAB substrate; Nichirei, Tokyo, Japan), and sections were counterstained with Mayer's hematoxylin. For double immunostaining, the same process from DAB staining was repeated, and immunoreactivity was visualized using Green chromogen (Vina Green Chromogen kit; Biocare Medical, Pacheco, CA, USA). Sections were counterstained with Mayer's hematoxylin.

Table 1. Antibodies used in immunohistochemistry.

Primary Antibody	Immunized Animal	Antigen Retrieval	Dilution	Supplier
SHH	Rabbit	Heated in 0.01 mol/L citrate buffer for 3 min	×100	Abcam (Tokyo, Japan)
PTCH	Goat	Heated in 0.01 mol/L citrate buffer for 3 min	×100	Abcam (Tokyo, Japan)
CD31	Mouse	Pressurized with 0.01 mol/L citrate buffer for 8 min in microwave oven	×100	Novocastra (Newcastle upon Tyne, UK)
SDF-1	Rabbit	Heated in 0.01 mol/L citrate buffer for 3 min	×200	Abcam (Tokyo, Japan)
CXCR4	Rabbit	Pressurized with 0.01 mol/L citrate buffer for 8 min in microwave oven	×300	Abcam (Tokyo, Japan)
CD68	Rabbit	Heated in 0.01 mol/L citrate buffer for 3 min	×200	Santa Cruz Biotechnology (Dallas, USA)
CD11b	Rabbit	Pressurized with 0.01 mol/L citrate buffer for 8 min in microwave oven	×500	Abcam (Tokyo, Japan)
α-SMA	Rabbit	Heated in 0.01 mol/L citrate buffer for 3 min	×200	Abcam (Tokyo, Japan)

SHH and PTCH expression in OSCC were evaluated independently by the 2 pathologists. The results were scored from 0 to 3 based on the intensity of the staining at the membrane or in the cytoplasm: 0, no reactivity; +1, weak; +2, moderate; and 3, marked.

4.3. Double-Fluorescent IHC Staining

Double-fluorescent IHC for PTCH-CD31, PTCH-CD11b, PTCH-CXCR4, and PTCH-α-SMA was performed using PTCH monoclonal antibodies (goat IgG) (Abcam, Tokyo, Japan). The secondary antibodies applied are detailed in Table 2. Antibodies were diluted with Can Get Signal A (TOYOBO, Osaka, Japan). After antigen retrieval, sections were treated with Block Ace (DS Pharma Bio-medical, Osaka, Japan) for 20 min at room temperature. Specimens were incubated with primary antibodies at 4 °C overnight and then incubated with secondary antibodies (1:200) for 1 h at room temperature.

After the reaction, the specimens were stained with 1 mg/mL DAPI (Dojindo Laboratories, Kumamoto, Japan).

Table 2. Antibodies used in double-fluorescent immunohistochemistry.

Second Antibody	Immunized Animal	Fluorescent Dye	Supplier
Anti-Rabbit IgG	Donkey	Alexa Flour 568	Thermo Fisher (Tokyo, Japan)
Anti-Goat IgG	Donkey	Alexa Flour 488	Thermo Fisher (Tokyo, Japan)
	Donkey	Alexa Flour 568	Thermo Fisher (Tokyo, Japan)
Anti-Mouse IgG	Donkey	Alexa Flour 568	Thermo Fisher (Tokyo, Japan)

4.4. Cell Counting

Sections were examined under a microscope at 400× magnification. Ten areas were randomly chosen in each sample. The number of positively labeled cells was counted manually, the average was obtained, and the ED type and EX type were compared.

4.5. Statistical Analysis

All values are the mean ± standard deviation. Statistical analysis was performed using one-way analysis of variance and Tukey's tests. A p value <0.05 was considered significant. All calculations were made using PASW Statistics 18 (SPSS Inc., Chicago, IL, USA).

Author Contributions: Conceptualization, K.T. and T.S.; Immunohistochemistry Experiment, K.T., C.A., H.K., S.Y., O.H., and M.W.O.; Statistic Analysis, S.S.; Draft Preparation, K.T., H.T., K.N. and H.N.; Funding Acquisition, J.M.

Funding: This study was jointly funded by the Japan Society for Promotion of Science (JSPS) KAKENHI Grant-in-Aid for Scientific Research (16K11722, 18K09789, 18K17224, 19K19159, 19K24094 and 19K19160).

Conflicts of Interest: The authors declare no conflict of interest.

Abbreviations

SHH	Sonic hedgehog
PTCH	Patched-1
SDF-1	Stroma cell-derived factor-1
CXCR4	C-X-C motif chemokine receptor 4

References

1. Stewart, B.W.; Wild, C.P. *World Cancer Report 2014*; IARC Nonserial Publication: Lyon, France, 2014; pp. 422–423.
2. El-Naggar, A.K.; Chan, J.K.C.; Grandis, J.R.; Takata, T.; Slootweg, P.J. Tumours of the oral cavity and mobile tongue. In *WHO Classification of Head and Neck Tumours*, 4th ed.; Reibel, J., Gale, N., Hille, J., Hunt, L., Lingen, M., Muller, S., Sloan, P., Tilakaratne, W.N., Williams, M.D., Odell, E.W., et al., Eds.; World Health Organization: Geneva, Switzerland, 2017; pp. 105–131.
3. Yong, H.; Meng-Ying, H.; Li-Fang, Z.; Cong-Chong, Y.; Mei-Ling, Z.; Qiong, W.; Wei, Z.; Yang-Yu, Z.; Dong-Miao, W.; Zeng-Qi, X.; et al. Tumor-associated macrophages correlate with the clinicopathological features and poor outcomes via inducing epithelial to mesenchymal transition in oral squamous cell carcinoma. *J. Exp. Clin. Cancer Res.* **2016**, *35*, 12.
4. Maria, N.M.R.P.; Karen, C.; Fernanda, G.S.; Maria, A.Z.F. Role of tumour-associated macrophages in oral squamous cells carcinoma progression: An update on current knowledge. *Diagn. Pathol.* **2017**, *12*, 32.
5. Imelda, S.; Nadège, K.; Géraldine, D.; Justine, B.; Jérôme, R.L.; Quentin, M.; Charles, P.; Fabrice, J.; Sven, S. High infiltration of CD68+ macrophages is associated with poor prognoses of head and neck squamous cell carcinoma patients and is influenced by human papillomavirus. *Oncotarget* **2018**, *9*, 11046–11059.

6. Colegio, O.R.; Chu, N.Q.; Szabo, A.L.; Chu, T.; Rhebergen, A.M.; Jairam, V.; Cyrus, N.; Brokowski, C.E.; Eisenbarth, S.C.; Phillips, G.M.; et al. Functional polarization of tumour-associated macrophages by tumour-derived lactic acid. *Nature* **2014**, *513*, 559. [CrossRef]

7. Traweek, S.T.; Kandalaft, P.L.; Mehta, P.; Battifora, H. The Human Hematopoietic Progenitor Cell Antigen (CD 3 4) in Vascular Neoplasia. *Am. J. Clin. Pathol.* **1991**, *96*, 25–31. [CrossRef] [PubMed]

8. Yoshida, S.; Kawai, H.; Eguchi, T.; Sukegawa, S.; Oo, M.W.; Anqi, C.; Takabatake, K.; Nakano, K.; Okamoto, K.; Nagatsuka, H. Tumor Angiogenic Inhibition Triggered Necrosis (TAITN) in Oral Cancer. *Cells* **2019**, *8*, 761. [CrossRef] [PubMed]

9. Teicher, B.A.; Fricker, S.P. CXCL12 (SDF-1)/CXCR4 pathway in cancer. *Clin. Cancer Res.* **2010**, *1*, 2927–2931. [CrossRef] [PubMed]

10. Zhang, J.; Chen, J.; Wo, D.; Yan, H.; Liu, P.; Ma, E.; Li, L.; Zheng, L.; Chen, D.; Yu, Z.; et al. LRP6 Ectodomain Prevents SDF-1/CXCR4-Induced Breast Cancer Metastasis to Lung. *Clin. Cancer Res.* **2019**, *25*, 4832–4845. [CrossRef] [PubMed]

11. Skalli, O.; Ropraz, P.; Trzeciak, A.; Benzonana, G.; Gillessen, D.; Gabbiani, G. A monoclonal antibody against alpha-smooth muscle actin: A new probe for smooth muscle differentiation. *J. Cell Biol.* **1986**, *103*, 2787–2796. [CrossRef]

12. Desmoulière, A.; Geinoz, A.; Gabbiani, F.; Gabbiani, G. Transforming growth factor-beta 1 induces alpha-smooth muscle actin expression in granulation tissue myofibroblasts and in quiescent and growing cultured fibroblasts. *J. Cell Biol.* **1999**, *122*, 103–111. [CrossRef]

13. Bhowmick, N.A.; Neilson, E.G.; Moses, H.L. Stromal fibroblasts in cancer initiation and progression. *Nature* **2004**, *4*, 332–337. [CrossRef] [PubMed]

14. De Wever, O.; Mareel, M. Role of tissue stroma in cancer cell invasion. *J. Pathol.* **2003**, *200*, 429–447. [CrossRef] [PubMed]

15. Lim, K.P.; Cirillo, N.; Hassona, Y.; Wei, W.; Thurlow, J.K.; Cheong, S.C.; Pitiyage, G.; Parkinson, E.K.; Prime, S.S. Fibroblast gene expression profile reflects the stage of tumour progression in oral squamous cell carcinoma. *J. Pathol.* **2011**, *223*, 459–469. [CrossRef] [PubMed]

16. Chuang, P.T.; McMahon, A.P. Vertebrate Hedgehog signalling modulated by induction of a Hedgehog-binding protein. *Nature* **1999**, *397*, 617–621. [CrossRef]

17. Pepicelli, C.V.; Lewis, P.M.; McMahon, A.P. Sonic hedgehog regulates branching morphogenesis in the mammalian lung. *Curr. Biol.* **1998**, *8*, 1083–1086. [CrossRef]

18. Ramalho-Santos, M.; Melton, D.A.; McMahon, A.P. Hedgehog signals regulate multiple aspects of gastrointestinal development. *Development* **2000**, *127*, 2763–2772.

19. Ingham, P.W.; McMahon, A.P. Hedgehog signaling in animal development: Paradigms and principles. *Genes Dev.* **2001**, *15*, 3059–3087. [CrossRef]

20. Hebrok, M.; Kim, S.K.; St Jacques, B.; McMahon, A.P.; Melton, D.A. Regulation of pancreas development by hedgehog signaling. Hedgehog induction of murine vasculogenesis is mediated by Foxf1 and Bmp4. *Development* **2000**, *127*, 4905–4913.

21. Astorga, J.; Carlsson, P. Hedgehog induction of murine vasculogenesis is mediated by Foxf1 and Bmp4. *Development* **2007**, *134*, 3753–3761. [CrossRef]

22. Candice, C.; Sarah, G.; Pierre-Louis, H.; Marie-Ange, R. Role of Hedgehog Signaling in Vasculature Development, Differentiation, and Maintenance. *Int. J. Mol. Sci.* **2019**, *20*, 3076.

23. Kusano, K.F.; Pola, R.; Murayama, T.; Curry, C.; Kawamoto, A.; Iwakura, A.; Shintani, S.; Ii, M.; Asai, J.; Tkebuchava, T.; et al. Sonic hedgehog myocardial gene therapy: Tissue repair through transient reconstitution of embryonic signaling. *Nat. Med.* **2005**, *11*, 1197–1204. [CrossRef] [PubMed]

24. Shimo, T.; Gentili, C.; Iwamoto, M.; Wu, C.; Koyama, E.; Pacifici, M. Indian hedgehog and syndecans-3 coregulate chondrocyte proliferation and function during chick limb skeletogenesis. *Dev. Dyn.* **2004**, *229*, 607–617. [CrossRef] [PubMed]

25. Young, B.; Minugh-Purvis, N.; Shimo, T.; St-Jacques, B.; Iwamoto, M.; Enomoto-Iwamoto, M.; Koyama, E.; Pacifici, M. Indian and sonic hedgehogs regulate synchondrosis growth plate and cranial base development and function. *Dev. Biol.* **2006**, *299*, 272–282. [CrossRef] [PubMed]

26. Tara, L.L.; William, M. Hedgehog pathway as a drug target: Smoothened inhibitors in development. *Oncol. Targets Ther.* **2012**, *5*, 547–558.

27. Hochman, E.; Castiel, A.; Jacob-Hirsch, J.; Amariglio, N.; Izraeli, S. Molecular pathways regulating pro-migratory effects of Hedgehog signaling. *J. Biol. Chem.* **2006**, *281*, 33860–33870. [CrossRef]

28. Liao, X.; Siu, M.K.; Au, C.W.; Wong, E.S.; Chan, H.Y.; Ip, P.P.; Ngan, H.Y.; Cheung, A.N. Aberrant activation of hedgehog signaling pathway in ovarian cancers: Effect on prognosis, cell invasion and differentiation. *Carcinogenesis* **2009**, *30*, 131–140. [CrossRef]

29. Feldmann, G.; Dhara, S.; Fendrich, V.; Bedja, D.; Beaty, R.; Mullendore, M.; Karikari, C.; Alvarez, H.; Iacobuzio-Donahue, C.; Jimeno, A.; et al. Blockade of hedgehog signaling inhibits pancreatic cancer invasion and metastases: A new paradigm for combination therapy in solid cancers. *Cancer Res.* **2007**, *67*, 2187–2196. [CrossRef]

30. Souzaki, M.; Kubo, M.; Kai, M.; Kameda, C.; Tanaka, H.; Taguchi, T.; Tanaka, M.; Onishi, H.; Katano, M. Hedgehog signaling pathway mediates the progression of non-invasive breast cancer to invasive breast cancer. *Cancer Sci.* **2011**, *102*, 373–381. [CrossRef]

31. Thayer, S.P.; Magliano, M.P.; Heiser, P.W.; Nielsen, C.M.; Roberts, D.J.; Lauwers, G.Y.; Qi, Y.P.; Gysin, S.; Fernández-del, C.C.; Yajnik, V.; et al. Hedgehog is an early and late mediator of pancreatic cancer tumorigenesis. *Nature* **2003**, *425*, 851–856. [CrossRef]

32. Xu, F.G.; Ma, Q.Y.; Wang, Z. Blockade of hedgehog signaling pathway as a therapeutic strategy for pancreatic cancer. *Cancer Lett.* **2009**, *283*, 119–124. [CrossRef]

33. Huaitong, X.; Yuanyong, F.; Yueqin, T.; Peng, Z.; Wei, S.; Kai, S. Microvesicles releasing by oral cancer cells enhance endothelial cell angiogenesis via Shh/RhoA signaling pathway. *Cancer Biol. Ther.* **2017**, *18*, 783–791. [CrossRef] [PubMed]

34. Kuroda, H.; Kurio, N.; Shimo, T.; Matsumoto, K.; Masui, M.; Takabatake, K.; Okui, T.; Ibaragi, S.; Kunisada, Y.; Obata, K.; et al. Oral Squamous Cell Carcinoma-derived Sonic Hedgehog Promotes Angiogenesis. *Anticancer Res.* **2017**, *37*, 6731–6737. [PubMed]

35. Gonzalez, A.C.; Ferreira, M.; Ariel, T.; Reis, S.R.; Andrade, Z.; Peixoto, M.A. Immunohistochemical evaluation of hedgehog signalling in epithelial/mesenchymal interactions in squamous cell carcinoma transformation: A pilot study. *J. Oral Pathol. Med.* **2016**, *45*, 173–179. [CrossRef] [PubMed]

36. Fan, H.X.; Wang, S.; Zhao, H.; Liu, N.; Chen, D.; Sun, M.; Zheng, J.H. Sonic hedgehog signaling may promote invasion and metastasis of oral squamous cell carcinoma by activating MMP-9 and E-cadherin expression. *Med. Oncol.* **2014**, *31*, 41. [CrossRef]

37. Wang, Y.F.; Chang, C.J.; Lin, C.P.; Chang, S.Y.; Chu, P.Y.; Tai, S.K.; Li, W.Y.; Chao, K.S.; Chen, Y.J. Expression of hedgehog signaling molecules as a prognostic indicator of oral squamous cell carcinoma. *Head Neck* **2012**, *34*, 1556–1561. [CrossRef]

38. Paluszczak, J.; Wiśniewska, D.; Kostrzewska-Poczekaj, M.; Kiwerska, K.; Grénman, R.; Mielcarek-Kuchta, D.; Jarmuż-Szymczak, M. Prognostic significance of the methylation of Wnt pathway antagonists-CXXC4, DACT2, and the inhibitors of sonic hedgehog signaling-ZIC1, ZIC4, and HHIP in head and neck squamous cell carcinomas. *Clin. Oral Investig.* **2017**, *21*, 1777–1788. [CrossRef]

39. Würth, R.; Bajetto, A.; Harrison, J.K.; Barbieri, F.; Florio, T. CXCL12 modulation of CXCR4 and CXCR7 activity in human glioblastoma stem-like cells and regulation of the tumor microenvironment. *Front. Cell Neurosci.* **2014**, *8*, 144.

40. Schimanski, C.C.; Bahre, R.; Gockel, I.; Müller, A.; Frerichs, K.; Hörner, V.; Teufel, A.; Simiantonaki, N.; Biesterfeld, S.; Wehler, T.; et al. Dissemination of hepatocellular carcinoma is mediated via chemokine receptor CXCR4. *Br. J. Cancer* **2006**, *95*, 210–217. [CrossRef]

41. Sun, Y.; Mao, X.; Fan, C.; Liu, C.; Guo, A.; Guan, S.; Jin, Q.; Li, B.; Yao, F.; Jin, F. CXCL12-CXCR4 axis promotes the natural selection of breast cancer cell metastasis. *Tumour. Biol.* **2014**, *35*, 7765–7773. [CrossRef]

42. Goto, M.; Yoshida, T.; Yamamoto, Y.; Furukita, Y.; Inoue, S.; Fujiwara, S.; Kawakita, N.; Nishino, T.; Minato, T.; Yuasa, Y.; et al. CXCR4 Expression is Associated with Poor Prognosis in Patients with Esophageal Squamous Cell Carcinoma. *Ann. Surg. Oncol.* **2017**, *24*, 832–840. [CrossRef]

43. Lee, C.; Lee, J.; Choi, S.A.; Kim, S.K.; Wang, K.C.; Park, S.H.; Kim, S.H.; Lee, J.Y.; Phi, J.H. M1 macrophage recruitment correlates with worse outcome in SHH Medulloblastomas. *BMC Cancer* **2018**, *18*, 535. [CrossRef] [PubMed]

44. Franklin, R.A.; Liao, W.; Sarkar, A.; Kim, M.V.; Bivona, M.R.; Liu, K.; Pamer, E.G.; Li, M.O. The cellular and molecular origin of tumor-associated macrophages. *Science* **2014**, *344*, 921–925. [CrossRef] [PubMed]

45. Karhadkar, S.S.; Bova, G.S.; Abdallah, N.; Dhara, S.; Gardner, D.; Maitra, A.; Isaacs, J.T.; Berman, D.M.; Beachy, P.A. Hedgehog signalling in prostate regeneration, neoplasia and metastasis. *Nature* **2004**, *431*, 707–712. [CrossRef] [PubMed]

46. Park, K.S.; Martelotto, L.G.; Peifer, M.; Sos, M.L.; Karnezis, A.N.; Mahjoub, M.R.; Bernard, K.; Conklin, J.F.; Szczepny, A.; Yuan, J.; et al. A crucial requirement for Hedgehog signaling in small cell lung cancer. *Nat. Med.* **2011**, *17*, 1504–1508. [CrossRef] [PubMed]

47. Guo, L.Y.; Liu, P.; Wen, Y.; Cui, W.; Zhou, Y. Sonic hedgehog signaling pathway in primary liver cancer cells. *Asian Pac. J. Trop. Med.* **2014**, *7*, 735–738. [CrossRef]

48. Williamson, A.J.; Doscas, M.E.; Ye, J.; Heiden, K.B.; Xing, M.; Li, Y.; Prinz, R.A.; Xu, X. The sonic hedgehog signaling pathway stimulates anaplastic thyroid cancer cell motility and invasiveness by activating Akt and c-Met. *Oncotarget* **2016**, *7*, 10472–10485. [CrossRef]

49. Shin, K.; Lim, A.; Zhao, C.; Sahoo, D.; Pan, Y.; Spiekerkoetter, E.; Liao, J.C.; Beachy, P.A. Hedgehog signaling restrains bladder cancer progression by eliciting stromal production of urothelial differentiation factors. *Cancer Cell* **2014**, *26*, 521–533. [CrossRef]

50. Bhattacharya, R.; Kwon, J.; Ali, B.; Wang, E.; Patra, S.; Shridhar, V.; Mukherjee, P. Role of hedgehog signaling in ovarian cancer. *Clin. Cancer Res.* **2008**, *14*, 7659–7666. [CrossRef]

51. Varnat, F.; Duquet, A.; Malerba, M.; Zbinden, M.; Mas, C.; Gervaz, P.; Ruiz, A.A. Human colon cancer epithelial cells harbour active HEDGEHOG-GLI signalling that is essential for tumour growth, recurrence, metastasis and stem cell survival and expansion. *EMBO Mol. Med.* **2009**, *1*, 338–351. [CrossRef]

52. Gulino, A.; Ferretti, E.; De Smaele, E. Hedgehog signalling in colon cancer and stem cells. *EMBO Mol. Med.* **2009**, *1*, 300–302. [CrossRef]

53. Mukherjee, S.; Frolova, N.; Sadlonova, A.; Novak, Z.; Steg, A.; Page, G.P.; Welch, D.R.; Lobo-Ruppert, S.M.; Ruppert, J.M.; Johnson, M.R.; et al. Hedgehog signaling and response to cyclopamine differ in epithelial and stromal cells in benign breast and breast cancer. *Cancer Biol. Ther.* **2006**, *5*, 674–683. [CrossRef] [PubMed]

54. Ertao, Z.; Jianhui, C.; Chuangqi, C.; Changjiang, Q.; Sile, C.; Yulong, H.; Hui, W.; Shirong, C. Autocrine Sonic hedgehog signaling promotes gastric cancer proliferation through induction of phospholipase Cγ1 and the ERK1/2 pathway. *J. Exp. Clin. Cancer Res.* **2016**, *35*, 63. [CrossRef] [PubMed]

55. Liu, Z.; Xu, J.; He, J.; Zheng, Y.; Li, H.; Lu, Y.; Qian, J.; Lin, P.; Weber, D.M.; Yang, J.; et al. A critical role of autocrine sonic hedgehog signaling in human CD138+ myeloma cell survival and drug resistance. *Blood* **2014**, *124*, 2061–2071. [CrossRef] [PubMed]

56. Yoo, Y.A.; Kang, M.H.; Lee, H.J.; Kim, B.H.; Park, J.K.; Kim, H.K.; Kim, J.S.; Oh, S.C. Sonic hedgehog pathway promotes metastasis and lymphangiogenesis via activation of Akt, EMT, and MMP-9 pathway in gastric cancer. *Cancer Res.* **2011**, *71*, 7061–7070. [CrossRef]

57. Xu, X.; Zhou, Y.; Xie, C.; Wei, S.M.; Gan, H.; He, S.; Wang, F.; Xu, L.; Lu, J.; Dai, W.; et al. Genome-wide screening reveals an EMT molecular network mediated by Sonic hedgehog-Gli1 signaling in pancreatic cancer cells. *PLoS ONE* **2012**, *7*, e43119. [CrossRef]

58. Islam, S.S.; Mokhtari, R.B.; Noman, A.S.; Uddin, M.; Rahman, M.Z.; Azadi, M.A.; Zlotta, A.; van der Kwast, T.; Yeger, H.; Farhat, W.A. Sonic hedgehog (Shh) signaling promotes tumorigenicity and stemness via activation of epithelial-to-mesenchymal transition (EMT) in bladder cancer. *Mol. Carcinog.* **2016**, *55*, 537–551. [CrossRef]

59. Yamazaki, M.; Nakamura, K.; Mizukami, Y.; Ii, M.; Sasajima, J.; Sugiyama, Y.; Nishikawa, T.; Nakano, Y.; Yanagawa, N.; Sato, K.; et al. Sonic hedgehog derived from human pancreatic cancer cells augments angiogenic function of endothelial progenitor cells. *Cancer Sci.* **2008**, *99*, 1131–1138. [CrossRef]

60. Scotton, C.J.; Wilson, J.L.; Scott, K.; Stamp, G.; Wilbanks, G.D.; Fricker, S.; Bridger, G.; Balkwill, F.R. Multiple actions of the chemokine CXCL12 on epithelial tumor cells in human ovarian cancer. *Cancer Res.* **2002**, *62*, 5930–5938.

61. Koshiba, T.; Hosotani, R.; Miyamoto, Y.; Ida, J.; Tsuji, S.; Nakajima, S.; Kawaguchi, M.; Kobayashi, H.; Doi, R.; Hori, T.; et al. Expression of stromal cell-derived factor 1 and CXCR4 ligand receptor system in pancreatic cancer: A possible role for tumor progression. *Clin Cancer Res.* **2000**, *6*, 3530–3535.

62. Geminder, H.; Sagi-Assif, O.; Goldberg, L.; Meshel, T.; Rechavi, G.; Witz, I.P.; Ben-Baruch, A. A possible role for CXCR4 and its ligand, the CXC chemokine stromal cell-derived factor-1, in the development of bone marrow metastases in neuroblastoma. *J. Immunol.* **2001**, *167*, 4747–4757. [CrossRef]

63. Rempel, S.A.; Dudas, S.; Ge, S.; Gutiérrez, J.A. Identification and localization of the cytokine SDF1 and its receptor, CXC chemokine receptor 4, to regions of necrosis and angiogenesis in human glioblastoma. *Clin Cancer Res.* **2000**, *6*, 102–111. [PubMed]

64. Tachibana, K.; Hirota, S.; Iizasa, H.; Yoshida, H.; Kawabata, K.; Kataoka, Y.; Kitamura, Y.; Matsushima, K.; Yoshida, N.; Nishikawa, S.; et al. The chemokine receptor CXCR4 is essential for vascularization of the gastrointestinal tract. *Nature* **1998**, *393*, 591–594. [CrossRef] [PubMed]

65. Orimo, A.; Gupta, P.B.; Sgroi, D.C.; Arenzana-Seisdedos, F.; Delaunay, T.; Naeem, R.; Carey, V.J.; Richardson, A.L.; Weinberg, R.A. Stromal fibroblasts present in invasive human breast carcinomas promote tumor growth and angiogenesis through elevated SDF-1/CXCL12 secretion. *Cell* **2005**, *121*, 335–348. [CrossRef] [PubMed]

66. Eiró, N.; Pidal, I.; Fernandez-Garcia, B.; Junquera, S.; Lamelas, M.L.; del Casar, J.M.; González, L.O.; López-Muñiz, A.; Vizoso, F.J. Impact of CD68/(CD3+CD20) ratio at the invasive front of primary tumors on distant metastasis development in breast cancer. *PLoS ONE* **2012**, *7*, e52796. [CrossRef] [PubMed]

67. Shimo, T.; Matsumoto, K.; Takabatake, K.; Aoyama, E.; Takebe, Y.; Ibaragi, S.; Okui, T.; Kurio, N.; Takada, H.; Obata, K.; et al. The Role of Sonic Hedgehog Signaling in Osteoclastogenesis and Jaw Bone Destruction. *PLoS ONE* **2016**, *11*, e0151731. [CrossRef] [PubMed]

68. Komohara, Y.; Jinushi, M.; Takeya, M. Clinical significance of macrophage heterogeneity in human malignant tumors. *Cancer Sci.* **2014**, *105*, 1–8. [CrossRef]

69. Bailey, J.M.; Swanson, B.J.; Hamada, T.; Eggers, J.P.; Singh, P.K.; Caffery, T.; Ouellette, M.M.; Hollingsworth, M.A. Sonic hedgehog promotes desmoplasia in pancreatic cancer. *Clin. Cancer Res.* **2008**, *14*, 5995–6004. [CrossRef]

70. Walter, K.; Omura, N.; Hong, S.M.; Griffith, M.; Vincent, A.; Borges, M.; Goggins, M. Overexpression of smoothened activates the sonic hedgehog signaling pathway in pancreatic cancer-associated fibroblasts. *Clin. Cancer Res.* **2010**, *16*, 1781–1789. [CrossRef]

71. Aghi, M.; Chiocca, E.A. Contribution of bone marrow-derived cells to blood vessels in ischemic tissues and tumors. *Mol. Ther.* **2005**, *12*, 994–1005. [CrossRef]

72. Lyden, D.; Hattori, K.; Dias, S.; Costa, C.; Blaikie, P.; Butros, L.; Chadburn, A.; Heissig, B.; Marks, W.; Witte, L. Impaired recruitment of bone marrow-derived endothelial and hematopoietic precursor cells blocks tumor angiogenesis and growth. *Nat. Med.* **2001**, *7*, 1194–1201. [CrossRef]

MDPI
St. Alban-Anlage 66
4052 Basel
Switzerland
Tel. +41 61 683 77 34
Fax +41 61 302 89 18
www.mdpi.com

International Journal of Molecular Sciences Editorial Office
E-mail: ijms@mdpi.com
www.mdpi.com/journal/ijms